"The Earth Constitution Solution is a guide for current and future action, and Dr. Glen Martin outlines the world structures needed to ensure the rights and common good of all humanity. This book is a vital presentation of the avenues to transcend the current geopolitical, conflictual structures toward a world society in which individuals can develop their full potential."

> ~ **Rene Wadlow**, President of the Association of World Citizens and Professor Emeritus at Graduate Institute of Development Studies in Geneva, Switzerland

"Martin's *Design for a Living Planet* is an amazing piece of work – an irrefutable argument for an *Earth Constitution* that should be compulsory reading for all the presidents and prime ministers of the world – preferably as a preparation for a new international convention on a world constitution to replace the obsolete U.N. Charter. Such a convention should be held before the world's injustices, inequalities, nationalisms, religions and climate change all combine to create the conditions that trigger the next world war."

> ~ **Ian Hackett,** Co-Founder of the World Federalist Party, former Chair of the UK Association of World Federalists, Director of the One World Trust, and Head of the International School of London

"The concept of a world organization to ensure peace and welfare is an ancient one. The *Upanishads* speak of the human race as a family and see a spark of divinity in every human being ... [yet] nationalism turns out to be a much more powerful and potent emotion than universalism. ... Martin has written a book entitled *Earth Constitution Solution* in which he has outlined a concrete proposal for achieving this great ideal. My good wishes are with Martin and other friends who are committed to an idea whose time is bound to come sooner or later."

> ~ **D. Karan Singh Ji,** Associate of the World Constitution and Parliament Association, former Cabinet Member and longest serving Representative in the Raj Sabba (Upper House of the Parliament of India)

"Design for a Living Planet is an essential guidebook for everyone seeking to transcend tribalism and unite humanity through democratic world federation. The world administration advocated by this book and the *Earth Constitution* is the most complete depiction of what united humanity could look like."

> ~ **Eston McKeague,** President of the Young World Federalists

"Design for a Living Planet has arrived just when our world needs it the most. Every page of this extremely informative, profound and illuminating book offers us seminal wisdom that will guide our generation into a new Age of Holism. Ratifying the *Earth Constitution* will help lead the way into this magnificent new era for our beloved Planet Earth, all beings in our vast ecosystem and future generations that deserve a beautiful, thriving, and life-affirming home."

~ **Diane Williams,** Founder of The Source of Synergy Foundation and the U.N. NGO Committee on Spirituality, Values, and Global Concerns (NY), and co-author of *Our Moment of Choice*

"Earth *is* a living planet. At the current rate of destruction, we will soon minimize the life cycle of Earth and its people. Hence, *Design for a Living Planet* can be taken as a hub for the wheel of holistic life. The *Earth Constitution* has been mooted by Dr. Martin; it is now for us to implement it without any apprehension, whatsoever. Moreover, the War Pentagon should mandatorily be replaced with the Peace Pentagon."

~ **Colonel Tejendra Pal Tyagi, Veer Chakra,** Executive President of the World Constitution and Parliament Association in India, and National President of Rashtriya Sainik Sanstha in India (National Vanguard of ex-servicemen and patriotic citizens)

"The Earth Constitution Solution has the potential to become a classic, as it proclaims the 'Big Idea' using academic theory and practice at the highest level of excellence and thoroughness. In arguing for the *Earth Constitution,* Professor Martin's logic rivals the famous *Federalist Papers,* drafted to gain support for the U.S. Constitution. There's plenty here for intellectuals, the average citizen, save-the-world activists, and politicians from around the world."

~ **Roger Kotila, Ph.D.,** President of the Democratic World Federalists, Board Member at the Center for U.N. Constitutional Research (CUNCR Brussels), and Editor of *Earth Federation News & Views*

"La *Constitución para la Federación de la Tierra* es un instrumento concreto para acercarse a una utopía alcanzable, y *Design for a Living Planet* es una guía necesaria para cualquiera que crea que hay un hermoso futuro para la humanidad. Para aquellos apasionados por la paz mundial, la justicia, la equidad planetaria y la sostenibilidad ambiental, el Dr. Martin presenta una solución holística en un lenguaje accesible para todos."

~ **Leopoldo A. Cook,** Board Member of the Earth Constitution Institute and Vice President of the World Constitution and Parliament Association in Latin America and the Caribbean

"Many great thinkers throughout history have argued that only a democratically elected global government – complete with a Constitution, Bill of Rights, and the other structures necessary to make it effective – can solve humanity's most pressing problems and ensure global peace and prosperity. Glen T. Martin is a superb spokesperson for these ideas, and their philosophical, scientific and moral foundation."

> ~ **Bob Flax, Ph.D.,** Executive Director of Citizens for Global Solutions
> and professor of Transformative Social Change at Saybrook University

"Dr. Martin conveys a great message: If thinkers, philosophers, and activists work with sincerity, honesty, commitment, and LOVE for all humanity, a new human civilization is possible. The *Earth Constitution* should be studied by universities around the world so that coming generations will be empowered to build a cooperative economy and justice-centered, global political system."

> ~ **E. P. Menon,** Board Member of the Earth Constitution Institute, global
> peace activist, and author of *Foot Prints on Friendly Roads,* which chronicles
> his 8,000 mile "peace walk" around the world

"Glen Martin delivers practical, intelligent, and comprehensive arguments for the adoption of the *Earth Constitution.* His *Design for a Living Planet* deserves to be number one on the reading list of anyone concerned with the freedom and well-being of humanity."

> ~ **Eugenia Almand,** Secretary General of the World Constitution and
> Parliament Association

"Carl Jung said the way to wholeness is comprised of inevitable detours and even wrong turnings but a snake-like path that ultimately unites opposites. I think the journey to an *Earth Constitution* will be much the same. Let the conversation begin!"

> ~ **Kurt Johnson, Ph.D.,** Founder of the Interspiritual Network, co-host of
> The Convergence series on Voice America, and author of *Fine Lines,*
> *The Coming Interspiritual Age,* and *Our Moment of Choice*

"Glen Martin's *The Earth Constitution Solution: Design for a Living Planet* is a hopeful vision that is a deeply inspiring, precise roadmap for a better future. This new work by Martin is an excellent *must-read* for anyone who cares about the future of our planet."

> ~ **H.E. Rev. Patrick McCollum,** World peace and planetary sustainability
> activist and Founder of the McCollum Foundation for Peace

"Martin's *Design for a Living Planet* is a compendium of the most advanced thinking concerning sustainable economics, ecological sanity, a holistic world-view, and 'planetary consciousness.' As human beings we must become *humane beings,* motivated by compassion and imagination as well as by reason. ... The legal and humane wisdom of the *Earth Constitution* can lead us to a world without war, where cultures can flourish and nations continue, but in which sovereignty is exchanged for a transnational security that preserves the peace, prosperity, and environmental health of all life-forms on Earth."

~ **Richard D. Sharp, Ph.D.,** Professor Emeritus in Literature at Catholic University and Fulbright Fellow to India

"The Earth Constitution Solution is an important contribution to the most critcal story of our time, the movement toward genuine Earth Unity."

~ **Ben Bowler,** Executive Director of Unity Earth

"A growing chorus of voices is calling for a transformation in global governance that prioritizes the idea of 'WE' and the 'Common Good' to ensure that humanity will have a livable future. Glen Martin is a powerful and controversial voice. Sharply critical of current systems and strategies, Martin nevertheless presents compelling alternative approaches to achieve a world that works for all."

~ **Rick Ulfik,** Founder of We, The World and the WE Campaign at WE.net

"The *Earth Constitution* – like the *Magna Carta*, the *Declaration of the Rights of Man and of the Citizen*, and the *Universal Declaration of Human Rights* – provides a framework for peaceful and just human interaction. Transcending those historic legal documents, the *Earth Constitution* explains how we, as world citizens, can sustainably interact with Earth and begin to govern our world as one human family. In his latest book, Dr. Martin presents clear and compelling arguments for humanity to immediately engage this paradigm shift in law and politics.

~ **David Gallup, Esq.,** President of World Service Authority and Convener of the World Court of Human Rights Coalition

"Our entangled world has become a single community of fate. But the proto-country of Earth is a failed state with no overarching constitution to bind a democratic polity. A pragmatic visionary, Glen Martin has given us a roadmap for the long transition from chaos and crises to a flourishing and unified commonwealth."

~ **Paul Raskin, Ph.D.,** Founder and President of Tellus Institute, and author of *Journey to Earthland: The Great Transition to Planetary Civilization* and *Great Transition: The Promise and Lure of the Times Ahead*

PREVIOUS BOOKS BY GLEN T. MARTIN

Global Democracy and Human Self-Transcendence: The Power of the Future for Planetary Transformation (2018)

One World Renaissance: Holistic Planetary Transformation Through a Global Social Contract (2016)

Anatomy of a Sustainable World: Our Choice Between Climate Change or System Change (2013)

The Earth Federation Movement: Founding a Social Contract for the People of Earth. History, Documents, Philosophical Foundations (2011)

Constitution for the Federation of Earth: With Historical Introduction, Commentary, and Conclusion (2010)

Triumph of Civilization: Democracy, Nonviolence, and the Piloting of Spaceship Earth (2010)

Emerging World Law: Basic Documents and Decisions of the World Constituent Assemblies and the Provisional World Parliament (co-edited with Eugenia Almand, 2009)

Ascent to Freedom: Practical and Philosophical Foundations of Democratic World Law (2008)

World Revolution Through World Law: Basic Documents of the Emerging Earth Federation (2006)

Millennium Dawn: The Philosophy of Planetary Crisis and Human Liberation (2005)

From Nietzsche to Wittgenstein: The Problem of Truth and Nihilism in the Modern World (1989)

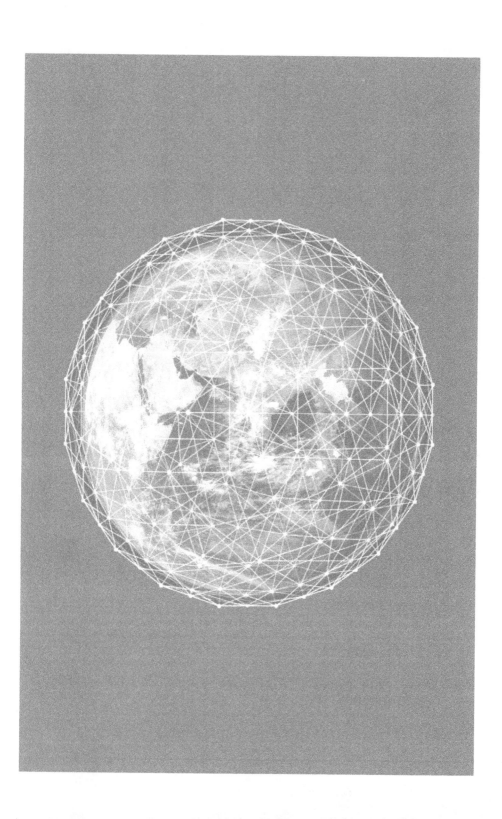

THE EARTH CONSTITUTION SOLUTION

Design for a Living Planet

Glen T. Martin, Ph.D.

Foreword by Ellen H. Brown, J.D.
Edited by Rev. Laura M. George, J.D.

Peace Pentagon Press
Independence, Virginia

Published by Peace Pentagon Press
An imprint of Oracle Institute Press, LLC

The Peace Pentagon
88 Oracle Way
Independence, VA 24348
www.TheOracleInstitute.org

Copyright © 2021 by Glen T. Martin

Publisher's Cataloging-in-Publication Data

Names: Martin, Glen T., 1944-, author. | Brown, Ellen H., foreword
    author. | George, Laura M., editor.
Title: Earth constitution solution : design for a living planet /
    Glen T. Martin, Ph.D.; foreword by Ellen H. Brown, J.D.;
    edited by Rev. Laura M. George, J.D.
Description: Includes bibliographical references and index. |
    Independence, VA: Peace Pentagon Press, an imprint of
    Oracle Institute Press, LLC, 2021.
Identifiers: LCCN: 2021932312
    ISBN: 978-1-937465-28-5 (pbk.) | 978-1-937465-29-2 (ebook) |
    978-1-937465-30-8 (MOBI) | 978-1-937465-31-5 (epub)
Subjects: LCSH International organization--Environmental aspects. |
    International law. | International cooperation. | International relations. |
    Sustainable development--Political aspects. | Climatic changes--
    Political aspects. | BISAC NATURE / Environmental Conservation
    & Protection | POLITICAL SCIENCE / Globalization | SCIENCE /
    Global Warming & Climate Change | BUSINESS & ECONOMICS /
    Environmental Economics | SOCIAL SCIENCE / Future Studies
Classification: LCC JZ1318 .M3763 2021 | DDC 341.2--dc23

Cover by www.KateInk.com
Interior Design by www.EMTippettsBookDesigns.com

CONTENTS

FOREWORD

In 2020, the world reached a state of existential crisis and "lockdowns" brought national economies to a standstill. Much concern has been expressed about the "Great Reset" proposed by the World Economic Forum, which posits a top-down global government run as an authoritarian technocracy. In *The Earth Constitution Solution: Design for a Living Planet,* Professor Glen T. Martin offers a much-needed alternative global model that recognizes the rights and freedoms of individuals and nations working together as a symbiotic whole.

I have been aware of Dr. Martin's critical work for a number of years, since first meeting him at the annual conferences of the American Monetary Institute in Chicago. His presentations at these meetings spoke about the *Constitution for the Federation of Earth* in relation to the need for global public banking and monetary policy. Reflecting that need, I added in the 2008 update to my book *The Web of Debt: The Shocking Truth About Our Money System and How We Can Break Free*:

> This sort of model has been proposed by an organization called the World Constitution and Parliament Association, which postulates an Earth Federation working for equal prosperity and well-being for all the Earth's citizens. The global funding body would be authorized not only to advance credit to nations but to issue money directly. ... These funds would then be disbursed as needed for the Common Wealth of Earth.

The "Common Wealth of Earth" is Glen Martin's passion. He understands that any solution to our most fundamental human problems must include all of us. No strictly national, regional, or parochial religious solution is capable of dealing with the immense threats that confront human existence today.

I became more familiar with Glen's visionary work when I was invited to participate in a series of "Building the New World Conferences" in 2008 and 2015, which he organized at Radford University in Virginia, where he is a Professor of Philosophy. He also is a Board Member with The Oracle Institute in Virginia, which sponsored a third, "virtual" Building the New World Conference in 2020. Once again, I was invited to participate, and I talked about the debt trap and the need for public banking in the "Economics and Business" sector at these conferences. Glen presented on the need for planetary solutions in the "Justice and Governance" sector. In his books, lectures and articles, Glen begins with a penetrating critique of the current world system, rooted in an outmoded paradigm of "modernity" formulated on 17th and 18th century assumptions that have long since been invalidated by 20th and 21st century revolutions in scientific theory. He shows how this outdated paradigm still determines our institutions and thought patterns, making it impossible to effectively deal with global problems like war, injustice, or environmental crises.

Glen also understands and incorporates the great spiritual traditions of the world, including those of Buddhism, Hinduism, Christianity, Islam, the meditative traditions, eastern and western spirituality, and Integral Theory. His vision of a redeemed world system includes an understanding of the dimensions of intellectual and spiritual growth necessary for humanity to free itself and become truly alive and fulfilled, able to live in peace, harmony, satisfaction, and justice on our mutually shared home planet.

As indicated in the title to his 2018 book *Global Democracy and Human Self-Transcendence*, Glen sees both institutional transformation and spiritual self-transcendence as interdependent and necessary features for creating a viable future for ourselves and future generations. This holistic focus makes Glen's concrete program for moving into the future unique and critical reading. He coherently brings together political and social critique, philosophical synthesis, spiritual insight, and practical, transformative action in his vision for a better world.

All of these features are captured in this new book, *The Earth Constitution Solution: Design for a Living Planet*. Chapter One presents an overview of the evolutionary process from the Big Bang to the present, integrating this process with the dynamics of human spiritual and moral growth. Chapters Two and Three lay bare the failed "modern" paradigm that includes sovereign militarized nation-states and unrestrained global capitalism. These chapters reveal the systemic flaws that have led to our present endangered condition and explain the contemporary holistic paradigm that has emerged from the sciences during the 20th and 21st centuries. It is this holistic paradigm that needs to be integrated into our thinking and our institutions to restore harmony with the biosphere and realize the fullness of life that is possible for ourselves and future generations.

Chapters Four and Five review the dimensions of our planetary environmental crisis as explained by top environmentalists and economists. In Chapter Six, Glen analyzes the United Nations "Sustainable Development Goals," which seek to guide the world toward sustainability by 2030. He shows the structural reasons why these goals cannot succeed. He argues that the U.N. is premised on the outdated "modern" paradigm of militarized sovereign nation-states that structurally defeats our need to bring humanity into harmony with our planetary biosphere. The holism embodied in the *Earth Constitution* does not abolish the U.N., but embraces its work in a democratically designed institutional framework.

Chapter Seven details the *Earth Constitution* itself, showing how it creates a redeemed and transformed world system premised on peace, justice, freedom, and sustainability – what a "design for a living planet" should look like. Chapter Eight concludes the book by bringing all these themes together in a synthetic vision in which politics, economics, science, culture, and spiritual growth fit together in the framework of the *Earth Constitution* and the holistic world view on which it is premised.

This book helps to guide us out of our seemingly hopeless human condition in the early 21st century by embracing the entirety of our situation and illuminating the ways in which the parts can be integrated into the whole through a comprehensive constitution, specifically designed for our living planet. The book provides not only a compelling vision of a sustainable future that is possible, but also a practical blueprint for achieving it.

~ Ellen H. Brown, J.D.
December 2020

Ellen H. Brown, J.D. is an attorney and the founder of the Public Banking Institute, which assists municipalities in setting up banks that serve as depositories for local funds and service local community needs. She is the author of more than a dozen books and hundreds of articles. In her bestselling book *Web of Debt*, Ellen exposed the Federal Reserve as a private banking cartel that has usurped the people's power to create money. The book also explains how "We the People" can take back this power. In her latest book, *The Public Bank Solution*, Ellen explains how public banks are designed to operate in the public interest. She also co-hosts a radio program called "It's Our Money," and she is a Fellow of the Democracy Collaborative.

PREFACE

As a philosopher, I have long been especially interested in moral and cognitive growth and in human spiritual development in relation to oppressive planetary systems of exploitation and domination. Since 2005, when I wrote *Millennium Dawn: The Philosophy of Global Crisis and Human Liberation*, my various books have been articulating, in one or another form, a synergistic vision of the various aspects of our human situation that require transformation in order for a deep and abiding human liberation to occur.

The present book is different because it focuses on climate change. However, this book draws on the same immense range of study and reflection involved in the theory and practice of human liberation. Climate change is an imminent threat to our existence, therefore this text explores fundamental themes related to this issue. However, it also includes a synthetic vision of interrelated transformations necessary for human liberation and self-actualization.

For this reason, certain themes merit greater exploration to make them more fully intelligible. For example, Chapter One speaks of our human existential situation as a whole, and brings in the process of growth toward a transpersonal "cosmic consciousness." To do this topic justice would require an entire book in its own right (*Millennium Dawn* has several chapters related to this theme). However, in light of the goal of this book – addressing our climate crisis – the treatment of some themes is necessarily limited. I am not attempting to provide a complete theory or account of such dimensions of human life, though I will indicate to the reader what must be taken into account if we are to deal comprehensively and holistically with our climate crisis.

This book is about establishing and preserving a *living planet*. It is very much about climate science and sustainable economics. But it also is about cosmic evolution as understood by contemporary science, theoretical possibilities for human growth and transformation, and the historic paradigm-shift from fragmentation to holism. As an integral aspect of this broad vision, I discuss the necessity of democratic world law under the *Constitution for the Federation of Earth*. Thus, this book attempts to compact these immense themes into a fairly small tome.

As we shall see, any design for a living planet requires the application of climate science, yet much more is needed than climate science alone. Indeed, this challenge requires serious spiritual and moral awakening on the part of humanity. *Our living planet requires human beings who are themselves fully alive, who are living life fully.* Today, many humans appear half dead. Ultimately, we will be able to successfully address climate crisis only if we transform ourselves as well as our actions and institutions on this planet. The two go together. Only whole human beings can bring our civilization into harmony with our biosphere.

This book is not about epidemiology, nor biological weapons research, nor the current international chaos induced by the present global pandemic of the Covid-19 virus. However, these phenomena underscore the great difficulties that humanity faces in creating an ethical and regenerative world. The crises of today parallel the ideas in this book. The range of issues that the pandemic raises bears a striking similarity to the issues surrounding the climate crisis.

Consequently, this book analyzes various vulnerabilities of our present world system. These disasters highlight the inability of our present system to anticipate or prepare for other disasters that we know are coming. The present pandemic also points to the economic and political failures of our current system – a world that is fragmented and out of touch with the reality of our human situation. Basically, we live in a world full of anarchy, competing corporations, mutual hostility, and other atomistic rivalries.

The world as we know it, and that we now experience openly in the face of global pandemic, has little capacity to coordinate, to cooperate, to organize in the face of disaster, or to build a transformed world of peace, justice, and sustainability. One commentator from Italy (an early epicenter of the pandemic) describes the growing chaos:

> The whole world says it's a war. And for the first time in history it seems to be everyone against everyone, without any alliances. Each nation thinks for itself, using every means to guarantee the winning weapons against the virus: swabs, masks, respirators. So the United States managed to buy in Brescia half a million kits to detect the infection. And they shipped them to Memphis in a military aircraft. ... In these hours, there are world auctions to buy every rising stock of masks and respirators at increasing prices: an economic challenge, in which the strongest wins. Like in war. But waged without any alliances.[1]

The world has institutionalized political and economic systems that reflect a mirror-image of war, whether due to massive military confrontations, low intensity warfare, terrorism, or economic rivalry. In the face of global threat or disaster, our regressive political and economic institutions reveal themselves as

---

[1] http://www.newsilkstrategies.com/international-relations/us-scarfs-up-supplies-urgently-needed-in-epidemic-stricken-italy

what they really are: totally inadequate to provide social harmony, legal equality, ecological resilience ... or even health-care protection to humanity.

In truth, the fragmentation of humanity amounts to a war-system. This designation applies whether we consider: (i) the threat of nuclear war and global apocalypse that has been with us since the 1950s, (ii) the threat of climate collapse, which has emerged as a very real specter within the past 50 years, or (iii) the present global pandemic, which threatens the health of the human race.

While the world likely will recover from this virus, the threat of climate change and the growing threat of nuclear war from a renewed nuclear arms race are far more serious because we cannot recover from either climate collapse or all-out nuclear war. The pandemic symbolizes the heartless cruelty of the neoliberal economic system that reduces everything to the bottom line for private accumulation of wealth. Meanwhile, most of the world has no medical backup, no reserves, not even sufficient hospital beds due to a for-profit system that milks the masses. In sum, preparations for a global pandemic – what many knew was a real possibility – are simply deemed "uneconomical."

Do we have a future on planet Earth? Not if we continue with irrational neoliberal absurdities. Nor do we have a future if we continue with the system of militarized sovereign nation-states developing their biological and nuclear weapons of mass destruction. Today, some countries are even exploring how to place these weapons-systems in orbit around Earth (Johnson 2006). We do not have a future unless we can eliminate the burning of fossil fuels as our main source of energy within the next few years.

In my 2018 book *Global Democracy and Human Self-Transcendence,* I explored the existential, structural, and spiritual significance of our future. Please realize that the future lives as an authentic dimension of our present consciousness, and it must be taken seriously because it is a key to transforming our world into a just and resilient society for all. I argued that the spirituality of mindfulness, of awareness of the depths of the present moment, is clearly required for a dynamic spiritual awakening and transformation of our planet. That spirituality is, however, fully compatible with what I called the "utopian horizon" that is present to us in everyday awareness.

This utopian horizon, implicit in the awareness of every normal human being, reveals our vast potential to become who we really are and what we are really meant to be: intelligent, gentle, loving, rational, cooperative persons living on our planetary home and focusing on the adventure of intellectual, scientific, moral, and spiritual growth. As we shall see, the new economists of sustainability insist that our economic goals must become *qualitative* rather than quantitative. This insight contains the seed of our deeper "ontological vocation," which calls us to embrace this exact goal: dynamic, qualitative growth. Psychologists such as Abraham Maslow (2014) and Eric Fromm (1996) called this potential within us all "being values," as opposed to materialistic "having values."

From a different (but fundamentally related) point of view, spiritual philosopher Raimon Panikkar calls our openness to the future "faith." "Faith," he writes, "is a constitutive dimension of our being: an openness to the more, the unknown, transcendence, the infinite: openness to the given. Faith is an awareness that we are still on the way, incomplete, unfulfilled, not yet *totally* realized ...." (2013, 306). The difficulty of the task before us should not be minimized.

With faith and commitment, we are capable of actualizing "being values." We must have faith in the open possibilities of self-realization and self-transcendence. But we must find ways to unite all humanity under this banner. Practicing these principles for ourselves or with our local community simply will not suffice to give humanity a future. As many ecological thinkers have declared, a global transformation of consciousness is necessary. We will examine some of these ideas shortly.

In this volume, I have not repeated my extended analysis concerning our utopian horizon that I made in *Global Democracy*. But I am hoping the reader will discern the practical power of the future as one of the assumptions behind the present book's focus on regenerative ecology. I show that thinking and living differently must necessarily include a new world system, like the one envisioned by the *Earth Constitution*. In my view, ratifying the *Earth Constitution* is the most effective, practical, and timely way to initiate the process of converting humanity from *having* to *being* values.

Developing both our critical reasoning powers and our capacity for love will mark the difference, Erich Fromm declares, between living as "*awake* or *dreaming* 'utopians'" (1996, 173). We need to be "awake utopians" he contends. As I showed in *Global Democracy,* our utopian horizon exists as a real and necessary feature of our humanity. We should be focusing on what we could truly become, and we have the means to do this in a practical, realistic, scientific, and rational manner. Clearly, we need to embrace a "practical utopia" that can guide us into a transformed future.

The *Earth Constitution,* as a practical and concrete set of arrangements that can be studied and implemented, bridges that gap and makes us aware, intelligent, and thoughtful utopians. This book, therefore, integrates the themes of cosmic evolution, scientific holism, climate science, economic rationality, and moral growth with the goal of ratifying the *Earth Constitution.* It reveals interconnections and inner harmonies between our current debased human situation and the pregnant promise of the *Earth Constitution*. Recently, I wrote an article published on-line in response to the Covid-19 pandemic:

> But after this current disaster, what will the world have "moved on" to? What will it move on to after the global pandemic and global economic collapse of 2020? Will it be more of the same? More war preparations, more bio-terror research, more chance of accidents that create new pandemics? Will we return to continuing collapse of the climate while the world spends its precious resources on war and militarism and weapons research? Will we return to more government and mass-media lies and propaganda covering up a global culture of corruption and fragmentation? Isn't it clearly past time that the world begins to realize that we have no credible future at all unless we unite as a global community under democratic world law directed to the common good of everyone?

This book shows how and why we can manifest a glorious future. Chapter One describes our human situation in relation to the process of cosmic evolution. Chapters Two and Three show in some detail the dynamics of the paradigm-shift from early-modern Newtonian science to contemporary holistic science. Chapters Four and Five investigate the details of climate crisis based on the thoughts of the most advanced climate scientists and ecological economists. Chapter Six shows why the U.N. system cannot possibly address our climate crisis, and Chapters Seven and Eight show the ways in which the paradigm shift required of humankind is embodied in the *Constitution for the Federation of Earth*. The *Earth Constitution* itself is included as an Appendix.

Please reflect on all these issues as you read this book. And if you see the connections and the fundamental truth of these matters, please act on them. Become an "aware utopian," a truly "practical utopian" committed to crafting a beneficent future of peace, justice, freedom, and sustainability for our precious planet Earth, all its living creatures, and future generations. Either we realize a significant portion of our higher human potential in the near future, or we perish from Earth. Let not the whole of human history and the aspirations of thousands of generations become merely "a sound and fury, signifying nothing."

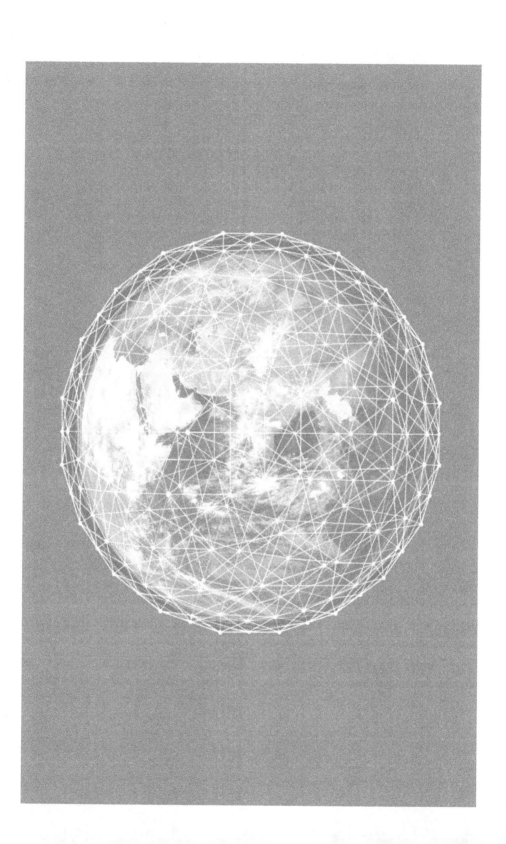

CHAPTER ONE
Cosmic Evolution and Living Design

We have not been seeing our Spaceship Earth as an integrally-designed machine which to be successful must be comprehended and serviced in total.

R. Buckminster Fuller

1.1 Design and Dimensions of Sustainability

What would it mean to see our Spaceship Earth as an integrally designed machine? Or, to speak in more contemporary terms, as an *integrally designed organism*, as a living whole, today often identified by the name of the Greek goddess Gaia? We will see that advanced Earth-systems thinkers have been speaking about Earth as a living, integrated whole for well over half a century, and that the holism of the cosmos has been recognized by advanced thinkers for more than a century. The Newtonian, static materialist model of the universe has been replaced throughout the sciences by an open, evolving, energy-based model.

The dynamic evolving holism of the cosmos, developing over billions of years, ascends to the ever-higher levels of organization seen in living creatures, beginning with their simple forms of life-consciousness and ultimately blossoming in self-aware mind. Our self-aware minds can imagine, nourish, project, and actualize transformed futures – futures that transcend past realities and establish new and better worlds. In our capacity for creative vision and transformation lies genuine hope. We can perceive the beauty of an ecologically sustainable vision for our planetary biosphere, and we can act to make it happen.

Achieving genuine sustainability, in which human beings may flourish upon our planet in perpetuity without stealing the ability to do so from future generations, requires that we treat our planet as a living whole. It requires that we organize human activities on our planet to harmonize with the ecological networks that sustain all higher forms of life on Earth. We will see that our current activities are doing the opposite. They are rapidly destroying the biospheric systems that are needed to support future generations.

Consider the subtitle of this book: *Design for a Living Planet*. Today, systems theorists discuss our planet as a living system that is rebelling against our overconsumption and desecration. They discuss the systems by which nature self-organizes all aspects of the cosmos. They discuss human economic, political, and social systems. Their general insight is that we need to be thinking about system design if we want to secure a decent human future on Earth. In this book, I will show that an adequate design for a sustainable Earth must include the following elements:

1. **Internalize Holism.** We must become deeply aware of the emergent holistic paradigm that derives not only from contemporary science but resonates with traditional wisdom about the cosmos and our place within it. This paradigm includes understanding the interdependence and interrelationship of all things in the universe, including the geosphere, biosphere, and noosphere of Earth. This awareness must inform all our design efforts.

2. **Establish Ecological Economics.** We must build an economy that benefits everyone while living within the parameters of the laws of entropy, renewability, and steady-state prosperity. We will examine what several ecological economists have to say about sustainable economics.

3. **Promote Spiritual Growth.** We must design a world system that promotes cognitive, moral, and spiritual growth that nurtures those aspects of human nature that transcend selfishness, egocentrism, fear and hate, while promoting those aspects that include love, compassion, dialogue, mutual respect, and concern.

4. **Institutionalize Planetary Applied Science.** We must create institutions – at the regional, national, and planetary level – that study and monitor Earth as a whole and that can respond in a timely fashion to maintain stability. Climate scientists must be empowered to manage our environment for health and stability. As Fuller says, this means seeing our spaceship as a whole. The good news is that we now have the expertise and technology to implement regenerative and sustainable practices.

5. **Give Earth a Brain.** We must generate a world noosphere (mind-sphere) that includes a coordinating center, a brain, and a common consciousness that is representative of humanity and capable of protecting stability, justice, and equity. This brain center also must be capable of planning

and acting for a stable, prosperous, and resilient future. To date, the world has no such coordinating self-awareness singularly focused on creating a decent and secure future for humanity.

My earnest belief is that the *Constitution for the Federation of Earth*, which is reprinted in the Appendix, provides our best bet for achieving all of these design features – each one necessary to achieve sustainability within the narrow time frame still available to humanity before climate change becomes unstoppable and our planet becomes uninhabitable. We will examine the details of climate change and its consequences, including what most climate experts and environmental leaders say about the current condition of Earth and its likely future if we fail to convert rapidly to a sustainable world system.

In sum, these five design features for sustainability clarify the goals of this book. They all are interrelated and interdependent, such that they cannot be treated separately. Therefore, please keep them in mind as you continue reading. My hope is that it will become clear why the *Earth Constitution* is central to achieving all these sustainability design goals.

1.2 Mind and Cosmic Holism

Scientists, biologists, and ecologists agree that life on our planet is a single, holistic biosphere. Over its 4.6 billion years of existence, Earth has evolved into dynamic unity, filled with diverse and integrated geological features, including its core, mantle, surface, oceans, rivers, landmasses, glaciers, polar caps, atmosphere, etc. (see Lenton 2016). These integrated features are famously known as the "geosphere." The geosphere dynamically integrates with our planet's web of life that encompasses the Earth's forests, grasslands, algae, phytoplankton, corals, bacteria, and living creatures in endless variety. Together, the geosphere and web of life constitute a dynamic living planet or "biosphere" (Hazen 2012).

Climate scientists have uncovered the myriad ways in which the web of life, over its 3.8 billion years of evolution, has harmoniously molded and modified the geosphere to develop and maintain life, making the planet a whole living system conducive to life and to its evolution into evermore complex forms. The most recent and most complex form of life is human life. Humanity is not only aware of itself, but aware of its relationships to all others and the world. As such, we possess an astonishingly creative potential beyond any other life form.

With human beings the vast underlying organizational intelligence of the evolutionary process has become self-conscious. With human beings a new level of connectivity and integration has appeared that surrounds and permeates both geosphere and biosphere. This is widely known as the "noosphere," from the Greek word for "mind" and the French word "sphere" (see Teilhard de Chardin 1975; Martin 2018, 22-3). "When we, or any other creature are truly thinking," writes philosopher Joel Kovel, "we are thinking in respect to the Whole; there is a sense in which it can also be said that the Whole is thinking through us" (2007, 114).

Mind is a self-aware manifestation of the dynamic, fractal-based, organizational intelligence found everywhere in nature. Mind now permeates the biosphere in energy wave-packets organized to carry *information* that fills every space on Earth's surface. This organized energy can be received and decoded by cell-phones, computers, radios, televisions, GPS systems, and other instruments that record and transmit the noosphere's vast content. Self-aware mind also produces the cities, roads, malls, electrical grids, industries, ships, and trains that cover the surface of our planet.

The holism of the cosmos evolved the geosphere, biosphere, and noosphere of Earth – an organized, networked, and integrated intelligence system. Today, humans possess the power to consciously evolve and co-create this matrix of life.

The holistic cosmos – informed by an integral, non-local, non-material coherence beyond anything we can conceive with practical imagination – gave birth to our Sun and planetary geosphere some 4.6 to 5 billion years ago. The geosphere has evolved the biosphere, the biosphere has evolved the noosphere, and they all continue to reciprocally evolve as a whole. Or, to put it differently, the holism of the cosmos has actualized itself throughout these three levels. However, with self-aware mind the evolutionary process also has created a *custodian, a steward, a protector, and a transformer –* a creature who can understand the universe with its holistic processes and who can harmonize (or not) with these processes, for better or for worse. "Holism," of course, does not mean complete and finished, but undivided. Holism constitutes the astonishingly deep coherence of the evolving cosmos and its offspring, life on Earth.

We are capable of blindly operating in a thousand fragmented ways that ultimately destroy the biosphere's integrated networks that sustain our lives. We are also capable, as Fuller puts it above, of comprehending the holism of our planet and "servicing" its integrated biospheric networks for health and sustainable living. His point is well taken, but the metaphor has changed. Our "spaceship" is not so much a vehicle that needs servicing but rather a living Gaia that must remain healthy, must be loved and cherished, and with which we must harmonize. We must understand how it operates and maintains its integrated and holistic mode of being. We also must ensure that humanity conforms to its processes and parameters.

However, understanding how Gaia works is not the end of the matter. We must recover the sense of human dignity, the awareness that we are more than merely rational animals who compete with one another through egoistic self-interest or collective nation-state identifications. There is a depth to human life that has been ignored by much of contemporary civilization. We are a keystone in this astonishing universe, not merely a custodian or protector of Gaia. Humans are *transformers* of life and *redeemers* of our planetary home into something precious, sacred, and valued for itself. Earth is our home and it frames and enriches our lives and that of future generations. This book is about how we manifest our human dignity in a concrete form.

Whichever metaphor we use, custodian or transformer, currently we humans are failing. E. O. Wilson alludes to the "four horsemen of the environmental apocalypse" (see Bormann, ed. 1991, 3-4). The first horseman is toxic waste, much of it in the form of invisible manmade chemicals that penetrate into every tissue of our bodies and every corner of Earth. The second is global warming, which is transforming our livable planet into a place of drought, desert, wildfire, and wasteland. The third horseman is ozone depletion that turns the naturally life-giving warmth of our Sun into toxic radiation, causing blindness in animals and cancer in humans. But the fourth horseman – just as in the *Book of Revelation* – is death itself, resulting from massive extinction of species and the very real possibility of extinction of our own human species.

Today, we are aware of many more factors than those captured in this image of the horsemen of the apocalypse. These factors include ocean acidification and rising sea levels, air pollution (atmospheric aerosol loading), biodiversity loss and loss of ecological functions, rapidly growing freshwater scarcity, deforestation and degenerative land conversion, nitrogen and phosphorus loading, acid rain, and over extraction of non-renewable resources to name a few. We will examine these phenomena and their absolutely serious consequences in detail later in this volume.

In short, we are allowing our home planet – the only home we will ever know – to die. We are failing to transform our thinking, our noosphere, to the holistic paradigm that has emerged across the board from the sciences over the past century. We have the capacity for "altruistic action" as David Sloan Wilson points out, if we *design* our relationships and governance systems properly (2015). We have the capacity for transformational action as David Korten points out, "to form networks of partnership, sharing, and mutual learning [that] … are integral not only to life, but as well to the whole of Creation" (2009, 23). But can it happen in time before the climate change feedback mechanisms grow beyond any possible human response?

Systems theorist and cosmologist Ervin Laszlo traces the scientific development of the concept of the deep "coherence" of all things in the universe, including all living organisms. "What happens to one cell or organ also happens in some way to all other cells and organs" (2007, 43). Laszlo writes:

> The German theoretician Marco Bischof summed up the insight emerging at the frontiers of the life sciences. "Quantum mechanics has established the primacy of the inseparable whole. For this reason ... the basis of the new biophysics must be the insight into the fundamental interconnectedness *within* the organism as well as *between* organisms, and that of the organism *with its environment.*" (2007, 49)

The holistic interconnectedness and coherence of Earth, including organisms within themselves and with their environment, have been perhaps the most fundamental discoveries of science. The only species we know of that *does not* coherently connect among its members and *does not* coherently connect with its environment is *Homo sapiens*. Tragically, humanity has yet to shift its consciousness from fragmentation to the holistic patterns of thinking called for by scientific breakthroughs. The miracle of cosmic life is awakening within us, but we must advance quickly, otherwise we will squander the time needed to become self-aware transformers of our shared destiny. Physicist Henry Stapp wirtes:

> Thus a radical shift in the physics-based conception of man from that of an isolated mechanical automaton to that of an integral participant in a non-local holistic process that gives form and meaning to the evolving universe is a seismic event of potentially momentous proportions. (Stapp 2011, 140)

Quantum physics has revealed the fact that the universe is informed by mind (intelligence) and that our minds are participants in a non-local cosmic process. We now understand that we have a unique role to play in the emergence, evolution, and protection of life on our planet. Yet, we have not allowed the implications of these discoveries to inform our lives and our activities. Human beings, using their minds, have colonized Earth with subjective self-awareness, dynamic cultures and languages, modes of production and distribution, scientific insights and methods, along with planet-wide political, economic, and bureaucratic institutional arrangements (Martin 2018, Chap. 1). However, this colonization has historically been a chaotic, haphazard, contingent process.

In large measure, we have colonized our planet at a relatively low level of self-awareness, barely conscious of nature's holism. As a result, we have given rise to climate change and on-going ecosystem collapse, as well as endless wars and the threat of nuclear holocaust. Fortunately, many of us are beginning to recognize ourselves as participants in "non-local holistic processes," as Stapp predicted. Our minds derive from and are now connecting with the integral foundations behind the entire evolutionary process. To truly think is to both think in terms of the Whole and recognize that the Whole is thinking in and through us. The significance of these insights are truly momentous, as Stapp observes, and they open up the horizon of the immense transformative possibilities described in this book and in my earlier texts, such as *Global Democracy and Human Self-Transcendence* (2018).

1.3 Our Human Existential Situation

The emergence of life on Earth some 3.8 billion years ago signaled the emergence of a phenomenon in which *form*, as an organized arrangement of molecules, became distinct from its surroundings. What we call "life" developed as form becoming distinct, highly organized, and self-perpetuating. One-celled creatures could preserve their form through taking in nourishment and excreting waste through a primitive *awareness* and *responsiveness* to their surroundings (Jonas 1984).

This developmental process was exceedingly slow for the first two billion years. Yet there had already begun what scientist Robert M. Hazen declares, "the astonishing coevolution of the geosphere and biosphere" (2012, 151). Then around 900 million years ago, life began to increase in complexity. Single cells merged to become multi-celled life forms, as our physical planet (the geosphere) continued to evolve through interaction with life (the biosphere). This ecologically interdependent network of living systems colonized the planet, thereby transforming and interacting with the geosphere itself (Hazen 2012; Redfern 2003; Maslin 2013).

Next, primitive consciousness arose, as a multitude of creatures began to proliferate and colonize the seas, lands, and atmosphere of Earth. Eventually, the interactions between the geosphere and biosphere produced the noosphere – true planetary consciousness. Only minutes ago on this immense time scale, human beings emerged as self-aware creatures and the most sophisticated development of form. By some scientific estimates there are 37.2 trillion cells in the human body, yet this vast complexity – divided into innumerable internal organs, tissues, processes, and systems – operates with astonishing unity, remarkable independence from local conditions, and the wonderful capacity for physical and mental action that we experience daily.

Despite the evolutionary impulse of unity in complexity, humans developed a unique form of consciousness that was no longer part of the vast oneness of life. The various levels of plant and animal consciousness remained, so to speak, within the womb of nature. We alone became self-aware and fell out of the primitive unity of consciousness within the biosphere and geosphere. We have the astonishing capacity to separate subject from object, personal perspective from objective reality. Mind is able to stand back from its experience and ask questions about how the world works, how the world is structured, and what constitutes reality. We became capable of asking: Who am I? What is the meaning of life and existence? How is my "I" related to the "I" of Being? We also are capable of intuiting the sacred foundations of all being and life, as mystics of every religion through history have testified.

Psychologist Erich Fromm points out that this "separation" from nature through self-awareness functions simultaneously as our greatest gift and our greatest curse. For Fromm, this separation is the root of depression, anxiety, fear, anger, conflict, and a great longing to "return to the womb" of nature, to a time before such anxieties and problems arose from self-awareness. He interprets the Garden of Eden story in the Bible in this way as well. We emerged from paradise (innocence) into self-awareness (knowledge of good and evil), and this led to the "fall" from paradise into our human history of struggle, conflict, war and separation, with one tragedy following another throughout the eons (1981, Chaps. 1-2).

As a psychoanalyst and social philosopher, Fromm saw 20th century man responding to heightened self-awareness in regressive, destructive, and unhealthy ways. In addition, he saw us attempting to satisfy needs for relatedness, rootedness, transcendence, identity, and a meaningful cognitive framework by seeking escape from or a reduction of self-awareness in a variety of regressive and destructive ways, from drugs or a narrow focus on personal life, to nationalistic fanaticism and fundamentalist religions. These extremes, ultimately rooted in poor self-awareness, constitute our fundamental human dilemma.

Consequently, we must face the problem of self-awareness squarely and rediscover how to be at home in the world. Yet today, in the face of our global crises, we must now respond in ways commensurate with our immense human potential to embrace life and the world, and to live in harmony with nature and one another. Fromm maintains that our higher potential for health and success requires developing "our reason and our love" to the point where we feel fully at home in the world. We can grow to the point where we live in harmony with the vast beauty, majesty, and intelligence of existence:

> Loving permits me to transcend my individualized existence and at the same time to experience myself as the bearer of the active powers which constitute the act of loving. What matters is the particular *quality* of loving, not the object. Love is in the experience of human solidarity with our fellow creatures, it is in the erotic love of man and woman, in the love of the mother for her child, and also in the love of oneself, as a human being; it is in the mystical experience of union. In the act of loving, I am one with All, and yet I am myself, a unique, separate, limited, mortal human being. Indeed, out of the very polarity between separateness and union, love is born and reborn. (1981, 3)

Of course, our self-aware separation from nature also allows us to "conquer" nature: to engineer, manipulate, create, and recreate. It allows us to "conquer" one another: to dominate, exploit, marginalize, and discriminate. It allows us to fragment the world into nations, races, classes, friends, and enemies. It makes possible endless wars and mutual self-destruction. It also allows us to love, bond with, consciously cooperate with, and organizationally unite human beings into families, communities, cities, regions, nations, and one planetary civilization. And our self-awareness opens something even deeper: We awaken to a sacred holism more fundamental than all this fragmentation, conflict, and chaos.

Spiritual thinker Raimon Panikkar says that rather than defining ourselves as "human" beings (a being that has distinctively human qualities), we should see ourselves as "humane" beings. "It is written that to give a glass of water may be enough for eternal life. ... Neither the past, nor the future, nor even the thought of eternity is paramount here, but just that moment, or simply, my life" (2013, 297, 300). Our immediate self-awareness awakens in moments of simplicity, of recognition of the other as myself, moments of innocence, beauty, and truth in the act of living.

Our reason can grow to understand the deep dynamics of life and the deep intelligence of the universe around us. And our love can grow to the point where we feel related and responsive to all the dimensions of existence. Our simplicity, self-awareness, and direct intuition (through the "third eye") can also deepen. Our awareness of the immediate depths, flow, and mystery of life can grow or dawn upon us in moments of awakening to the point where something entirely new shows up.

Reason, love, and immediate awareness form a concatenation of fundamental human potentials that together open us to the depths of existence. We begin to understand that human beings are not simply one contingent species whose self-extinction through nuclear holocaust or climate collapse matters little in the vastness of the universe. Reason, love, and intuition reveal, actualize, and connect us with the source of Being and the metaphysical depth of the All. Ultimately, at this level, language fails us. Nevertheless, for heuristic purposes let us follow the lead of John Hick, renowned scholar of religions, and use the word "Real" (2004). Hick affirms that "the ideal of love, compassion, generosity, and mercy has always been a basic factor in the recognition of someone as an authentic mediator of the Real" (2004, 326). The Real, for Hick, represents the ground of Being in the cosmos recognized by all the great world religions. The Real is manifested in the coherence and integrated harmony of the whole manifested in every part and every creature.

> *Reason, love, and intuition enable us to actualize our human potential and connect with the Source of Being. However, if we refuse to harmonize with Earth and forego cosmic maturity, our extinction is almost certain. Continued fragmentation will cause a metaphysical tragedy.*

The fullness and glory of existence can manifest itself in the joy of living and being as expressed by those who have developed cosmic consciousness, by those who mediate through their lives the depths of the Real. Human beings can awaken to themselves as living manifestations of the Real. Love and transformative action for holism flow forth from those aware of our human mind as "a non-local holistic process that gives form and meaning to the evolving universe." The integration of love, reason, and intuition makes us capable of creative transformation. These capacities themselves are metaphysical, manifesting the Real. We begin to understand reason, love, and intuition as gifts, as expressions of the divine ground of Being itself. The extinction of human life would be a metaphysical tragedy, not just a human one. What we are reflects the depths of reality. We are infinitely more than meets the eye.

We can be in harmony with the cosmos, the biosphere and one another, or we can frustrate that harmony through immature egoistic identifications with group, cult, nation, race, or some other fragmentation of the whole. Our capacity for such fragmentation arises from a relatively recent development in the evolution of consciousness over the past few millennia. At the same time, this capacity can be seen as a mere prelude to another level of consciousness in which love, reason, and intuition have reunited us with one another, with our home planet, and with the cosmos that gave birth to us.

Out of the perhaps two million years that the genus *homo* has been on the planet, our single universal species named *Homo sapiens* (all nearly genetically identical with one another) emerged to colonize the globe. The first signs of an emerging self-awareness began only about 20,000 to 40,000 years ago in the Paleolithic cave paintings that were created in various regions of the world during this period. This early "magical consciousness" widely transformed into a "mythological consciousness" some 12,000 years ago, giving rise to the great river valley civilizations along the Nile, the Yellow, the Indus, and the Tigris-Euphrates Rivers in the very ancient world. The era of mythological consciousness lasted perhaps 10,000 years. The fullness of our present self-aware condition only emerged during the famous Axial Period that occurred some 2,500 years ago (Jaspers 1953; Swimme and Berry 1992; Hick 2004).

During the Axial Period, specifically between about 800 and 200 BCE, an emergent self-aware consciousness was born more or less simultaneously all over the world. This emergent awareness included great transformative figures such as the Buddha and the author of the *Bhagavad Gita* in India, Lao Tzu and Confucius in China, Zoroaster in Persia, a number of Hebrew Prophets in the Near East such as Amos, Jeremiah, Ezekiel, and Isaiah, and ancient Greek philosophers such as Plato and Aristotle in the Mediterranean world (Martin 2005, Chap. One). Jesus and Muhammed came after this Axial Period. However, both these later teachers embodied this same emergent self-awareness with its post-mythological orientation.

Generally speaking, the emergent self-awareness of human beings since the Axial Period has not been complete enough to reintegrate with the intelligence everywhere embodied in the non-self-aware consciousness of the life forms that encompass Earth. Many in the ancient world, such as Plato and Aristotle, recognized the intelligence (*logos, nous*) in the universe (Martin 2008). Yet today, we are tasked to grow toward a self-aware maturity that recognizes cosmic intelligence and to ask the question of whether we are living in harmony with that intelligence or in conflict with it.

The noosphere developed out of the biosphere as the biosphere did from the geosphere, all mutually interacting and evolving together. The self-aware noosphere over these past 2,500 years colonized the entire planet, operating without much knowledge of Earth's ecosystems or the coherence of the biosphere which supports all life. We have only discovered the immense history of our planet's evolution, and our own codependence with the geosphere and biosphere during the past century or so.

We have not yet assimilated our holistic interdependence with the biosphere and geosphere. Especially since the rise of science in the 17th century through the present, we have operated with vast fragmentation and focused on disparate egoistic projects, initiatives, racisms, nationalities, religious dogmatisms, wars, and conquests. This process of fragmentation along with the development of powerful technologies has resulted in the crises humanity is now facing. In the light of our new understanding of our human situation, we now are tasked to consciously unite as a common humanity responsible for the quality of life everywhere on the planet.

The self-aware noosphere has given human beings *power over* the biosphere and geosphere that no other creatures possess. The separation of subject from object that emerged during the Axial Period allowed us to conceive of the world and other living creatures as objects subject to our domination and manipulation. Our obsession with this power and our limited awareness of the intelligence patterns inherent in nature that support and sustain life have led us into conflict with the biosphere, with one another, and even with the geosphere.

Starting with the Industrial Revolution, our power over nature has increased to such immense proportions that we now have disrupted the ecological support systems that make higher forms of life possible. We have begun to understand that much of the misinformation that permeates our globe has been based on paradigms, assumptions, and perspectives that are not in harmony with the biosphere's immanent organizational intelligence. Yet, as many contemporary thinkers have pointed out, we know that our love and our reason could transform civilization into patterns of harmony and sustainability. We have this immense human potential that largely remains unactualized.

The dominant economic model of the past several centuries coincided with this industrial power *over* nature and reduced all human activities to materialist premises, as economist E. F. Schumacher points out. What was "uneconomic" and could not be reduced to a market-price was ignored, and what was "economic" was fostered as "the single-minded pursuit of wealth," preventing civilization from solving "the most elementary problems of everyday existence." For our truly human problems require, "not endless material growth, but wisdom" (1973, 30-33).

Today, we are realizing that human civilization is a *manifested mind* that has colonized our entire planet, largely in disharmony with the biosphere. Mind has not yet understood the deeper intelligence at the heart of the evolutionary life process, an intelligence that had emerged from the evolutionary process taking place in the cosmos itself, ever since the Big Bang phenomenon some 13.8 billion years ago (Swimme and Berry, 1992). This intelligence ultimately includes both love and intuition. We lack wisdom about what is valuable in life, about what cannot be quantified nor satisfied by ever more material goods within a so-called "higher standard of living."

Sadly, most humans fail to perceive the coherence of life and civilization using their reason, love and intuition. Subject and object, separated from one another a mere 2,500 years ago, have not yet united in a fuller perception of wholeness. In truth, however, subject and object form two dimensions within the wholeness of consciousness, inseparably coherent and connected with one another. Human life cannot be reduced to industrialized materialism. Our ascent to self-awareness within the universe has led thinkers to declare that we are made in the "image of God" or that we are expressions of "Buddha nature." There is something very special about this ascent that has been ignored by modern civilization.

Today, the dominant modes of human civilization are destroying both the biosphere and the geosphere that have evolved over 4.6 billion years on this planet. We are tearing apart the integrated evolutionary networks that have supported life in evermore complex and intricate forms. If we do not want to cause the rapid extinction of most higher life forms on Earth – including human life – we must convert the paradigms, assumptions, and perspectives that now dominate the self-aware noosphere to a *planetary consciousness* in harmony with the organizational intelligence at the heart of the biosphere. Such a planetary consciousness would necessarily require wisdom, that is, non-economic, truly human, values and awareness.

For a sustainable civilization to occur, the noosphere must transform its paradigms, assumptions, and perspectives to evolve the *holosphere* – a *holarchical consciousness* that harmoniously integrates the noosphere, biosphere, and geosphere. Love, as Fromm points out, does not obliterate the separateness of the subject and object of love (1981, 3). It combines the two in a unity that embraces and protects that diversity. Human beings must unite on our home planet within a world system that unites and embraces this vast diversity. As we will see, humanity can indeed substantially institutionalize the love-reason relationship within our political and economic arrangements.

1.4 Emergence of Cosmic Consciousness and the Earth Federation Movement

Some thinkers have called the deep paradigm or higher awareness that harmonizes with the biosphere "integral science" (Goerner, et al., 2008). Others have called it a new "ecozoic awareness" (Swimme and Berry 1992). Still others have called it "biosphere consciousness" (Rifkin 2019). With this new level of awareness comes the capacity for "conscious evolution," a term often associated with Barbara Marx Hubbard and her book by that title. She writes:

> We see that planet Earth is herself a whole system. We are being integrated into one interactive, interfeeling body by the same force of evolution that drew atom to atom and cell to cell. Every tendency in us toward greater wholeness, unity, and interconnectedness is reinforced by nature's tendency toward holism. Integration is inherent in the process of evolution. Unity does not mean homogeneity, however. Unity increases diversity. We are becoming ever more connected as a planet while we seek further individuality for our cultures, ethnic groups, and our selves. (1998, 48-49)

The tendency in nature, as Teilhard de Chardin pointed out, is integration moving through ever higher levels of complexity-consciousness. This process is working within human civilization as well, bringing us to higher levels of integration. The noosphere synergistically integrates as a holosphere. Diversity is increased and protected through the process of integration. The many diverse parts are absolutely essential to the whole, while the whole is more than just the sum of its parts. However, human civilization has not yet united as a whole. The way to true unity in diversity remains elusive.

Global systems thinker Ervin Laszlo calls awareness of this process as it emerges in us "planetary consciousness" (2007, 139). Our emerging planetary consciousness corresponds to planet-wide, nearly instantaneous communication systems and global interdependence: "Social systems, like systems in nature, form 'holarchies.' These are multi-level flexibly coordinated structures that act as wholes despite their complexity. There are many levels, and yet there is integration. ... The world communicates practically instantaneously and becomes ... a global village. More exactly, it becomes a global holarchy" (2002, 51-2).

> Organization in nature comes to resemble a holarchic pyramid, with many relatively simple systems at the bottom and a few complex systems at the top. Between them all natural systems take intermediate positions: they link the levels below and above them. They are wholes in regard to their parts, and parts with respect to higher-level wholes. (Ibid.)

Scholars such as Laszlo represent a growing chorus of voices calling for a planetary consciousness shift from egocentric and anthropocentric modes of thought to planet centered, holistic, and integral modes of thought in harmony with the intelligence of our planetary biosphere. Such a transformed culture and consciousness would harmonize human civilization with the biosphere, and it would be sustainable, protecting higher forms of life and future generations. Karl Marx had already called for awareness of our "species-being." We must begin thinking like a species and, beyond that, we must begin thinking and experiencing like the cosmic intelligence that animates us and all other living things.

The noosphere converges into a holosphere and the holosphere into a holarchy. Humankind and the cosmic evolutionary process on planet Earth develop not only a brain, but a self-aware brain, a center from which we may coordinate and govern ourselves. Hence, we ought to organize the noosphere for global governance of the whole of humanity. Human beings as self-transcending embodiments of the divine cosmic principle actualize their potential to mirror the fractal-based structure of the universe that links higher and lower levels of organization in dynamic solidarity (Harris 1977, Chap. 3). We need to organize humanity holarchically and we need to see this project as somehow sacred and fundamental to the sanctity of life.

What would it look like if we federated Earth into a holarchy – from the lowest community levels, through regional levels, national levels, all the way to the global level? Such an emergent Earth Federation government would represent the worldcentric self-awareness of humanity. Human intelligence, now representing the sovereign authority of the people of Earth and reflecting the holism of humanity, could effectively address the elimination of war, the protection of universal rights, and the problems of establishing a sustainable world system.

Using the model of human development affirmed by psychologists and planetary thinkers such as Gene Gebser, Pierre Teilhard de Chardin, Erich Fromm, Raimon Panikkar, Ervin Laszlo, Ken Wilber, Abraham Maslow, Lawrence Kohlberg, Carol Gilligan, and Jürgen Habermas (cf. Martin 2018, 10-15), we can imagine the evolution of selfhood from immaturity to cosmically aware maturity as following the basic pattern outlined in the chart below.

In the process of cognitive, moral, and spiritual growth, the "I" comes to realize that the entire cosmos is presupposed by both the subject and the object, since these are united within what Immanuel Kant called the "transcendental unity of apperception." The referent "I" – our thinking, feeling, experiencing, dreaming sense of self – changes as we grow, as the worldcentric and cosmocentric levels become ever more transpersonal. Once we reach the transpersonal level, the unity of apperception is no longer confused with the ego or psychological self. The deeper transpersonal selfhood opens beyond the psychological self to awareness in which subject and object become ever more fully united and experienced as one. Often, this level of awareness is called "cosmic consciousness."

Figure 1 **Referent "I" Transforms** **as Development Ascends**
Cosmocentric *I live as a self-aware expression of the cosmic intelligence* (transpersonal and unconditional love)
The limits of language begin to appear. All experience and language itself is informed by the deep mystery and flow of being, of which I am a self-aware expression.
Worldcentric *I live as a self-aware expression of humanity* (love of humanity)
I realize that what is most true about me is reflected in the universality embedded within all languages: our human dignity, human rights.
Ethnocentric *I live as a member of my country, my church, or my community* (love of community)
My language reflects my culture and my background. I recognize that there are other cultures and other languages, but they are alien to mine, less reflective of the apparent truth embodied by my community.
Egocentric *I live to satisfy myself with pleasure, money, security, or power* (love of self)
Language here is often thought of as literal, as if the "I" refers to something atomistically real about myself.

As I clarified in *Global Democracy and Human Self-Transcendence,* the process of self-transformation and transcendence is built into the very foundations of human temporality. We live within a present, in which both past and present project themselves toward a transformed future. At higher levels of growth, we begin to understand that subject and object are related and, ultimately, are one. Myself and the other are one, as Mahatma Gandhi and many others have declared. Philosopher Errol E. Harris draws the same conclusion from the point of view of a scientifically informed rationalism:

> Through the physicochemical conditions prevailing on the Earth, the whole physical universe is registered in the biosphere and in each and every organism within it. ... The whole is the same throughout the scale of its self-specifications, each of which is a distinct manifestation of its immanent totality. ... The subject and its experience are one, and its experience is of an external world that is actually the same subject in becoming. ... The full philosophical realization of this dialectical relation is thus the identity of subject and object. (2000a, 286-87)

This paragraph underscores that as the "I" ascends to a cosmocentric consciousness, the non-local holistic consciousness informs me that I am "immanent totality." Becoming ever more aware of this true reality defines the process of growth toward cosmic consciousness. One begins to clearly feel oneself as a concrete embodiment of the whole, a particularized incarnation of the ever creatively flowing, cosmic-divine mystery. This ever-increasing oneness with all humanity, the biosphere, and the cosmos provides the bliss of knowing our immense human dignity.

Medieval Christian thinker St. Thomas Aquinas recognized the "I" as a unity of consciousness directly related to the whole of existence. In *De Anima* Thomas wrote: "It appertains to the nature of the intellect to be one, that is, not to be directed to anything. Its only nature is that it can be everything" (in Panikkar 2013, 382). The intellect, coalescing in the "I" of the knower, comprehends that it can be everything. Raimon Panikkar expresses this insight another way:

> Intelligence is our highest gift. Man awakens as man by an act of consciousness. Consciousness is the ultimate and irreducible bulwark of our openness to reality, the inseparable companion of Man. We know things because things are knowable; they are knowable because between things and our intellect there is a primordial relationship, which we have not established. ... We can be aware of the Whole that embraces us and that as a microcosm we are an icon of the entire reality. (2013, 146, 212)

For Panikkar, this consciousness includes the awareness of a primordial relationship between the world and ourselves. He does not call this "cosmic" but "cosmotheandric," a consciousness that includes self, world, and God. Similarly, in his classic work entitled *Cosmic Consciousness,* Richard Maurice Bucke describes three distinct levels of consciousness beginning with the higher animals. Higher animals have "simple consciousness," an immediate awareness of their surroundings but no consciousness of having a self that is aware of both itself and its surroundings.

The second level arises with the "self-consciousness" of human beings. We think and know, and we know that we think and know. Subject and object have become distinct, yet intimately related. From the Axial Age to the present, human self-awareness has allowed us to colonize and conquer the entire Earth. The third level is "cosmic consciousness":

> Cosmic Consciousness is a third form which is as far above Self Consciousness as is that above Simple Consciousness. With this form, of course, both simple and self consciousness persist (as simple consciousness persists when self consciousness is acquired), but added to them is the new faculty .... The prime characteristic of cosmic consciousness is, as its name implies, a consciousness of the cosmos, that is, of the life and order of the universe. (1974, 2)

It is as if a new race of men is emerging, Bucke declares, who "will occupy and possess the Earth," for they alone will be reflections and embodiments of the universal evolutionary cosmic process that has emerged in and through us. To become sustainable, human beings must "live as self-aware expressions of the cosmic intelligence." It is this immense potential within us that must ultimately be actualized if we are to overcome climate collapse and achieve a flourishing sustainable civilization. Harmony with the biosphere will come about only when human civilization is in harmony with the cosmic intelligence at the heart of the biosphere.

One must be sensitive, of course, to Bucke's mode of expression here, which sounds elitist and might appear to imply some form of domination. What informs awareness in so-called "cosmic consciousness" is the opening up and transformation of the sense of "I" to the point where the "I" of domination substantially disappears. Whether personal or cultural, domination is much more likely to appear on the lower egocentric and ethnocentric levels. For people entering universal and transpersonal levels, love merges with the impersonal compassion named, for example, by the *karuna* of Buddhism or the *agape* of Christianity. Love is willing to suffer in self-sacrifice (*tapasya*), as Mahatma Gandhi explained, so that others may be spared suffering.

Cognitive scientist Steven Pinker shows at length that thought and language are not the same (1995, Chap. 7). "I" is a place-marker in the grammatical universe. As philosopher Ludwig Wittgenstein puts it, "I is not the name of a person" (1958, 410). Buddhism has long understood these truths as evidenced by its famous *anatta* (no-self) doctrine.

Drawing on Buddhism, philosopher Nolan Pliny Jacobson identifies the egocentric confusion of the self and the word "I" as both immature and neurotic. Through growth we come to understand the possibility "that the Self may be an instrument of impoverishment and alienation from reality and life, and that richness and fulfillment of life and health can be realized only on condition that the control of this Self can be loosened and transcended" (1974, 89). Here we have the process of growth in a nutshell.

The false sense of "I" at the egocentric and ethnocentric levels fragments reality into incommensurable parts. For example: I (here) and you (over there). Borders protect "my" territory and "my" reality from what is other, what is not me and mine. Your very existence is a threat and danger to me. As a result, the unbroken wholeness and flow of creative existence is irrevocably bifurcated. The world becomes atomized chaos and our common human dignity is obscured.

Neither harmony with one another nor with our planetary biosphere will happen through a haphazard process of concerned writers exhorting us to achieve "biospheric consciousness" or "planetary consciousness." Such pleas have been going on for nearly a century with little effect. Similarly, harmony with the biosphere will not occur via rebellions like Extinction Rebellion (www. Rebellion.global) which attempt to block the business as usual practices that are leading us toward destruction. Courageous activists at Greenpeace also utilize an aggressive style, but their success is questionable.

Attempts at disrupting business as usual may be imperative at one level. As Naomi Klein pointed out so well in her book *This Changes Everything* (2014), resistance groups around the planet have had a modicum of success interrupting insane practices. However, without a vision of how to change the global political and economic system to sustainability, rebellion alone will not suffice. We need *transformative action* directed toward what I call "deep sustainability" (2020, Chap. 3). Meaningful action requires a new social-political-economic system that can effectively produce and sustain deep changes in both human behavior and human consciousness, while supporting practices necessary for our survival and the harmonious perpetuation of the biosphere.

In sum, the solution to the environmental crisis is necessarily connected with human moral growth and spiritual maturity. Laszlo affirms that "to integrate, harmonize, and unify all things, and at the same time embrace all things in oneness and love, is the *Telos* of all existence. … *It is why we are here*" (2017, 47). This idea of the "*Telos* of all existence" means that we have a uniquely creative role in the emergent evolutionary cosmos. The issue is how to unlock humanity's dignity and our collective destiny as co-creators of a truly transformed future.

What we need is transformative action based on systems theory and holistic practice. Theory and practice are intertwined in our quest to actualize a revolutionary vision of a redeemed humanity living in harmony with one another and with Earth. We don't know exactly what such a civilization will look like, as real and imperative as the vision may be, though now is the time to engage in practices that will inform theory, and vice versa. We do this by engaging in a dynamic learning and growing process guided by the broad vision of holism, harmony, and sustainability – a vision that is inherently within us as self-aware embodiments of the Whole.

This will only happen if we manage to transcend the fragmented forms of thinking that today dominate our planet. We also must transcend the outdated institutions that perpetuate and embody archaic forms of thinking. Human beings need to unite under a collective sense of our human selfhood, bound together by a common vision of planetary civilization. We need to adopt a *global* system that makes lasting transformation possible, thereby transforming not only our world systems, but ourselves as well.

As we will see, one key to transformative praxis that brings about a new world, a human renaissance, is love. Love is not simply a subjective feature arbitrarily added to our knowledge and understanding of our existence. Rather, love involves an active engagement of our whole being, our deeper selfhood, that already lives with a primal connection to the cosmos and the ground of Being. Hence, rebellion, important as it is, will not get us to our goal. For true transformation, we need love of life and love of one another.

1.5 World Systems and Life-Promoting Designs

Overcoming institutional fragmentation and the distorted selfhood of human beings associated with these institutions means synthesizing the principle of unity in diversity and bringing human consciousness to worldcentric and cosmocentric levels. Conscious evolution for ourselves as individuals or within our local social groups will not suffice. It must become a worldwide and universal process if we are to succeed.

The world needs to become a community dedicated to self-actualization of our higher human potential. There are direct relationships between the way we treat our world, the way we relate to others, and the way we understand ourselves. We need to be thinking about how we can design our world to promote not only sustainability but also human self-realization. How can we treat others as true brothers and sisters living within our common home Earth?

Social thinker Peter Gabel critiques our liberal but overly individualistic social paradigm, and he argues that it denies what is most real about us: that we are "mutually constituting social beings perpetually knitting each other together through the inter-experience of mutual recognition into a fabric of interconnectedness that is social through and through" (2013, 6). Our consciousness embodies this call to mutual recognition and loving connection with one another. But the design of our world – our legal systems, militarized nation-state systems, lethal economic system of absolute winners and losers – kills this connection and distorts our perception. Our personal, local, ethnic, national, and religious identities disconnect us from one another.

Fortunately, our collective consciousness contains the longing and the imperative for mutual recognition, planetary justice, and the "Truth," as Mahatma Gandhi put it, that we are all brothers and sisters within one human community. For Gabel, "law must maintain its connection to justice by following an ethical intuition anchoring the present to the future, an intuition of what we are in our being but are not yet in reality" (ibid., 19). At the worldcentric and cosmocentric levels, we become "in our being" manifest social creatures, capable of mutual recognition and love, whose diversity is constituted through and inseparable from our unity. Law must allow for this vision and this growth. Spirituality and ethics draw us toward a transformed future, and we must design our institutions, including our legal institutions, to make this growth toward maturity possible.

Consequently, we need to design a world system premised on holarchical unity in diversity. Such a world system will naturally promote a deep and rapid transformation of human consciousness to worldcentric and cosmocentric levels. Our political and economic institutions need to harmonize with the ecological patterns and processes that represent the deep intelligence of the biosphere. Only a holarchically organized world system that both promotes and represents biospheric consciousness can achieve this goal.

Such a world system would not only promote growth in consciousness but would necessarily continually monitor the health of our planet and prompt us to adjust as necessary. New global institutions would consciously act to keep the ecosystems of our planet in balance. Our present fragmented world system clearly produces consequences destructive of the biosphere. We must establish a holarchical planetary system that produces the ecological and cosmocentric consequences that are designed into the system from the very beginning.

Moreover, such a novel Earth system must be capable of regularly modifying its design to produce optimal sustainability consequences. This is part of what transformative praxis would look like. It would systematically transform our planetary economic and political institutions in the direction of ever-increasing holism while monitoring planetary ecosystems and producing useful feedback loops that allow for timely and continuous adjustments. We must be able to modify our theories of sustainability in the light of feedback from our experimental economic and political initiatives. We must be able to incorporate innovative technological developments into our theories and integrate these with new practices and changing conditions.

As Jay W. Forrester affirms, "Systems of information-feedback control are fundamental to all life and human endeavor, from the slow pace of biological evolution to the launching of the latest space satellite. ... Everything we do as individuals, as an industry, or as a society is done in the context of an information feedback system" (in Meadows 2008, 25). Genuine human biospheric consciousness would result in economic and political systems that reflect the holarchical structures of nature, with information feedback systems making possible continuous adjustments for integration, balance, and harmony.

We cannot credibly expect to attain a planetary, holospheric consciousness while maintaining the regressive, fragmented economic and political institutions that currently dominate life on Earth. Economist Kate Raworth recently underlined this exact fundamental truth: success in overcoming our planetary crisis will come from the way we *design our systems*. She writes, "Economics, it turns out, is not a matter of discovering laws. It is essentially a question of design. And ... the last two hundred years of industrial activity have been based upon a linear industrial system whose design is inherently degenerative. ... From a systems-thinking perspective ... far greater leverage comes from changing the paradigm that gives rise to the system's goals" (2017, 180, 182).

Our planetary noosphere can become harmonious with the biosphere if we implement a process of planetary transformative action. Many thinkers have pointed out that the biosphere operates as a network of holarchies – cells grouping together into larger units that again group together, until they become a living ecological network that forms a single planetary system.

"All natural systems," Laszlo writes, "are wholes in regard to their parts, and parts with respect to higher-level wholes" (see Capra 2004; Goerner, et al. 2008). They are therefore holarchies. A biospheric consciousness would mirror the laterally dispersed, interconnected, and ascending levels of organized energy patterns that make our planet a living planet. Transformative praxis empowered by holarchical civilizational systems that make space for the holistic expression of human love is by far our best bet for putting human beings into a deep sustainable relationship with our planet.

This requires that we organize life on Earth into a networked holarchy from local sustainable communities to ascending, cooperating, ever-larger communities in which these networked parts empower the whole and the whole in turn sustains and empowers the networked parts. As Buckminster Fuller states regarding Spaceship Earth, it "must be comprehended and serviced in total" (1972, 47). Translated into the language of Gaia: Our living planet must be cared for and harmonized with as a living whole.

We therefore need to join human beings into a planetary community that is interdependent and interlinked at every level. This will require deep changes in global design toward a culture of rational love – a culture of kindness, peace, justice, collaboration, compassion, and mutual recognition of our common humanity. These higher human qualities emerge as we grow from egocentric toward worldcentric and cosmocentric forms of awareness. Such changes in culture and consciousness are possible, if we implement the necessary institutional design changes.

1.6 Providing Earth with a Brain

Mirroring the holism of nature also requires fundamental changes in economic theory. Today, economists recognize that cooperation, collaboration, enhanced creativity, shared resources, open-source information, shared profits, and concern for the economic well-being of all participants are mandatory components of the analysis (Raworth 2008). Similarly, deep changes in government are needed. Collaborative democracy, concern for the common good of the whole of humanity, social justice, citizen participation, mutual respect, and war prevention are keys to achieving true unity in diversity (Martin 2008).

To date, the noosphere has encircled the planet without a biospheric consciousness. Our planetary mind has no center as yet, no selfhood, no true representation of the whole of humanity. Even the United Nations Charter is based on a regressive, militaristic, nation-state worldview. In human beings, the mind has a center and a sense of being a unitary self, what Immanuel Kant called "the transcendental unity of apperception." As noted above, all the thought-processes, reasoning, emotions, desires and imaginative ideas of an average human mind are associated with a unitary center of experience. However, this center need not be egoistic. It can be worldcentric and cosmocentric.

The noosphere appears to operate within a realm of chaos, falsehoods, mythologies, prejudices, ideologies, and distorted perceptions. There are a few truths, a few principles, a modicum of wisdom and integrity living in the interstices of this chaos. As a whole, the noosphere can perhaps be compared to the mind of an egoistic ten-year-old child, a mind filled with emotions, desires, fears, hopes, pleasures, and pains. It has no real center, no centralized human self-awareness. However, the child rapidly grows. Before we know it, the child may become a loving, worldcentric adult. The task before us is to find ways to rapidly transform this egocentric mode of existence into a holistic consciousness that can embrace genuine sustainability.

Everyone says, "I think, I feel, I want, I hope, I love, I imagine," etc. The diversity of experience is integrated within the unity of consciousness, indicated normally by the word "I." The world and myself are related, knowable on one level and intelligible, yet somehow at the same time inseparable as aspects of a living whole. Above, Erich Fromm posited selfhood as the self-aware locus of love that can unite the whole of existence while retaining its separate identity and personal integrity, living as a distinct center of creative responsiveness while at the same time being one with all. Subject and object are one at the primary level, yet at the same time two, interacting and flowing together, a seeming paradox that the philosopher Hegel made fundamental to his dialectical method (cf. Harris 1987; Frank 2020, 177-78).

On the other hand, Earth's noosphere presently exhibits the chaos of mental phenomena with no center, no integrative reasoning power, no capacity for unitary self-assessment or for planning for a coherent and intelligible future. We must discern our interdependence, our oneness with one another, a oneness that at the same time generates respect for our marvelous diversity. The world must unite to become a self-aware economic and political system that embraces unity in diversity.

So far, the world has not experienced a transformation comparable to what started happening to individual persons some 2,500 years ago during the Axial Age – a shift in self-awareness that may include global consciousness. I believe we are in the midst of a second Axial transformation, with many more people reaching planetary consciousness (Martin 2018, 22-27). But the majority still lack this awareness because our institutions are nowhere near a planetary consciousness. The result is a planet in grave danger of self-destructing.

The holism of humanity depends on our institutions both representing and fostering planetary selfhood. Networked arteries of commerce, communications, and transportation can ascend from a multiplicity of capillaries into central arteries that integrate the required planetary sustainability, harmony, and integrity. Without a united humanity integrating its activities with the health of Gaia, we will not succeed in creating a sustainable civilization in harmony with our Earth.

The noosphere can therefore only properly reflect the integrity of the biosphere through a centralized, worldwide, and democratic government. A democratic "World Parliament" predicated on the oneness of humanity in peace, justice, and sustainability is the obvious answer. Already, many people identify themselves as "World Citizens." However, this phenomenon has not yet resulted in people realizing that a networked humanity of world citizens depends on a cooperative process of uniting humanity around effective governance of our planet as a whole. Once the people of the world unite to create a universal selfhood and realize that they deserve holistic representation of their collective consciousness, democratic world government will become a "Tipping Point" issue.

This Tipping Point will be reached when people insist on global governance that represents the common good of the whole of humanity, democratically chosen from every locality on our planet. Today, the demand is growing for global governance, as it did after World War I and World War II when the United Nations was formed. But we still are dominated by extreme undemocratic economic and political forces, many of which pull in different directions and most of which remain uncoordinated, including the World Trade Organization, World Bank, International Monetary Fund, multinational corporations, U.N. conventions, and so-called "international laws." These powerful entities and interests dominate the masses via nation-states such as China, Russia and the U.S., and via corporatocracy. As a result, our world – replete with horrific bio-weapons research labs, hypersonic weapons, and nuclear arsenals ready for Armageddon – lurches disconnectedly into a fragile, chaotic, and deeply uncertain future.

A World Parliament would coordinate our collective planetary mind-phenomenon and create a networked humanity ready to a focus on protecting Earth and the common good of all. Hence, it is incumbent on us to assist the noosphere in evolving a holarchically networked humanity so that we may form a democratic World Parliament. Such a parliament would be the central nervous system of humanity and integrate the collective consciousness. The people of Earth would then have a collective will, a collective selfhood to express their desire to be a global community premised on peace, justice, and sustainability. Uniting politically and economically would simultaneously inspire conscious growth and further transformation.

The theory of fractals exploded in the 1970s when French mathematician Benoit Mandelbrot first used the computer to generate recurring patterns at ever smaller levels. His work revealed the astonishing similarity of the tiniest parts of systems to the larger wholes to which they belong (cf. Lokanga 2018, 104-108). This patterning occurs throughout the universe and makes possible, for example, the amazing functioning of the human lungs and the blood circulating in our veins and arteries. Our planet as a whole still lacks this optimum structure, though holistic and healthy patterns are beginning to coalesce.

> *A World Parliament would network humanity, allowing us to protect Earth and the common good of all. It also would allow us to express the collective consciousness of the entire planet.*

World democratic government would be designed as a fractal – a representation of unity in diversity elaborated in parliamentary, judicial, administrative, and conflict resolution functions. Moreover, these patterns of democratic governance would repeat at ever smaller levels from the national to regional to local levels. At present we have the chaos of a fractal pattern that has yet to unite and coordinate all its parts. It's as if the blood were trying to circulate without a whole living body to animate and enliven.

We have seen a number of contemporary thinkers, such as Barbara Marx Hubbard, emphasize *conscious* evolution. As self-aware creatures, we now are capable of understanding the structures of the cosmos and the direction of "complexity-consciousness" of life on Earth. We therefore have the responsibility to consciously evolve to higher levels of maturity, awareness, and spiritual illumination. As Laszlo writes, love is the elixir we need to reach this Tipping Point:

> What does it take to recover the intuitive feeling that we are part of it, that we are connected? ... It takes love, the deep, embracing feeling of love. Love is the recognition that the other is not other. The world is not beyond or outside of me: it is inside me just the same way as I am inside the world. ... I don't see even the remotest possibility of creating a sustainable and flourishing world on this planet unless we embrace this embracing love. (2014, 79)

Conscious evolution through a growing and transforming love also is a theme in Hubbard's book *Conscious Evolution*, which is full of inspiring broad statements about holistic education, governance, science, art, and culture. Yet, neither Hubbard nor Laszlo have provided a concrete plan for uniting Earth within a loving holarchical system. Nor do we find helpful or practical suggestions from environmentalists or economists. Most fail to recognize the ways in which a democratic World Parliament could and should unite humanity to deal effectively with climate change and social ills like the current Covid-19 pandemic.

Thus, urging people to evolve more "consciously" or more "lovingly" is not very helpful. Describing the progressive integration of our "global village" is fine, but where do we go from here? Why are there no concrete plans for transforming our broken and outdated global institutions in the direction of synergy, integration, and holism?

For most of the 20th century, great thinkers called for organizing the planet beyond war, fear, and fragmentation. What we ended up with, instead, is the United Nations, an organization that perpetuates the fragmentation of the world into militarized sovereign nation-states and that fails to prevent war, protect human rights, or establish a sustainable world system. Why have the voices of such profound minds like Albert Einstein and Mahatma Gandhi been ignored? Why has this fundamental dimension of planetary history been ignored and marginalized by so many concerned people?

Clearly, planet Earth is being governed by chaotic and conflicting forces. The noosphere is scattered, disconnected, and careening like a headless horseman toward the abyss. It lacks the wholeness within which fractal patterning of constituent parts may circulate, animate, and help guide our future. The present global pandemic reveals and symbolizes our immense disconnectedness, incoherence, and vulnerability. We have yet to understand that we need to democratically govern ourselves as a planetary civilization or face the nightmare of continued global chaos and possible extinction.

In sum, our Earth does not yet have a brain. It has no design for peace, justice, and sustainability. It has no carefully designed universal health-protection system that can ensure such pandemics never happen again. We face multidimensional planetary crises of immense proportions and we have no brain guiding us!

Thankfully, there does exist a plan for institutionalizing the collective consciousness – a brilliant plan. This plan embraces the love that unites all human beings and its design is holarchical, directed toward peace, justice, and sustainability. This plan was crafted by hundreds of world citizens who worked together for nearly thirty years, and it provides a beautifully organized world system designed to produce and sustain a global transformation that benefits all. The plan is called the *Constitution for the Federation of Earth*, and it was completed in 1991. It is an actual template for conscious evolution and for providing Earth with its own brain capable of organizing and synthesizing the noosphere (Martin 2010a; www.EarthConstitution.world).

By now it should be clear that humanity will not accomplish a "Paradigm Shift" without greater cooperation, collaboration, and organization. The *Earth Constitution* provides the deeper synergy needed to network the planet. It also contains the spiritual component needed to anchor our evolutionary progress. Human spirituality can no longer confine itself to participation in one religious movement or another, as my friend Swami Agnivesh declared:

> Transformation is not just any change. It is even more than a change
> for the better. It is a radical change that empowers the fulfillment
> of potentialities that remain hidden and untapped. The scope of
> transformation goes beyond that of reform. Reform is content with
> specific improvements, whereas transformation calls for the shifting of
> the very foundation on which a society or religious system is based. It
> is a total and comprehensive agenda. (2015, 38)

The transformation envisioned in this book constitutes "the shifting of the very foundation" of ourselves and our planet. We are talking about reorganizing and networking the foundation of human civilization, a shift that will draw upon all spiritual traditions, whether several millennia old or New Age. For the *Earth Constitution* calls for a new regime of sustainability, politics, economics – all of which can only be changed if we embrace a renewed spirituality that "empowers the fulfillment of potentialities that remain hidden and untapped."

Raimon Panikkar writes that "the world cries out for a radical change that cannot be merely theoretical, without a grounding in praxis." The *Earth Constitution* provides both a practical and elegant model for transformative praxis. It also spurs us to action here and now, at this auspicious and timely moment in our collective history. Panikkar continues with the insight that we must solve the problem of nation-state "ideology." The nation-state, he says, is not merely a political problem; it requires a "new vision of human life" that is animated by love and solves the problems of distribution of wealth and of keeping the peace (2013, 358-59). The *Earth Constitution* expressly contains such a "new vision of human life" that is all-embracing, comprehensive, and equitable for all.

In the remainder of this book, I will describe our present crisis in more detail and show the multiple ways that the *Earth Constitution* concretely and effectively addresses the many threats we face. I also will show how the *Earth Constitution* will allow humanity to flourish. This is not a utopian dream. It is an absolute necessity if we are to go beyond extinction rebellion and effectuate transformative praxis.

Why am I so certain that the *Earth Constitution* offers the planetary design we seek? Because it provides global democratic law that transcends the assumption of atomism now tearing us apart. Its design is directed to the common good of all and makes possible – for the first time in history – the emergence of our highest collective consciousness. It also applies universal moral values and ethics to build institutions grounded in community, justice, sustainability, and peace. In sum, the *Earth Constitution* contains the fractal roadmap needed to synthesize local, regional, and national cultures into a planetary whole that is greater than the sum of its parts.

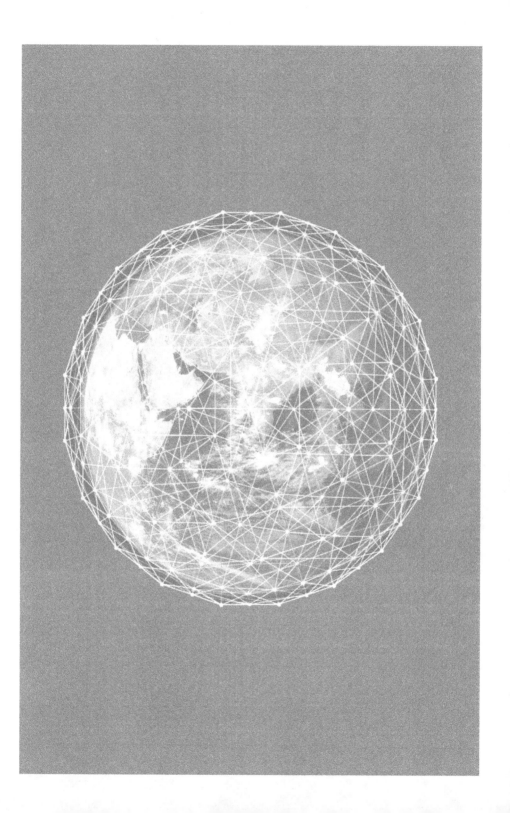

CHAPTER TWO
Paradigm Shift:
From Fragmentation to Holism

We are all citizens of a world community sharing a common peril. Is it inevitable that because of our passions and our inherent customs we should be condemned to destroy ourselves? No one has the right to withdraw from the world of action at a time when civilization faces its supreme test.

Albert Einstein

2.1 Ecology and the Science of Sustainability

Definitions of sustainability vary in emphasis and focus, but nearly all share a basic framework such as: "Meeting the needs of the present without destroying the ecosystem of Earth and depleting its resources so that future generations are unable to meet their needs." Simply stated, sustainability is the ability to flourish in the present without diminishing the ability of future generations to flourish. Only by recognizing our interdependence with the planetary biosphere, can we meet the needs of the present and preserve the life-prospects of future generations.

The concept of sustainability, therefore, includes holism – the defining feature of the 20th century revolutions in science. Specifically, ecology as a science focuses on the innate interdependence of organisms and their environment. At the very least, holism requires a new economics for the people of Earth. The emerging model comes from numerous non-mainstream economic thinkers, such as Lester R. Brown, Hazel Henderson, David Korten, Herman E. Daly, Richard Heinberg, Kate Raworth, and Jeremy Rifkin. In their book *Environmental Economics*, Turner, Pearce, and Bateman affirm that "conventional economics textbooks often convey a very misleading picture of the relationship between an economic system ... and the environment. ... Basically, simple economic models have ignored the economy-environment interrelationships altogether" (1993, 15).

Fundamentally, sustainability means that the economy cannot create goods and services to support human life without understanding that the economy is a subset of the biosphere. To date, economics has been practiced as if investment, production, and consumption were self-contained and able to grow and increase endlessly, without creating immense negative external consequences on society and our planetary ecosystem.

Sustainability also refers to the ability of human beings to integrate their societies and economies into the larger holism of ecological balance, which includes an ethical and social holism linked to the well-being of future generations. Economics will have to change to include transformed patterns of investment, production, engineering, design, transportation, distribution, consumption, waste generation, and disposal. It also will require changes in education, social media, and the way we think about life and one another.

Holism will generate transformed civil institutions, organizations, ways of working together, communicating, and doing business. It will necessarily mean the ending of militarism, which, in all its phases – from production to deployment to use of military organization and weaponry – is utterly unsustainable and wasteful (Sanders and Davis 2009). Politics will have to change as well, from a politics of wealth, power, and winners-take-all to a politics of cooperation, fairness, inclusiveness, and dialogue directed toward mutual understanding. In other words, holism demands authentic economic and political democracy, and democracy will have to be globalized.

In the future, non-renewable resources will be used at a bare minimum. All energy will have to be clean and renewable (sun, wind, tides, geothermal, etc.). Everything will have to be manufactured for maximum durability and reparability. Waste (both thermal and trash) will be reduced to a minimum, and whatever waste there is will have to be biodegradable and non-toxic. What little toxic waste is still produced will have to be detoxified. Forests will have to be replanted, watersheds repaired, and biodiversity protected. Worldwide agricultural lands will be cultivated with minimum soil loss and programs of soil restoration, while using natural insect repellants and natural fertilizers that maintain soil balance and productivity over time.

The implications are vast, for they require a transformation of human civilization throughout the entire range of our practices. In their book *The New Science of Sustainability: Building a Foundation for Great Change*, Goerner, Dyck, and Lagerroos elaborate on this point:

Sustainability began as a subset of environmentalism, as concern about how to manage the flow of natural resource inputs and human-generated outputs (pollutants and waste) in a way that could go on, if not forever, at least for a very long time. Yet, when people began to think about how to alter modern practices, they began to realize that the changes needed would require a major shift in the way we think about nature, agriculture, energy, economics, health, and, from there, ethics, democracy, and social justice. (2008, 31)

Economist Herman E. Daly's definition of sustainability as "development without growth beyond environmental carrying capacity, where development means qualitative improvement and growth means quantitative increase" carries with it vast ramifications for education, economics, culture, and government (1996, 9). In his book *Beyond Growth,* Daly demonstrates that the perpetual growth model of mainstream modern economics systematically ignores the carrying capacity, finite resources, and fragile ecosystem of Earth in its economic formulas. As Jeremy Rifkin puts this: "making the transition from a nonrenewable to a renewable base of energy represents a monumental task for the whole of civilization" (1989, 210).

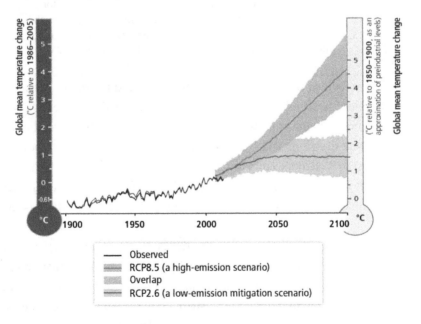

Figure 2.1
This chart shows the vast danger in temperature rise if we do not transform our planetary civilization to sustainability rapidly
Source: https://climate.nasa.gov/evidence/

Daly understands the pervasive holism of the new scientific paradigm and he applies this directly to sustainability or what he calls "the steady-state paradigm."

> Ecology is whole. It brings together the broken, analyzed, alienated, fragmented pieces of man's image of the world. Ecology is also a fad, but when the fad passes, the movement toward wholeness must continue. Unless the physical, the social, and the moral dimensions of our knowledge are integrated in a unified paradigm offering a vision of wholeness, no solutions to our problems are likely. (1996, 357)

In his 1996 book *The Web of Life: A New Scientific Understanding of Living Systems,* Fritjof Capra characterizes the ecological model in terms of five principles: interdependence, cyclical feedback loops (recycling), pervasive cooperation, flexibility, and diversity (298-304). All these concepts are interconnected and linked with one another. Interdependence, therefore, implies all the others. Capra writes:

> Interdependence – the mutual dependence of all life processes on one another – is the nature of all ecological relationships. The behavior of every living member of the ecosystem depends on the behavior of many others. The success of the whole community depends on the success of its individual members, while the success of each member depends on the success of the community as a whole. (1996, 298)

Can human beings think in terms of cooperation and interdependence rather than competition, hostility, fragmentation, and militarized defense? Of course, but the challenge demands that we make a complete and genuine paradigm-shift from our current divisions and fragmentation to holism. In his 2001 book *Eco-Economy,* Lester R. Brown calls this needed change "Copernican" – a shift as great as the movement from a geocentric to heliocentric worldview (21). Clearly, such a fundamental shift in worldview is an absolute necessity if there is to be a viable future at all.

In *Only One Earth,* Barbara Ward and Rene Dubos state that the transformation of human consciousness is indeed possible, affirming that "men can experience such transformations is not in doubt. From family to clan, from clan to nation, from nation to federation – such enlargements of allegiance have occurred without wiping out the earlier loves. Today, in human society, we can perhaps hope to survive in all our prized diversity provided we can achieve an ultimate loyalty to our single, beautiful, and vulnerable planet Earth" (1972, 220). But the issue is deeper than this, for it involves the way people's loyalties are shaped by dominant institutions. Lester Brown raises this issue in the following way:

> The central question is whether the accelerating change that is an integral part of the modern landscape is beginning to exceed the capacity of our social institutions to cope with change. Change is particularly difficult for institutions dealing with international or global issues that require a concerted, cooperative effort by many countries with contrasting cultures if they are to succeed. For example, sustaining the existing oceanic fish catch may be possible only if numerous agreements are reached among countries on the limits to fishing in individual oceanic fisheries. And can governments working together at the global level, move fast enough to stabilize climate before it disrupts economic progress? (2001, 20-21)

These statements by climate scientists underscore the failing ecosystem of our planet. The scientific evidence now is overwhelming, continuously mounting for at least the past fifty years. If the behavior of nations to date is any indication, the answer to Brown's questions is clearly "no." The system of sovereign nations will not be able to take the necessary measures to save the ecosystem of Earth. Yet in his book, Brown himself appears unable to imagine any alternative other than the unlikely cooperation of some 193 militarized nation-states in addressing our multiplicity of global environmental crises.

The same is true of the environmental economics proposed by Turner, Pearce, and Bateman. Though excellent in other respects, their book treats sustainable economics as if it were a matter of adding up some 193 separate national accountings. The authors state that "the current system of national accounts used in many countries fails, in almost all cases, to treat natural capital as assets which play a vital part in providing a flow of continuous output/income over time. ... Extended national accounts ... are required in order to improve policy signals relating to SD" [sustainable development] (1993, 56).

In short, all these scholars describe an economic accounting system that must treat the economy as a subset of the natural environment, and they assume that 193 independent nation-states will somehow come together to craft the needed policies. They are correct that global economic holism is the answer. So why does it appear impossible for governments to think in terms of planetary democracy? Let us review our human history with this question in mind.

2.2 Axial Period and the Emergence of Democracy

The Axial Age was a worldwide phenomenon that gave birth to our present level of self-reflective consciousness. Since the Axial Period of history (roughly the

8th to the 2nd centuries BCE), civilization has passed through three macro-stages or paradigms, along with many sub-stages. In *Triumph of Civilization,* I called the three central paradigms: Age of Static Holism (characterizing many of the ancient civilizations through the medieval period), Age of Fragmentation (characterizing the early-modern paradigm emerging from Europe and spreading worldwide after the Renaissance), and Age of Evolutionary Holism (the global transition taking place today).

The latter age has only recently emerged out of the discoveries of 20th century science and has not yet substantially influenced the world's dominant institutions, namely nation-states and global capitalism. However, by understanding our human situation in relation to the cosmos during this age of holism, it is but a very short step to comprehending the necessity of planetary federation if our institutions are to reflect the holism of nature and our common human civilization.

None of this is, so to speak, "ancient history." Based on our one to two million year human timeline, Plato and Buddha lived just yesterday, and they worked just two seconds ago on the planetary scale in what anthropologist Loren Eisley (1959) called the "immense journey" of development for the past 3.6 billion years (1959). And Plato and Buddha lived as our virtual contemporaries on the scale of the age of the universe, which exploded in its "primal flaring forth" some 13.8 billion years ago (Swimme and Berry 1992).

It was during the Axial period of human history that philosophy, ethics, and advanced religion first emerged as an activity of reason attempting to consciously comprehend our human situation. By then, human self-consciousness had emerged to the point where reflection was natural. This reflection was applied to ethics, law, social order, the natural world, transcendent reality, and nearly every other fundamental mode of human existence. Today, we would say that the general tenor of the philosophies and religious worldviews coming out of the Axial period was holistic.

In many parts of the world this ancient form of philosophical holism persisted nearly to the present, with the notable exception of Western society from the 16th century on, when the philosophy of mechanism, atomism, and fragmentation was introduced. This Western worldview has since spread globally in the form of capitalism and the nation-state system. My thesis provides for the continuing emergence of human self-consciousness since the Axial Period, such that we now are experiencing a paradigm shift – a decisive breakthrough, though not yet fully comprehended by all. Nevertheless, we collectively are beginning to understand the nature of holism, including the inseparability of parts and the dynamic order of our physical universe, biology, and human life.

This new holism is evolutionary on the cosmic level and revolutionary on the human level. The universe and human life are no longer characterized by a static holism but by an *evolutionary holism* in which one form of life develops through dynamic interaction with other life forms and with the wholes of which they are a part. The wholes or fields exist within ever-larger fields, providing a *telos* for the interaction of all life forms. Human consciousness has evolved in the same way, with thought itself moving toward an ever-greater, dynamic evolutionary holism.

It is palpably obvious to anthropological science that human self-consciousness and reason have emerged out of the process of evolution. For most of our one to two million years there were few signs of self-awareness expressed in culture or artifacts. It was only during the Axial Period of the first millennium BCE that our species as a whole began to distinguish between our subjective responses to things and the objective way that things exist independently of those responses (cf. Swimme and Berry 1992; Armstrong 2007; Hick 2004; Jaspers 1953).

Science no longer characterizes the universe by a static holism, but by an evolutionary holism in which life and consciousness change and develop through dynamic interaction with other life forms, via new ideas, and by relational interaction with the united wholes of which they are an integral part.

People were now capable of acting according to abstract ethical principles, not mere social custom and conformity directed toward preserving the mythologically grounded community. Oral traditions that had remained unchanged for centuries were suddenly open to examination from religious, ethical, and natural philosophers who looked for principles behind custom and nature, no longer just simply accepting the status quo. Often considered the three earliest Pre-Socratic philosophers were Thales, Anaximander, and Anaximenes of Miletus in Asia Minor. Thales was teacher of Anaximander who was teacher of Anaximenes, and it is significant that each pupil *disagreed* with his teacher. The quest for knowledge of the cosmos grounded in objective principles had begun.

Individuals were developing the ability of thinking for themselves and acting autonomously. For this reason, they started to accept responsibility for their behavior according to the great ethical and religious teachers of this period, and they based their morality on abstract principles such as the Golden Rule, not mere custom and social conformity (see Hick 2004, Chap. 17). In addition, religion was becoming concerned with individual responsibility and salvation, not simply maintaining the unchanging human and cosmic order. Human consciousness had opened up to its own possibilities in relation to the ground of Being (a.k.a. ultimate reality, Brahmin, Tao, Buddha Nature, or God). Human beings began to transform their lives and the world according to a radical potential for new relationships arising from the depths of this new reality.

In *An Interpretation of Religion – Human Responses to the Transcendent* (2004), John Hick studies this age in relation to the birth of the great religions. He writes concerning the mythic basis of the earlier, mythological age:

> This serves the social functions of preserving the unity of the tribe or people within a common world-view and at the same time of validating the community's claims upon the loyalty of its members. The underlying concern is conservative, a defense against chaos, meaninglessness, and breakdown of social cohesion. Religious activity is concerned to keep fragile human life on an even keel; but it is not concerned, as is post-axial religion, with its radical transformation. (23)

We are witnessing the emergence of individuality, Hick says, out of an earlier mode of human awareness not capable of clearly separating its individuality from the group and its general environment. Here "the symbolization of self and world are only very partially separate" (32):

> They were now able to hear and respond to a message relating to their own options and potentialities. Religious value no longer resided in total identification with the group but began to take the form of a personal openness to transcendence. And since the new religious messages of the axial age were addressed to individuals as such, rather than as cells in a social organism, these messages were in principle universal in scope. (30)

These several features of the emergent consciousness of human beings during the Axial Age form an interrelated part of a paradigm-shifting transformation, including the emerging ability to: separate one's personal subjectivity from objective principles; distinguish oneself from the social group; assess personal responsibility and personal potentialities; live according to universal, abstract ethical principles; and relate one's life to the ground of Being or God, that is, to the whole of existence (in relation to which the ground of Being is both immanent and transcendent). Human beings were emerging from their two-million-year evolutionary process as what I have called "rational freedom oriented toward wholeness" (*Ascent to Freedom,* 2008). This capacity can simply be termed "reason."

Heraclitus identified it as the *logos.* Human reason had attained the ability to reflect or mirror the *logos* – the objective order of things hidden at the heart of the world process. The human microcosm could rationally and objectively know the cosmic macrocosm. The older mythological consciousness that primarily acted in *response to the world*, that is, in a relation to the world as a "Thou," was coming to an end. The new orientation appeared to *know the world*, independent of the subjective feelings and responses of individuals or cultures.

The most fundamental point here is that the emergence of these qualities laid the foundation for democratic world law and, ultimately, sustainable civilization. Democracy was not simply one political system among others. It was an ethos and a set of values that correlated with the emergent and newly inherent characteristics of human beings during the Axial Period. These characteristics are still developing toward maturity within humanity. They are both qualities we inherit from the Axial Period and potentialities within us for ever greater realization.

Thus, the set of values that we call "democracy" arose during the Axial Period from a new definition of what it meant to be human beings: our sense of individuality and our human potential; the capacity for ethical responsibility according to abstract principles; the ability to formulate and self-consciously follow laws; the ability to reason with all that this implies for knowledge and freedom; the capacity to separate the merely subjective and personal from the objective; the ability to progressively formulate and know the principles of nature, philosophy, and science; the drive to love the democratic community with its ideal of unity in diversity; the ability to cooperate for the common good; and the desire to engage in communicative discourse (see Jürgen Habermas, 1984 and 1987, and other contemporary thinkers). These values require that society be organized in such a way that these human capacities are recognized, encouraged, built upon, and further developed.

The comprehension required for democratic world law has been steadily evolving since the emergence of these basic capacities that define our humanity (Martin 2016, Chap. 3). Today, they are integrated principles and define our potential for ever greater maturity. Since these capacities are now universal, they need to be institutionalized in a planetary civilization. As we saw in Chapter One, we need to acknowledge our worldcentric, planetary consciousness and recognize the equality and dignity of all persons. A truly democratic world system also means a sustainable civilization.

The comprehension required for federating Earth has been steadily emerging. A holistic and sustainable democratic civilization will arise once we collectively acknowledge our full ascent to authentic global maturity.

Economic and political relations under authentic democracy are not splintered into militarized antagonistic units, nor are they fragmented into exploiters and exploited. Democracy implies all dimensions of liberty, equality, and community on a planetary scale. A sustainable, holistic civilization will result from our ascent to authentic planetary maturity and democracy. But the relationship between maturity and institutions is reciprocal. If we ratify the *Earth Constitution,* this in turn will substantially increase planetary maturity and global biospheric consciousness.

Authentic democracy manifests the principle of unity in diversity because its holistic framework recognizes the equality, rights, and dignity of all members of society. However, after the Axial Age, modern physics and philosophy took a wrong turn and failed to assimilate any serious form of holism. Instead, the science that emerged from the 17th and 18th centuries sculpted a physical and social world order based on atomism and fragmentation, and it lacked any significant redeeming unity. This outdated Newtonian worldview continues to dominate our thinking today.

2.3   Age of Newtonian Physics and Philosophy

This separation of subject from object during the Axial Period eventually made possible the discovery of the scientific method during the 17th century. The observer created an idealized situation called an "experiment" to test a hypothesis under controlled conditions. The observer was a detached spectator who observed the results in mathematically quantifiable and systematically recorded terms.

The result was an explosion of knowledge about the physical world. Neither subjective feelings nor the observer's values were to interfere in the observations. Fact and value were considered two different dimensions of existence. In contemporary (non-Newtonian) physics, physicist Ediho Lokanga observes that there remain unexplained anomalies and theoretical questions, for example, about the relation of quantum theory to Einstein's relativity theory. Yet, the one thing all theories agree on is "the unbroken wholeness of the universe – everything in the universe is infinitely and internally connected to everything else" (2018, 40).

However, Lokanga declares, "it is very unfortunate that we continue to depict, describe, and interpret the laws of the universe as if we were still living in a purely material-mechanical system" (ibid., xvi-xvii).

In the *Principia Mathematica* (1687), Sir Isaac Newton synthesized the laws of planetary motion discovered by Johannes Kepler with the laws of earthly motion discovered by Galileo Galilei. It appeared as if science had uncovered the most fundamental workings of the universe. Newton posited an absolute space and time wherein the mechanics of bodies in motion operated. The universe was conceived as a vast machine whose parts (bodies) operated in *external relations* with one another and which were reducible to simpler parts down to the smallest atoms.

This worldview was characterized by the assumption of universal causal necessity. The universe was composed of "bodies in motion" that could be understood through analyzing them into their component atomistic parts. These bodies in motion filled the universe, moving like a gigantic machine, all of them in external relationships with one another. Events and developments in the machine should not (and could not) be understood in terms of any *telos* or purpose, but only as units being pushed from behind (so to speak) by efficient causality. The human mind could objectively observe the machine and formulate its "laws of motion," but mind was never observed in nature. This universe had no observable place for mind.

Also during the 17th century, René Descartes posited the mind as a "mental substance" entirely different from matter or "extended substance." Mind and matter confronted one another: one physical and the other non-physical, with no bridge between them. The question of how non-physical mind could control physical matter (which it obviously did in the human body) plunged Descartes into great conceptual difficulties. However, subject and object were now formally distinct, and the project of knowing the world "objectively" could proceed unhindered. With the connection between mind and matter now severed, Descartes in his *Sixth Meditation* ultimately had to depend on God as a *deus ex machina* to guarantee the veracity of the mind's relationship with the external world (1975, 185-199).

All these early-modern thinkers – from Spinoza, Leibniz, and Kant on the European continent to the empiricists, Locke, Berkeley, and Hume in England – were strongly influenced by the assumptions of the newly emerging science of physics. In his *Essay on Human Understanding*, John Locke attempted to construct human knowledge on the basis of "simple ideas" that entered into the "blank tablet" of the mind through the five senses. As an empiricist, he would only accept ideas of what he termed the primary and secondary qualities of bodies in the world that originated in sense experience. However, he soon realized that one never experiences the "substance" within which these qualities appear to inhere. He concluded that this substance underlying the qualities of the world was "something ... I know not what" (1978, 101).

Bishop George Berkeley, also an empiricist, followed the Lockean path to its logical conclusion: If substances were not observable, they could not be part of human knowledge and must be dropped as a fiction. "To be means to be perceived" (*esse est percipi*), he concluded. All things existed as collections of ideas and these collections of ideas were available to perceivers, including the ultimate perceiver, God, who guaranteed the solidity and stability of the world. Berkeley perceived that subject and object must be *internally* related in some way ("to be" must be internally related to a perceiver). However, Berkeley's Newtonian atomism and reductionism prevented his moving forward with this insight. Rather, as with Descartes, God had to be brought in as a *deus ex machina* to guarantee the existence of the world when no finite creatures were perceiving it (1957).

David Hume, drew even more radical conclusions based on his powerful, so-called "logical empiricism." The "simple ideas" of Locke become "simple impressions" for Hume, who rejected the vaunted "universal causal necessity" of Newtonian physics. Bodies in motion appeared to causally affect one another. A billiard ball strikes another ball, stops moving, and the second ball begins moving. However, Hume observed that in the relation of two events in which one is said to cause the other, we only *perceive* "constant conjunction," never causality itself, never necessity. Hence, causal necessity in nature must be dropped as a fiction.

Hume was bringing empiricism to its logical conclusions. Regarding the self that is said to observe these objects in the world, we do not empirically observe any such thing. We do not perceive any unitary, subsistent "self" that Descartes had assumed as the objective observer of the world. If we don't perceive it, it too must be dropped. What we do experience in our subjective lives – disparate desires, urges, and fantasies – led him to conclude that "reason is, and ought only, to be the slave of the passions" (1949, 127).

The result was a skepticism so pervasive that by the end of the 18th century, Immanuel Kant realized that something was wrong. Throughout the western tradition, reason had connected humanity with the highest values and aspirations, such as peace, justice, beauty, and truth. Could it be the case that such ideals were *merely subjective* as Hume maintained? Kant remarked that Hume's philosophy had awakened him from his "dogmatic slumber" to rethink the entire grounds of human knowledge.

Empiricism, as part of this early-modern set of assumptions, did not appear to be able to account for human knowledge. Instead, it demanded a "value-free science" without providing any place for "mind" within the universe. It required a rigorous separation of subject from object, assuming that this was the only route to successful, objectively grounded science. Kant undertook an entire rethinking of these assumptions. However, even this rethinking of the Cartesian starting point was hindered by the fact that Kant also assumed the fundamental "truth" of the Newtonian paradigm. Nevertheless, he understood that subject and object must be *internally related* in any coherent account of knowledge, and he developed his *Critique of Pure Reason* (1781) on this basis.

In Chapter One we encountered this unity of consciousness, the mysterious unity that we often call "I" and termed by Kant the "transcendental unity of apperception." This unity consciousness tends to be conflated by people at the egocentric level of consciousness with their psychological self. As we grow in awareness, this conflation diminishes. We move into more mature, transpersonal levels of consciousness where "I" becomes identified with the whole of humanity. Kant made a major step forward by pointing out the deep relationship between this "I" and the world.

Kant also understood that the "transcendental unity" of knowledge correlates to the assumption that the world comprises a single, systematic unity (CPR, sec. B 139). There is only one world, and this unity is implicit in the unity of apperception. Subject and object are fundamentally related. Mind, therefore, was not merely a collection of subjective fantasies driven by desires, as Hume had maintained. Rather, mind and reason were directly connected with the very intelligibility of the universe. With this realization, Kant foreshadowed the 20th century scientific paradigm that has discovered the fundamental unity of the universe and the holism by which diversity is integrated into that unity. Newtonian science appeared to demand the opposite.

To Newtonian scientists, the world appeared as a vast machine operating automatically and entirely disconnected from human values and aspirations. Hume drew this conclusion, and in the 19th century Nietzsche (1969) carried this argument to its ultimate implication: nihilism, where all human values are devalued, worthless, and merely subjective illusions. What we call the "objective world" is itself just another subjective point of view. Subjective perspectival relativism swallows everything. Nothing is true. Everything is an illusion.

On the other hand, Kant had grasped something about the relationship and ultimate identity of subject and object that escaped the Newtonians and their theoretical heir, Nietzsche. He saw that our experience of the world went beyond merely empirical sense experience. Reason – transcending even the forms of understanding through which we experienced nature – was intimately connected with the intelligibility of the world. The law of reason discerns relationships within ever-greater unities and requires us to seek these unities. Reason is irrevocably linked to the unity of the cosmos.

Kant deduced that this reasoning power was a necessary law; otherwise, we would have no reason at all. Without reason, we would not possess coherent comprehension, and devoid of reason, no sufficient criterion of empirical truth or any truth could exist. Therefore, in order to secure an empirical criterion for science, we have no option but to presuppose the systematic unity of nature as objectively valid and necessary (1965b, sec. B 679, 538). In sum, there is a transpersonal structure to human reason that links us to the coherent and intelligible universe around us. Subject and object are inseparably linked, and the revolutions in 20th century science have borne out Kant's deductions.

G. W. F. Hegel picked up Kant's principles in the 19th century (1967). Thereafter, these principles became fundamental to a new paradigm that recognized the inseparability of subject and object in the constitution of knowledge and the holism of nature. Even before the theory of evolution proposed by Darwin in 1859, Hegel discerned *emergent evolution* in the cosmos, within which human beings were a necessary component. He viewed human life as capable of dialectically evolving into truly new paradigms. Subject and object were necessary to one another, dialectically interacting through time and moving into a future characterized by ever greater universality, rationality, and freedom.

Thinkers like Kant and Hegel were discerning the flaws in the mechanistic and fragmented worldview of early-modern philosophy and science. However, it was too little too late – the basic premises of the early-modern paradigm had become fundamental to many philosophers, and now were embodied within worldwide institutions like capitalism and the system of sovereign nation-states, both of which emerged during the early-modern period. Fragmentation became institutionalized, obscuring the holism that had been promoted in different ways by Spinoza, Leibniz, Kant, and Hegel.

In the case of each empiricist thinker (Locke, Berkeley, and Hume), "simple" ideas or impressions were treated like Newtonian atoms, and the attempt was made to construct human knowledge from these simples, just as the world was understood as a machine constructed from material atoms in external relationships with one another. The world was understood in terms of universal efficient causality only, and the teleology fundamental to both ancient and contemporary thought was excluded. The world therefore appeared to operate like a vast machine through its own self-contained motions.

God (if recognized at all) was relegated to the role of an original creator who no longer played a part in the functioning of the machine. Mind found no place in the machine either, and scientists were supposed to approach their work "objectively," without letting their merely "subjective" values, emotions, or intuitions interfere. They also were not responsible for how their discoveries were used. The scientific invention of bombs was not intrinsically or morally different from the scientific invention of medicines to cure diseases.

Similarly, modern capitalism – developing since the Italian Renaissance and theoretically conceived by Adam Smith and others in the 18th century – treats human beings as if they were Newtonian mass-points. The multi-dimensional complexity of human beings is ignored in favor of a conception of persons as individual self-interested atoms seeking only to maximize their personal advantage, the so-called *homo economicus*. Just as the bodies (mass-points) of Newton were governed in their external relations with other bodies by the causal law of gravitation, so the capitalist paradigm imagines self-interested human units operating within assumed natural laws of supply and demand.

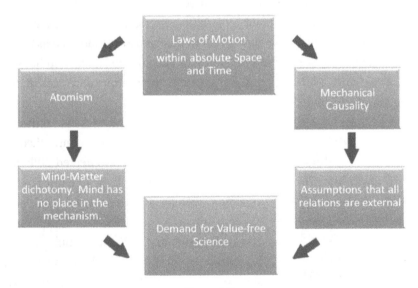

Figure 2.2
The nexus of concepts integral to the Newtonian Paradigm:
Mind has no place in the mechanical world assumed by Newtonian science,
creating the false demand that science must be value-free

Moreover, just as Newtonian bodies in motion were meshed together in the orbiting of the planets and the intricacy of natural processes on Earth, so the human atoms of capitalism were inserted into mathematical formulas producing theoretical benefits for the majority of humanity. Adam Smith first formulated this seemingly miraculous outcome as an "invisible hand." The meshing of innumerable greedy atoms in the "free" marketplace produces the greatest benefit to the greatest number of people.

In fact, this economic model had to be (and continues to be) forced on the world through brutal colonial and imperial wars. No nation or corporation has ever truly followed or believed in "free trade." Instead, they seek to control markets and eliminate competition. As a result, poverty and misery in the world have increased everywhere capitalism has extended its greedy hands, due in part to the legions of academic mandarins and philosophers of exploitation who continue to spew forth versions of this insidious ideology (cf. Smith 2005, Ch. 6).

Like the "value-free" scientists, "free trade" economists were viewed as objectively formulating the laws of economics and not responsible for the consequences. "Values," in any case, were deemed merely subjective and carried no binding universal authority. Then so-called "political scientists" followed suit, promoting such ideologies as "spheres of influence." The fragmentation of capitalism and the acceptance of nation-states, these "objective" models concluded, were based on the early-modern mechanistic model that assumed the self-interested and competitive orientation of each of the "atoms" in the system. Today, this faulty philosophy permits every corporation and each nation to operate out of self-interest.

Under the mechanistic view of the universe, nations operate out of hegemonic self-interest, just like individuals, and corporations operate out of capitalistic and competitive self-interest.

Thus, Newtonian physics is mirrored in nearly every social theory, economic theory, and political philosophy. It also infected 20th century theories of positivism, logical empiricism, and epistemic realism. It is reflected in the "political realism" theories put forward by thinkers like Hans Morganthau (2006, orig. pub. 1948) and Leo Strauss (1965), both of whom saw the nation-state system as merely a benign, amoral struggle of giant power-centers for ascendency and hegemony. These unfortunate historical developments explain how the economic self-interest of capitalism and the power-oriented operations of nations replaced traditional and rationally grounded values like peace and justice.

This "political realism" still dominates in the 21st century, especially through the foreign policy of the United States (cf. Engdahl 2009; Hardt and Negri 2000; Escobar 2006). Sadly, proponents of "world systems theory" –

thinkers like Immanuel Wallerstein (1983) and his many academic followers – appear to see no hope beyond the perpetual unbridled competition of the global imperial power-centers for control of markets, resources, and exploitable labor (cf. Shannon 1989).

2.4 Paradigm Shift of the Twenty-First Century

As we have seen, the distinction that emerged during the Axial Period between our subjective perceptions and the objective world became the basis for the ancient, medieval, and modern paradigms that impacted the development of human history. The 19th century sociologist Auguste Comte also saw history as progressive, and he formulated three large stages that he labeled the Theological, the Metaphysical, and the Positive. The Theological stage included both ancient and medieval thought. The Metaphysical stage included the 17th, 18th, and 19th century philosophies that encouraged principles such as democracy, human rights, and sovereign states as autonomous personalities. The Positive age involved the restriction of knowledge to what is confirmable by science.

During the Positive age of science, Comte believed that real progress was possible: from material progress, to improvements in human physical well-being, to intellectual progress in understanding scientific knowledge and its foundations, to moral progress that included increased common sympathy, benevolence, and sense of community (Blaine 2004, 16). Comte's relatively short view of development covered only the past 2,500 years or so, and it can be questioned on empirical grounds. His basic idea – that moral progress is possible and emerges from the process itself – can be placed within the context of the one to two million years of human existence. Only yesterday (on this time scale) did we begin to think in terms of moral principles, right and wrong, and society in relation to these issues. Today, we see that scientific and technical progress are not equivalent to progress in human well-being. Nor are they equivalent to progress in human moral consciousness.

Much of contemporary philosophy of law and government involves the Newtonian and Darwinian framework, thereby reducing human beings to insignificant creatures struggling for survival in a vast, impersonal universe. In their book *The Conscious Universe: Part and Whole in Modern Physical Theory* (1990), Menas Kafatos and Robert Nadeau validate the emergent 20th century paradigm of the inseparability of whole and part and make their point in the following way:

> And yet it is also demonstrably true that theoretical reason does over time refashion the terms of construction of human reality within particular linguistic and cultural contexts, and thereby alters the dynamics of practical reason. As many scholars have exhaustively demonstrated, the classical paradigm in physics has greatly influenced and conditioned our understanding and management of human systems of economic and political reality. Virtually all models of this reality treat human systems as if they consisted of atomized units which interact with one another in terms of laws for forces external to the units. (181)

The implications of the early-modern paradigm were discerned, perhaps most clearly, by the 19th century philosopher Friedrich Nietzsche, because he perceived a tendency toward disaster in the Newtonian science of his day. In the *Genealogy of Morals* Nietzsche expressed his dismay at the devaluing of human beings by the apparent implications of science: "*All* science (and by no means only astronomy, in the humiliating and degrading effect of which Kant made the noteworthy confession: "it destroys my importance"…), all science … has at present the object of dissuading man from his former respect for himself, as if this had been nothing but a piece of bizarre conceit" (1969, 155).

Progress to date is far from obvious, given that human beings remain mired in the outmoded Newtonian paradigm in terms of economics, nation-state political organization, and social theory. But the paradigm shift of the 21st century does point toward a new foundation for economics, government, psychology, and social theory. The new paradigm does not replace the older Newtonian paradigm with an incommensurable alternative, as Thomas Kuhn (1970) would have it, suggesting a relativism of paradigms. Rather, according to physicist Henry Stapp, the Newtonian view was "fundamentally incorrect" (Kitchener 1988, 56). Therefore, the new paradigm encompasses the older within a larger and more coherent framework revealed by 20th century breakthroughs in science. It replaces those assumptions that were incorrect and embraces those that were correct within a larger framework.

Philosophy itself derives from an examination of the *coherence* of experience. Strict empiricism is an epistemology that derives from the Newtonian paradigm assuming that knowledge is built up from discrete sensations organized into patterns by self-conscious observers who are separate from what is observed. However, such reductionist and atomistic empiricism cannot begin to account for 20th century developments in human knowledge and understanding of the holistic universe in which we live.

Observation, of course, plays a necessary role, and we are far removed from any rationalist reliance on "pure reason." Truth results from the coherence of empirical experience along with thought, and this coherence reveals a universe of interconnected wholeness encompassing a multiplicity of parts in a descending series of "fields" or systems (Harris 2000a, Chap. 16). It reveals a universe observably structured on the principle of *unity in diversity*, and it reveals a universe in which values – like mind itself – emerge from the cosmic process that has come to self-awareness in the human project.

Human beings emerged out of a cosmic process some 13.8 billion years in the making. This cosmic process appears as a comprehensive whole integrating a multiplicity of elements into ever-greater levels of complexity, resulting in ever-greater levels of unity in diversity. Indeed, human beings exemplify the highest level of integrated *unity in diversity* yet observed. We are aware that we are one species inhabiting planet Earth, and we are conscious of our vast multiplicity of unique differences among individuals, ethnicities, cultures, nations, and religions.

In short, our emergence out of the cosmic process has transformed the way we look at ourselves. Pioneering psychologists, physicists, and philosophers began articulating this new holistic paradigm during the 20th century. In the 21st century, this new paradigm has begun to transform general thinking, though not our outdated institutions. Henry Stapp compares the view of ourselves derived from "the classical view" of early modern philosophy and physics with what he takes to be the substantially correct view put forward by physicist Werner Heisenberg.

> This classical view of man and nature is still promulgated in the name of science. Thus, science is seen as demanding a perception of man as nothing more than a local cog in a mechanical universe, unconnected to any creative aspect of nature. For, according to the classical picture, every creative aspect of nature exhausted itself during the first instant.
> ...
> [T]he real world of classical physics is transformed into a world of potentialities, which condition, but do not control, the world of actual events. These events or acts create the actual form of the evolving universe by deciding between the possibilities created by the evolving potentialities. These creative acts stand outside space-time and presumably create all space-time relationships. Human mental acts belong to this world of creative acts, but do not exhaust it. (In Kitchener 1988, 56-57)

The classical view of nature exhausted creativity in the first instant because a mechanism is fixed by its structure and understood in terms of efficient causality rather than a creative *telos* guiding the emergence of what is genuinely new – the realization of potentialities that are not yet actualities. Stapp affirms that science has revealed potentiality as prior to actuality, that the universe is an emergent process of realizing higher potentialities, not a mechanical process of evolution pushed from behind by efficient causalities. Further, the human mind participates in the creative process at the heart of the universe itself, a perspective that coheres with the view of many psychologists such as Robert Lifton (1993), that a human being must be understood in terms of the primacy of his or her potentialities. In other words, new conditions in the universe give rise to new possibilities. Many of our highest potentialities are precisely in the dimensions of values and reveal that we are not locked into some sordid, irredeemable "human nature," but have the potential to grow cognitively, morally, and spirituality, that is, to grow in compassion, kindness, gentleness, respect, and mutual understanding.

Today, scientists are not locked into an objective stance that is value-free and detached from the consequences of what they discover. The highest potentialities of science are now unleashed. The shift to holism, Errol E. Harris points out, "abolishes the opposition of fact to value, for the criterion of truth and value is the same: coherent order and unity …. For what ought to be done is what promotes health, unity, and harmony, as well in the biosphere, as in human society" (2000b, 261-62). Mind permeates the whole as a manifestation of its organized energy flow. Mind comes to self-awareness in human beings, eventuating in values and discernment of the true, the good, and the beautiful.

Figure 2.3
Fanciful image of Cosmic Intelligence informing an evolving space-time-energy matrix
that gives birth to life, then mind, then values moving into an ever-expanding horizon
Source: Barbara Marx Hubbard (permission given to Oracle Institute Press)

In exactly this same manner, the *Earth Constitution* represents a milestone of human evolution. It provides the potentiality for us to actualize the whole. This very actualization gives rise to new possibilities for communication, cooperation, peace, and biospheric harmony. The Newtonian mechanical universe has been replaced by the emergent evolutionary universe: dynamic, full of synergistic potential and purpose.

Those who claim that democratic world law and a sustainable world system will take decades or centuries to evolve are still thinking in terms of the social relationships and causal mechanisms derived from the early-modern paradigm. The global crises that confront humanity put pressure upon human consciousness and simultaneously create potentialities that did not exist a mere fifty years ago. Rapid transformations of consciousness, attitudes, institutions, and relationships are very possible.

Moreover, the possibility of democratic world law is not locked into the causal relationships derived from outdated institutions and, therefore, may come to fruition very rapidly. The values of peace, justice, freedom, and sustainability are no longer merely subjective sentiments. They are objective demands arising from our cosmic roots. Below, we see that a civilization operating under democratic world law can be founded now. We need not wait for it to slowly evolve because its time has come.

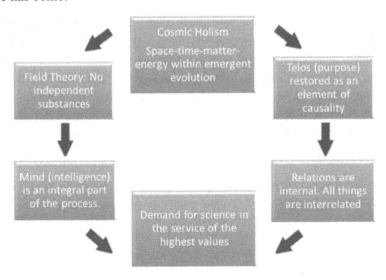

Figure 2.4
The nexus of concepts integral to the Holistic Paradigm:
The scientific paradigm-shift over the past century promotes
what is valuable in culture, human life, and civilization.

In his book *The Tao of Physics* (1975), physicist Fritjof Capra describes the interconnected holism of our universe as the central discovery of 20th and 21st century physics:

> Thus modern physics shows us once again – and this time at the macroscopic level – that material objects are not distinct entities, but are inseparably linked to their environment; that their properties can only be understood in terms of their interaction with the rest of the world. ... [T]his interaction reaches out to the universe at large, to the distant stars and galaxies. The basic unity of the cosmos manifests itself, therefore, not only in the world of the very small but also in the world of the very large; a fact which is increasingly acknowledged in modern astrophysics and cosmology. In the words of the astronomer Fred Hoyle: "Present-day developments in cosmology are coming to suggest rather insistently that everyday conditions could not persist but for the distant parts of the Universe, that all our ideas of space and geometry would become entirely invalid if the distant parts of the Universe were taken away. Our everyday experience even down to the smallest details seems to be so closely integrated to the grand-scale features of the Universe that it is well-nigh impossible to contemplate the two being separated." (209-210)

Once again, the holism of the universe is identified, and with it the dynamic interdependence between local and macroscopic conditions. Human beings are implicated in this holism, and it is relevant that the idea of an "Anthropic Principle" originated among physicists themselves. This is the idea that the emergence of human beings has been integral to the evolution of the universe from its very beginning. Philosopher and scientific cosmologist Milton K. Munitz, in his book *Cosmic Understanding: Philosophy and Science of the Universe* (1986), remarks on the "Anthropic Principle" developed by 20th century physicists, who discovered the astonishing connection between the exact physical parameters of the universe and the emergence of human life:

> It is not simply that certain properties of the universe throw light on human existence, but the reverse also holds. The universe and human life are coupled. If we are to understand either, we need to move in both directions: from the universe to man and from man to the universe, since they are mutually involved in a very special way. (236-237)

It is high time that philosophers of law, jurists, and legal theorists adopt the transformed paradigm bequeathed to us by 20th century science. Democratic law is a fundamental expression of our human reality. If the holism of human life on this planet clearly dawns upon us, democratic world law and sustainable civilization will soon follow. Conversely, any theory of law based on false foundations or mistaken premises can only spell disaster for human existence on this planet. As Munitz puts it, "the universe and human life are coupled." Clearly, we need planetary institutions premised on this connection.

In 21st century physics, everything in nature is integrated into wholes, including human beings, and the cosmos cannot be understood apart from the unity in diversity of this holism. Economic and political models must likewise be transformed to appropriate the new holism of unity in diversity. According to Capra, the new holistic model in physics is mirrored across the spectrum of sciences that now examine the world from the perspective of "self-organizing systems."

> The broadest implications of the systems approach are found today in a new theory of living systems, which originated in cybernetics in the 1940s and emerged in its main outlines over the last twenty years. ... As I mentioned before, living systems include individual organisms, social systems, and ecosystems, and thus the new theory can provide a common framework and language for a wide range of disciplines – biology, psychology, medicine, economics, ecology, and many others. (In Kitchener 1988, 149)

For Capra, self-organizing systems do not have their organization imposed upon them by external forces (the environment) as in the Newtonian model. Rather, they tend toward establishing their own order while interacting with the environment (ibid., 149-150). This approach means that nature includes within itself a *telos* or *nisus* (internal self-direction) for self-organization and wholeness. Ultimately, this approach also includes the human mind, no longer understood as a mysterious opposite of matter. Additionally, mind cannot be understood as an egoistic atom of rationalized self-interest, as imagined by the theorists of capitalism. Rather, mind is now integral to matter at every level:

> The organizing activity of living, self-organizing systems, finally, is cognition, or mental activity. This implies a radically new concept of mind, which was first proposed by Gregory Bateson (1979). Mental process is defined as the organizing activity of life. This means that all interactions of a living system with its environment are cognitive, or become inseparably connected. Mind, or more accurately, mental process is seen as being immanent in matter at all levels of life. (Ibid., 151)

Today, mind can be expressed on various organizational levels and is understood not to be a different kind of substance from its environing universe. In his book *Human Potentialities* (1975), psychologist Gardner Murphy concludes that the human mind may well express "cosmic potentialities, cosmic trends as yet unparalleled elsewhere in the knowable universe." Consequently, human beings should be defined in terms of our immense potentialities for self-realization and self-transcendence. Similarly, Raimon Panikkar concludes that "Man is much more than a mere part of the cosmos." For Panikkar, we are "the center of that consciousness that pervades everything .... We are more than passive inhabitants of the universe .... We can 'effect' or provoke the failure of the entire adventure of being. ... Man may not be the absolute king of creation, but certainly is its gardener" (2013, 349). The potentialities of the cosmos dovetail in us in ways that provide for explosive growth and genuine transformation. Yes, human beings are poised to ascend to higher levels of maturity, self-actualization, and self-transcendence.

Physicist Ediho Lokanga observes that "many researchers have argued that the three foundations of reality consist of matter, energy, and consciousness, and that each of these components enfolds the other two" (2018, 114). Consciousness pervades the universe, rendering the human brain a receiver and holographic synthesizer of consciousness, not its origin. The traditional dualism between mind and matter is no longer tenable in a universe now understood as composed of various forms of interconnected energies. Likewise, fact is no longer easily separable from value, since the holism of our situation interweaves human beings with the environing world. In his book on the theoretical grounds for a holographic view of the universe, Lokanga states that "we have come to realize that every individual point in the universe contains the whole universe itself ... every subatomic particle comprises a web of interconnections by which it becomes intertwined with other parts of the universe" (2018, xx). This interconnectedness, we now see, applies to human beings as well as all other things.

Peace and justice are now understood as objective values mirroring the unity in diversity of our cosmic and human background. Our human potentialities are now understood as "cosmic," and necessarily include the dimension of values. Values are no longer "merely subjective," but arise from the intelligence that informs the universe. Errol E. Harris explains that "genuine rational love, therefore, must extend to the entire human race," and that this love gives rise to the moral ideal of "the unity of the perfected human community" (1988, 162-63).

Consequently, 21st century holism requires us to act as cooperative parts of wholes in perpetual relationship to the wholes that encompass and sustain us. The new paradigm also compels us to establish institutions reflecting this holism. In a manner similar to all natural systems, planetary democracy will one day function as a "holarchy" as well. Local communities will interact democratically and economically, addressing local problems and issues within a federated world order. Larger regional social and political units – for example, cantons of China, pradesh of India, or states within the United States – also will function democratically, dealing with regional problems and issues. The parts will interact with the whole and fractally pattern the unity in diversity of the world at every level.

Nations naturally and holarchically include these smaller units and are themselves included within the Earth Federation that addresses planetary problems and issues. In this way, democratic processes are not only preserved but also enhanced. Protection of the rights of individuals and the federated units within the system are enforced through laws maximizing the equal freedom of everyone to develop his or her potential within a framework of the common good for the whole of humanity and our planetary ecosystem.

Furthermore, a world governed by a democratic constitution that integrates all nations, persons, and communities will provide purpose to the whole and positively influence the functioning of all parts. The parts and whole of such a global democracy will synergistically empower one another, as Buckminster Fuller pointed out in *Operating Manual for Spaceship Earth* (1972, 88). The present chaos of competition, greed, and self-interest will be transformed by this new paradigm, which Fuller calls "true democracy" (ibid., 77). Holism and planetary consciousness finally will replace fragmentation, and this actualization will mean the beginning of a sustainable planetary civilization.

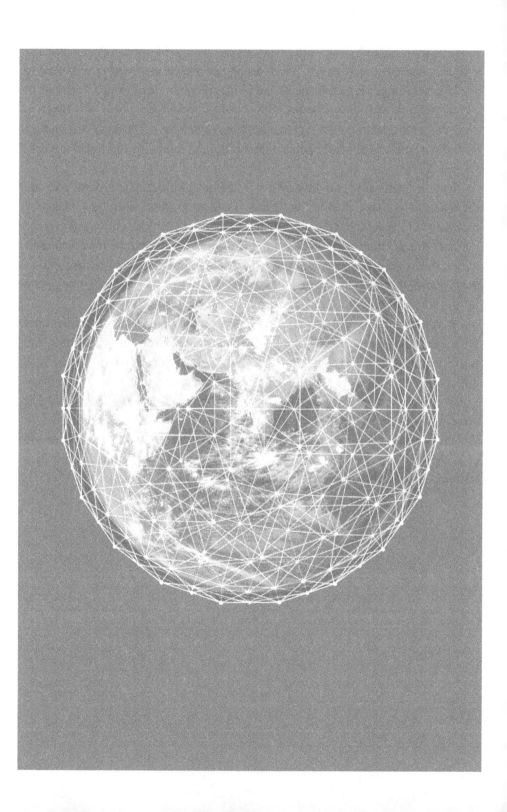

CHAPTER THREE
Political Collapse:
The Advent of Evolutionary Science

Our species is now at the most important turning point since the Agricultural Revolution. For the first time humanity has the knowledge to destroy itself quickly, and for the first time humanity has the knowledge to take its own evolution into its hands and change the way people comprehend and think.

Robert Ornstein and Paul Ehrlich

3.1 The Twisted Logic of the Nation-State System

In a world divided into nearly two hundred territorial units competing with one another politically, militarily and economically, there is a perpetual danger of total disaster. It is the logic of the nation-state system itself that lays the ground for disaster, not the corruption of peoples or leaders. The system creates corruption due to its implacable logic, which is inherently destructive of democracy, freedom, justice, and peace.

Many scholars point to the 1648 Peace of Westphalia at the end of the Thirty Years War in Europe as the origin of today's system of sovereign nation-states. While not mentioning the concept of "sovereignty," the document does define nations according to a principle in which each bounded territory is to have a single, unified government conducting all internal affairs of the nation. Secondly, the treaty states that each nation will be independent in its relation with other nations. Not surprisingly, this system of independent units mirrored the atomistic science of the 17th century, just as the system of independent companies and self-interested human economic units of capitalism mirrored the reductionism of the day.

However, some thinkers of that era, such as Thomas Hobbes in Britain and Baruch Spinoza in the Netherlands, pointed out that this constitutes a war system. Militarized independent entities recognizing no enforceable laws above themselves inevitably form a war system, even when they are not engaging in hostilities. Kant pointed this out again in the 18th century, and Hegel did the same in the 19th century. During the 20th century, Emery Reves made this explicit point in his book *The Anatomy of Peace* when he wrote, "*War takes place whenever and wherever non-integrated social units of equal sovereignty come into contact* (1946, 121, emphasis in original). Today, many 21st century scholars have made the same point, as we shall soon see.

In terms of lethality, 17th century armies were primarily riding on horseback or marching on foot, fighting with swords or muskets. Beginning with World War I, however, war became industrial scale extermination of the enemy and his life support systems. Since the late 1950s, it literally has become possible to exterminate all of humanity. Today, the situation is dire. Hypersonic missiles can carry nuclear weapons around the world in minutes, wiping out everyone and everything in a nuclear winter that would follow such a calamity. Moreover, the current Covid-19 global pandemic may have been a leak from one of the bioweapons labs encircling the globe. Whether nations utilize nuclear weapons or bioweapons, the twisted logic is the same.

In all seriousness, how stupid are human beings for failing to transcend this mindless, lethal, and counter-productive system of militarized sovereign nation-states? It is a truly absurd system and produces all manner of mayhem, not the least of which is the threat of wiping out humanity. Yet most people on our planet appear to accept these absurdities as if they were inevitable facets of human existence.

National sovereignty is by its very nature a root cause of war, because every national government regards security and defense against possible invasion or aggression as a first and most vital interest.

The first and most obvious danger inherent in the logic of nation-states is that of war, as thoughtful people have pointed out for centuries. War and militarism destroy all the values identified in Chapters One and Two, even when a war is promoted as a "defense of democracy" or some other noble sounding ideal. National sovereignty is by its very nature a root cause of war, because every national government regards security and defense against possible invasion or aggression as a first and most vital interest.

Most sovereign states, therefore, seek to maintain as powerful an arsenal of weaponry as they can afford and spend inordinate amounts of their wealth on armaments, which other states view as a potential threat. The result is an ever-increasing arms race in the attempt to maintain a balance of power, which generates international tensions and crises that constantly tend to explode into armed conflict (Harris 2000). Today, the United States and Russia are in yet another absurd arms race for both hypersonic weapons and upgraded nuclear arsenals. They are spending ludicrous amounts of money on preparations for an Armageddon, when all would be better served by directing resources toward saving our planetary environment.

As Errol E. Harris points out in his essay "Are We Unteachable?," the United Nations cannot prevent war because Article 2 of its Charter commits it to uphold the sovereign status of its members, and Chapter VII provides – as its means of "peace-keeping" – only warlike measures "by land, sea, and air" (in Martin, ed. 2005b). As long as states remain sovereign and independent, world peace cannot be established because the U.N. is premised on the same flawed logic of nation-states that belie the noble ideals enshrined in its Charter.

Perhaps even more insidious than nuclear madness is the proliferation of bioweapons research labs throughout many nations around the world. These bio-arsenals have genocidal implications because bioweapons are more difficult to monitor, detect, and contain. Most of these labs are not open to inspections, while others are part of military installations, such as Fort Dietrich in the U.S., also not open to observation. Many cliff-hanger movies have plots with terrorists attempting to steal the material necessary to build a nuclear weapon, but real bio-terror labs are much more vulnerable to simple accidents or unintentional escape of deadly pathogens. The global pandemic may well be linked to bioweapons research, as a number of informed thinkers have pointed out.

This absurd war system is indelibly tied to the system of sovereign nation-states. If we reduced sovereign power by joining these nations and their leaders all under the rule of enforceable world law, then the danger of global destruction would substantially disappear. For this reason alone, we should be doing everything in our power to ratify the *Earth Constitution*.

Linked to this system of sovereign nations are the global conglomerates, which seek to accumulate wealth regardless of the common good and outside of independent oversight of their activities. An economics predicated on the secret machinations of shareholder self-interest will oppress those human values that interfere with its primary directive – to accumulate and reinvest private wealth. The most obvious example is the military industrial complex, comprised of giant corporations with a vested interest in the propagation of war or, at a minimum, perpetuating the war system within the governments of their respective nation-states.

Even more terrifying is the fact that many multinational corporations operate autonomously, outside the confines of any individual nation, and hence beyond the possibility of effective regulation by any and all governments. They use the anarchic system of a multiplicity of governments to their own advantage, regardless of the impact on the people of Earth. Indeed, some corporations wield more power than some nations, with assets exceeding the nation-states within which they operate. Such businesses certainly influence governments, which essentially are colonized into the service of private, profit-making institutions. Thereafter, governments become unaccountable to the common good or the welfare of the community they are supposed to be governing. Anarchy, greed, exploitation, and cynical self-interest then become normalized. In the poor nations of the world, bribery is rampant and criminal deeds can be purchased by powerful corporations. Even murder is ignored, as in the case of oil companies killing native peoples defending against the rape of their ancestral lands in the Nigerian Delta and Ecuador.

The system of sovereign nation-states gives institutional support to the lowest aspects of human nature: blind territoriality, a warped belief in national "exceptionalism," the likelihood of surrendering civil rights and individuality into a collectivism of state worship, and a tendency to blind obedience to state authorities and their military mandarins. This in turn produces a rich environment for fascism to grow. Fascism requires enemies to bind the nation into a unit submerged in the power of the state. It requires the complicity of corporations who link with state power as the source of obscene profit and wealth, and it crushes diversity and freedom, submerging the citizenry in obsessive "patriotism" that asks no questions and eventually rejects the uncertainty of democracy.

To be clear, nations are more than mere territories. They are a mythological construct of epic and archetypal proportion. Consider, for example, the profundity of "mother Russia," or the "children of light" fighting the "forces of darkness" in the United States. Consider the common phrase "American exceptionalism," despite the fact that the U.S. rarely ranks on the top ten lists of any meaningful global measure. Sheldon Wolin points out that such a "myth presents a narrative of exploits, not an argument or a demonstration. It does not make the world intelligible, only dramatic" (2008, 10). Yet countries everywhere use mythology to dramatize their imagined unique roles in world history.

After World War II, the U.S. embraced the myth of leading the "free world" in a fight against the "evils of communism." This narrative powered an empire hell-bent on warring, torturing, dominating, and exploiting the nation-state system for its own interests. By the 1990s, the U.S. took credit for the collapse of the

Soviet Union. A victory over evil! Meanwhile in the Soviet Union, the collapse was cataclysmic, leading years later to the Russia we know today dominated by Vladimir Putin and his fellow oligarchs. *H. W. Bush + Oligarchs - 9/11*

But did America defeat evil when the Soviet Union fell? Of course not. What enemy would the children of light need to conquer next? After a desperate search to perpetuate the U.S. myth, a miraculously convenient event occurred: 9/11. Thereafter, the "children of light" found their new "force of darkness": global terrorism. Like communist conspiracies, the new enemy was everywhere and nowhere, conspiring in secret, creating sleeper cells at home and abroad, waiting to strike, wanting to destroy, determined to obliterate the peace and prosperity of the empire and its free market.

As long as territorial instincts and national fantasies are exacerbated by dividing the world into nearly two hundred autonomous units, we will live in danger of obsessive nationalism and fascism. Earth is not the private property of the people who happen to be born at a particular place, nor does God magically gift certain nations with a special role in human history. Indeed, any honest history course covering the modern period will reveal the havoc caused by the nation-state system. Only a united world, premised on the principle of unity in diversity, can overcome these threats ... and more.

Another danger of nation-state logic is that mass media tend to operate in the service of the nation with which they identify. Mass media dedicated to the freedom of information and the creation of an educated and informed public are essential to the values of democracy, freedom, justice, and peace. Yet today, the mass media of each nation-state tend to reflect the dominant values and mythological fantasies of that state and its system of power. The "news" is disseminated to the population of each country from the point of view of the national establishment and its dominant governmental and business elites, who often own or control the mass media. The result is a clouded mirror that fails to reflect the destructive logic of its host country, thereby promoting toxic nationalism, imperialism, international misunderstanding, war, fascist tendencies, and injustice.

As scholars have pointed out, the people currently living within the global imperial center are the most propagandized, manipulated, and brainwashed population in the world (Chomsky 1989). The people within the U.S. experience job loss, outrageous health-care and insurance costs, environmental degradation, decaying infrastructure, repression of civil liberties, efforts to privatize social security, growth of the prison industry, draconian laws targeting minorities and the poor, and growing social disintegration. At the same time, when U.S. troops invade or bomb other countries, yellow ribbons appear on autos and in windows announcing: "Support our Troops." It does not matter if the troops invaded another country, if they are committing mass murder of civilians, if they are engaged in torture or repression. The mantra is always: "Support our Troops." The propaganda system in the service of the ruling class is mightily effective, to the point where exploited and marginalized people in the U.S. often support the very system that diminishes their lives.

The mass media focus on the "aggression" or "atrocities" of official enemies. We are told these are our enemies based upon secret evidence. The "evil" actions of these enemies are then published for public consumption according to misinformation created by the CIA or their own government (see Valentine 2017; Blum 2005). Yet the public is told they must trust their government as it attempts to protect them from horrible enemies in a dangerous world, while mass media uncritically repeat the formulas, photos, selection of certain information, and ideological framework provided by government and the ruling powers.

For example, millions of people were slaughtered in Vietnam and the American people were trained to only mourn 65,000 of these deaths. U.S. troops invaded Iraq and slaughtered its people, and the mass media spoke of the courageous Iraqis who resisted the invasion of their country as "insurgents" or "rebels." As many as 200,000 people were slaughtered in the first two years of the U.S. invasion of Iraq, yet most Americans mourned only our soldiers – just 1,800 of the lost lives. Terror groups were clandestinely armed in Libya and Syria by the U.S. and its allies with the object of removing national governments, and the American people and mass media sat idly by, repeating the propaganda about "fighting terrorism" and "bringing democracy."

In addition, the system creates a near inevitability that powerful states will act to dominate and exploit weaker states. Inherent in the modern world system are the aspirations to empire and hegemony of the strong over the weak. The entire history of modernity bears this out. Imperialism, colonialism, slavery, economic exploitation, spheres of influence, and secret pacts and intrigues comprise the history of sovereign nation-states. Ambitious rulers and compliant populations are not only at fault. The logic of the system itself recruits despotic rulers and promotes blind patriotism in trusting or naïve populations. In truth, the sovereign system is premised not on ethical principles, but on territorial power blocks, which act in service to mythologies that trump authentic principles at every turn.

Military personnel in brutal dictatorships are trained by the imperial troops to repress their own people, as happened in Indonesia, East Timor, Columbia, El Salvador, Guatemala, the Philippines and elsewhere, while the U.S. remains silent about the connection, the massacres, and the implications of its "foreign policy." Are the American people so hopelessly corrupt that they accept the horrific acts of their military and care only about their own forces? I refuse to believe it. The problem is not the American people, but the sovereign nation-state that institutionalizes a system of blind loyalty to the country where one happens to be born. The people of Germany supported Nazi troops in the same way. They chanted *Deuschland über alles!* and organized civic groups to "support our troops" in the field.

In this manner, imperial nation-states cultivate and promote a deadly patriotism in order to maintain blind popular support for their systems of domination and exploitation. In the U.S., the slogan is "God bless America." The fact that this translates into the destruction and domination of other peoples worldwide is of no concern, since in the popular mind, America has been chosen by God (and is closely linked to Israeli Zionism). Presently, the horrible suffering of the Venezuelan people due to the U.S. blockade of their nation is of little concern to the American public. I say again: The problem is not the people of the United States. The problem is the world system of sovereign states where such blind loyalty is promoted, thereby destroying democracy, freedom, justice, and peace.

Lastly, the military oppression of other peoples by the imperial centers integrates with economic exploitation. The wealthy imperial countries seek control of all natural resources: oil, diamonds, gold, manganese, forests, and coal. They want "financial stability" for their investments, and they understand that dictatorships and authoritarian regimes that will accommodate the multinational corporations, so long as the tiny wealthy elite in their own countries are richly compensated (Chomsky 1993; Blum 2005).

Tragically, these enumerated flaws in the logic of the nation-state system make a world of democracy, freedom, justice, community, and peace nearly impossible. These ideals developed during the Enlightenment period of the 18th century, and they form the greatest legacy of modernity as well as the highest hope for our future on this planet. The principles of

> *Democracy, freedom, and justice possess an inner logic that is fundamentally at odds with the logic of the nation-state system.*

democracy have an inner logic that is fundamentally at odds with the logic of the nation-state system.

Democracy will exist only when there is world democracy. The twisted logic of the sovereign nation-state system gives us the dangerous world described above. Under these conditions, authentic democracy within nations is nearly impossible. Governments must necessarily keep their foreign and military policies secret from their own citizens, thereby defeating civilian oversight that is the very essence of democracy. Economic decisions within a nation necessarily impact people in other countries, and decisions made by other countries are similarly capricious and unpredictable.

Hence, the only real freedom must be world freedom, guaranteed to every citizen of Earth. Currently, governments maintain elaborate immigration protection systems, systems of visa restrictions, and security arrangements because of the threat of terrorism, subversion, digital warfare, or machinations of other "rogue" nation-states and "terrorists." None of this would be necessary in a world where gross injustices have been eliminated and every nation and all people have equal rights. Only democratic world government can secure real freedom for the peoples of Earth.

Justice, too, can only be accomplished through world justice. As long as some nations have substantially more wealth, power, and influence than others, this will never be a just world. The logic of nation-states ensures that there will never be economic, environmental, juridical, or democratic justice among the diverse peoples of Earth. The current system will always marginalize the weak in the service of the powerful. Only world government under the *Earth Constitution,* which contains the clear mandate to treat all people equally, can succeed at this task.

Peace can only come to humanity as world peace. All consequences of the nation-state system – militarism, unregulated corporate power, the tendency to fascism and blind patriotism, the distortion of news and the media, and imperialism – work to destroy the possibility of peace. Fundamentally, war is a criminal activity, no matter what fantasies its perpetrators may use as justification. The production of weapons of war is a criminal activity. Only when these activities are recognized for what they are and abolished by enforceable world law will there be peace on Earth, and no force on Earth can create world peace except democratic world government.

Throughout the four-hundred-year history of nation-states, one or more of these logical consequences of the sovereign system have coalesced to perpetually destroy democracy, freedom, justice, and peace in our world. Today, with the technology of weapons of mass destruction brought nearly to perfection, this system portends the self-destruction of Earth and our common human project. Ironically, democracy is considered an unstable form of government by the U.S., whose foreign policy has routinely subverted, overthrown, or suppressed dozens of countries since the ascendancy of its global dominance at the end of World

War II (Blum 2005; Parenti 2011) In sum, a fragmented world system cannot support unity in diversity.

Only the sovereignty of all the people who live on Earth, as institutionalized in the *Earth Constitution*, can achieve the goals inherent in democracy: peace, equity, freedom, and prosperity. Real revolution can be evolutionary and need not be violent; whereas violent revolutions usually establish a new government on the same flawed assumptions as the old. Real revolution involves thinking outside the box of outmoded institutions and solving the governance crisis by establishing a new, more elegant paradigm. The holistic approach expressed in the *Earth Constitution* cuts through these immense problems and creates a decent life for all the diverse peoples of our planet.

The assumptions behind the system of sovereign nation-states create imperialism, and it is this flawed system that impels countries to act in destructive ways. A new world system based on democratically legislated world law will transform our planet. We must demilitarize, unite under a single constitution, and solve our problems through authentic democratic

> *Real revolution involves solving the current crises by establishing a federated government model for the entire planet.*

processes. The only legitimate route to world revolution is through democratic world law.

3.2   The Current World System

We have seen that there are two gigantic and dominant global institutions that must be transformed if we are to adopt a truly holistic planetary orientation: The first institution is global corporate capitalism, and the second is the global system of sovereign nation-states. Both are interdependent, interrelated, and inseparable from one another in the modern world system. It is quite astonishing when economists claim to articulate a complete theory of sustainability and the need for a major transformation of our fragmented institutions seem unable to think beyond the clearly unconnected and unworkable system of sovereign nation-states.

For centuries, many scholars have described the development of this modern world system, along with its internal dynamics and processes. By "modernity," we mean that period of world history since the European Renaissance during which sovereign nation-states and the multi-national conglomerates developed as the dominant world institutions. The accounts of historians such as Arnold Toynbee, sociologists such as Max Weber, economists such as Thorstein Veblen and Karl Marx, and philosophers such as Jürgen Habermas have regularly revolved around the interrelation of these two dominant institutions of the modern world, which have been interdependent for centuries.

Thus, these dominant features of the current system are quite clear. While there is active debate as to the developmental periodization of the world system, as well as over the cyclical features of the system (i.e., how often and of what severity must depressions and recessions recur: Chase-Dunn 1998), the interrelation of the dominant institutions has been described at length in volume after volume (see Parenti 1995; Smith 2005; Petras and Veltmeyer 2005; Klein 2007). Yet the work of these thinkers has not led to a general understanding among the mainstream population. These domineering institutions are so pervasive that they form the invisible background of life as if they were natural phenomena – like the air we breathe or the water many of us take for granted – and their self-justifying propaganda is everywhere.

What is not controversial and forms part of the awareness of most educated people is the understanding that the sovereign nation-state evolved out of the medieval feudal political system just as modern capitalism evolved from feudal economic relations. These institutions have operated as an inseparable set of interrelated systems, inextricable and interdependent. Their destructive consequences have operated in tandem since the very beginning. Social scientist Christopher Chase-Dunn expresses this accurate observation:

> The state and the interstate system are not separate from capitalism, but rather are the main institutional supports of capitalist production relations. The system of unequally powerful and competing nation-states is part of the competitive struggle of capitalism, and thus wars and geopolitics are a systematic part of capitalist dynamics, not exogenous forces. (1998, 61)

Consequently, the multiplicity of negative consequences caused by our world system originate in the systemic nexus of these two institutions that are inseparably linked and determine the social, political, economic, and environmental conditions in which planet Earth is today mired. Hardt and Negri analyze the relation between global expansion of capitalism and nation-state imperialism. It was the need for new markets and new ways of investing surplus value under the capitalist imperative to "expand or die" that led the ruling classes of the imperial nations into foreign imperial and postcolonial conflicts (2000, 221-239).

The dysfunctional macro-features of the modern world system have developed through a number of overlapping and interrelated periods, often identified by thinkers under such headings as the era of the conquistadors (or so-called primitive accumulation), the era of mercantilism, the era of colonization, the era of slavery, the era of decolonization and neocolonialism, etc. The immense crises that today threaten to tear our world asunder are the result of the largely unspoken and unexamined premises under which we have operated for the past several centuries. Today, these assumptions are hopelessly outdated and inadequate for our planetary needs.

Nevertheless, quality work has been done by the thinkers of the 19th and 20th centuries who attempted to extrapolate the observable and probable consequences of this system. Every political, economic, and social assumption carries within it probable consequences, now proven based on the factual conditions we find all about us in today's world. The internal logic of the world system produces demonstrable and dire political, economic, and social consequences, such as commodification of existence, wide-spread poverty, perpetual wars, militarism, human rights violations, and environmental destruction (see Wolin 2008).

World Systems Chart

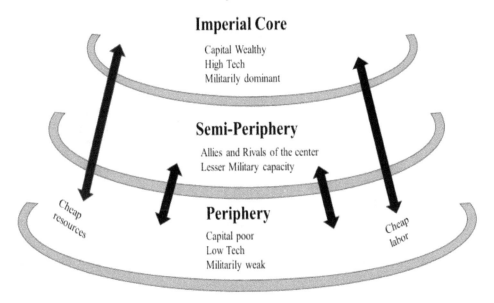

Figure 3.1
World Systems Theory reveals that the current system as a whole is designed for exploitation of the poor by the rich and powerful nation-states

Chris Williams, in his 2010 book *Ecology and Socialism: Solutions to Capitalist Economic Crisis*, affirms that "capitalism is thus systematically driven toward the ruination of the planet and we underestimate how committed the system is to planetary ecocide at our peril. As stated above, ecological devastation is just as intrinsic to the operation of capitalism as is the exploitation of the vast majority of humans in the interests of a tiny minority, imperialism, and war" (232). Yet, Williams ignores the fact that capitalism cannot engage in imperialism and war on its own. For capitalism to work, it needs synthesis with militarized nation-states. Hence, a fundamental change in capitalism will require a concomitant change in the system of sovereign nation-states.

In addition, capitalism is inherently anti-ecological just as the nation-state is anti-ecological. The nexus between countries and capital allows the imperial capital centers to expand, in perpetuity theoretically, with surplus value reinvested. It also requires the constant appropriation of resources, institutions, and peoples into capitalist production for the creation of ever-more surplus value. This economic system cannot continue in its present form if human beings are to achieve a sustainable civilization for Earth.

Futurist and environmental economist Jeremy Rifkin correctly states that the concept of "private property" will have to change radically within a low-entropy, sustainable culture. While personal consumer goods and real estate may remain protected, large tracts of land and major resources will have to be deemed public property and directed toward the common good of all (1989, 245). Yet the public trust will have to include more than real estate. It will necessarily have to protect the oceans of the world, the atmosphere around the world, the ozone layer, and the vital underground natural resources of the world. What kind of global institutions will be built once we accept that the global commons must be protected for all the people of Earth? We shall soon see how the *Earth Constitution* accomplishes this goal directly.

In his essay on "Environmental Ethics," Shridath Ramphal centers on a fundamentally correct evaluation of our situation. He writes, "Today the concepts of neighbor and neighborliness are being enlarged and refined in our interdependent world but have yet to be brought under the rule of law. What global interdependence means is that we all need each other in some measure: for prosperity, for subsistence, for survival. Our planet offers no sanctuaries. There are no shelters that insulate anyone, anywhere, from disease, from the effects of poverty, from military holocaust, from environmental collapse" (in Mayur 1996, 172). In *The Great Turning*, David Korten rightly declares that now is the time for a paradigm shift, away from this system of domination toward a new model of being human:

> We now have the means to end the five-thousand-year era of empire that has reproduced hierarchies of domination at all levels of human organization. A global cultural and spiritual awakening is building momentum toward the birthing of a new era of Earth Community based on a radically democratic partnership model of organizing human relationships. This awakening gives us cause for hope. (2006, 21)

The dual hierarchy of the sovereign nation-state intertwined with systemic capitalist greed has come to the end of the line. Thankfully, the rule of law and the new global community – as clearly defined in upcoming chapters – does not have to empower capitalistic companies and nations, nor their relentless quest for expansion on a finite planet. Rather, we have the insight to structure world law and economics based on a new system of sustainability, cooperation, and equity.

3.3 Truth, Love, and Our Common Humanity

We have seen that holism comprises the most fundamental understanding of our cosmic world order emerging from the 20th century revolutions in science. This holism can transform our entire way of thinking and living on Earth. It can become the basis for a new economics, a new ethics, and a new understanding of human social and political life. Yet these new understandings simply represent the fulfillment of certain civilizational fundamentals – like democracy – that go back to the ancient world. This paradigm shift in human thinking has not yet taken root in our ethical, social, or institutional life. We remain trapped in the older paradigms predicated on fragmentation and division, and our immense suicidal problems of the 21st century stem from this fragmentation.

But it is more than this, for we are under the illusion that our distinctions and divisions describe some fragmented "reality" that we believe is there for all to see. Physicist David Bohm points out that these distinctions (i.e., nation, race, caste, culture, ethnicity, etc.) are *generated by us* and do not "correspond" to any so-called external realities (1980, 9). Hence, "men who are guided by such a fragmentary self-world view cannot, in the long run, do other than to try in their actions to break themselves and their world into pieces, corresponding to their general mode of thinking." Any time we create "groupings of people," whether political, economic, or religious, we create separations and divisions from the rest of the world, and "this cannot work" because "the members are really connected with the whole" (ibid., 19-20).

In nearly every field – quantum theory, cosmology, ecology, systems theory, social science, and psychology – part and whole now are understood to be inseparable from one another. The very meaning, structure, and function of the parts has become incomprehensible apart from the wholes within wholes in which the parts are embedded and in which their nature, evolution, and functioning must be understood. Yet our thinking remains mired in divisions, separations, and fragments that appear incommensurable with one another. The result is collective and personal egoism, war, conflict, economic exploitation, destruction of nature, and destruction of one another.

Similarly, science has revealed that individuals are part of an interrelated matrix of energetic societies: fields that relate individuals to one another in a multiplicity of ways and distinguish them as distinct beings within the fields. In other words, individuals are not contradictory to one another in the sense that "A" and "not A" are mutually exclusive. They are complimentary to one another as instances of a more encompassing set of social fields (Harris 2000a; Martin 2008: Ch. 3).

Thus, holism requires that we enlarge our thinking to encompass the manifold fields within which we are embedded. No longer is "A" incommensurable with "not A." A clear view of reality requires that I discern the ways in which "not A" is compatible, complementary, and collaborating with "A." The other person is inseparable from my existence, since the fields within which we both are embedded make possible the existence of my and the other's existence. Consequently, the other does not contradict me in an irreconcilable manner but becomes complementary to me as another essential part within a more encompassing whole. The outmoded logic of sets is superseded by the dialectical logic of wholes (Harris 1987).

Indeed, the other person remains a center of moral freedom that cannot be reduced to any scientific or behavioristic set of compulsions or bio-chemical reactions. The absolute dignity of the other derives from this fact, as will become clear below (cf. Levinas 1969). However, other persons and I interpenetrate and overlap in a vast multiplicity of ways that unify us as human beings within our common moral and civilizational project. Today, we realize that this civilizational project includes the precious Earth on which we dwell – its beauty, its ecological integrity, its fragile biosphere, and its proper, holistic governance.

In his book *The Systems View of the World: A Holistic Vision for Our Time*, Ervin Laszlo calls this structure of our world "holarchy." The holism of individuals flourishing within the fields that sustain and make possible their individual existence is reflected in a nested hierarchy of wholes within still greater wholes, from the sub-atomic level to the level of the cosmos. He writes:

> A holarchically (rather than hierarchically) integrated system is not a passive system, committed to the *status quo*. It is a dynamic and adaptive entity, reflecting in its own functioning the patterns of change

over all levels of the system. ... The holistic vision of nature is one of harmony and dynamic balance. Progress is triggered from below without determination from above, and is thus both definite and open-ended. To be "with it" one must adapt, and that means moving along. There is freedom in choosing one's path of progress, yet this freedom is bounded by the limits of compatibility with the dynamic structure of the whole in which one finds oneself. (2002, 58)

Human beings are integral parts, not only of the holism of the cosmos and the ecosystem of the Earth, but of the planetary society encompassing the Earth. However, just as we have not yet harmonized our civilization with the delicately balanced biosphere that sustains all life on our planet, so we have not harmonized our social life to the holism of planetary society. We remain trapped in systems of fragmentation that are destroying the biosphere and continue to destroy planetary society through war, patterns of exploitation, linguistic forms of deceit, organized violence, and perpetual conflict.

The *Earth Constitution* provides civilization with an organic structure, one that is a natural outcome of the holism of the cosmos which manifests through the evolution of the geosphere, biosphere, and noosphere. To actualize the holism at the heart of our common humanity and human civilization, we need democratic world law. Such an actualization of our unity in diversity will be a necessary factor, if not the central factor, in creating a sustainable civilization.

> *Just as we have not yet harmonized our civilization with the delicately balanced biosphere that sustains all life on our planet, so we have not harmonized our social life to the holism of planetary society.*

Fragmented systems go hand in hand with fragmented patterns of thought. The holistic view of the cosmos and human life emerging from 20th century science has to be assimilated if we seek a true paradigm shift. The ways we think and organize our political, economic, and cultural lives must shift first. As Laszlo suggests, holarchically organized systems are influenced from below, not determined from above.

However, influences are *reciprocal* in any truly holistic system. In a true whole, the parts function as integral elements in the functioning and maintenance of the entire system. They fractally pattern and mirror the whole in reciprocal harmony and synergy. This is what Buckminster Fuller meant when he declared that the synergy of the united whole was so much greater than that of sovereign parts attempting to work together (1972, 88). The work of philosopher Errol

E. Harris (1987), also demonstrates the reciprocal nature of relationships. The principle of wholeness does not diminish individuality, but rather actualizes genuine individuality such that the complementary functioning of the parts integrates and maintains the integrity of the whole.

The Earth Constitution brings the holarchical systems of nature into human civilization

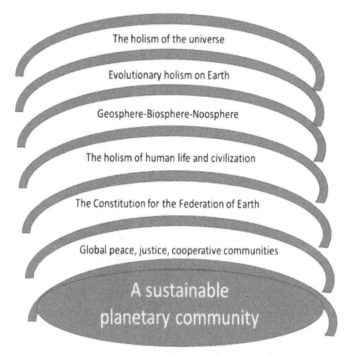

The holism of the universe

Evolutionary holism on Earth

Geosphere-Biosphere-Noosphere

The holism of human life and civilization

The Constitution for the Federation of Earth

Global peace, justice, cooperative communities

A sustainable planetary community

Figure 3.2
The Holism of the Universe and Evolution integrating the Earth Constitution

When applied to human life and ethics, thinkers who understand the principle of holism advocate linking our individual lives with all human beings, since our oneness with all other persons is inseparable from our uniqueness as individuals. Our ability to link our lives in this way means discovering our own fundamental humanity. As we become ever more fully human, we begin to realize that nothing and no one separates me from others. Indeed, we appear to have a key role in the universe itself. Here is what physicist Henry P. Stapp concludes concerning the implications of quantum physics:

> The radical change swept away the meaningless billiard-ball universe, and replaced it with a universe in which we human beings, by means of our value-based intentional efforts, can make a difference first in our own behaviors, thence in the social matrix in which we are imbedded, and eventually in the entire physical reality that sustains our streams of conscious experiences. (2011, 7)

A fragmented sense of self begins to give way to a deeper sense of self that lives from the universality of its own humanity. Our value-based intentionality transforms us, the social world, and the deeper reality of human experience. Yet, so long as we define ourselves in terms of the antiquated system of militarized sovereign nation-states, we will continue the "meaningless billiard-ball universe" of conflict, blind soulless forces, and death.

Today our world struggles with a global pandemic that many believe emerged from one of the bioweapons research labs that are active in many countries around the world. Any lab researching bioweapons is illegal under international law, yet there are dozens of such labs around the world experimenting with plagues that could wipe out humanity (whether released accidentally or intentionally). Like the nuclear weapons regime that hides in the belly of the nation-state beast and could wipe out humanity at any time in thermonuclear holocaust, the utter absurdity of a world system that allows such bioweapons research to go on should be apparent to all.

We have seen both Erich Fromm and Ervin Laszlo define our human calling as actualizing the potential for reason and love, to which I added a third dimension of intuition or "immediate, direct awareness." Higher potentials of selfhood are progressively actualized. This process of universal actualization within ourselves is affirmed by many leading psychologists and philosophers. On the other hand, nation states do not "love." They are political entities that systemically seek national interest, national security, national power, national secrecy, etc.

Fromm affirms that the purpose of a healthy life involves "a constant striving to develop one's powers of life and reason to a point at which a new harmony with the world is attained; it means striving for humility, to see one's identity with all beings, and to give up the illusion of a separate, indestructible ego" (1962, 156). Psychologist Robert J. Lifton writes, "One moves toward becoming what the early Karl Marx called a 'species-being,' a fully human being. Once established, the species identification itself contributes to centering and grounding. In no way eliminated, prior identifications are, rather, brought into new alignment within a more inclusive sense of self" (1993, 231).

Similarly, psychologist and spiritual teacher Richard De Martino affirms that "to the degree to which I can rid myself of this filter and can experience my self as the universal man, that is, to the degree to which repressedness diminishes, I am in touch with the deepest sources within myself, and that means with all of humanity" (1960, 127). Spiritual teacher Jiddu Krishnamurti declares that "if you don't know how your own mind works you cannot actually understand what society is, because your mind is part of society; it is society. ... Your mind is humanity, and when you perceive this, you will have immense compassion" (1989, 83-86). For such psychologists and spiritual teachers, careful attention to the workings of our own consciousness and our common human situation inevitably illuminates our identity with all humanity.

If we truly see ourselves as aspects of one humanity, we no longer can justify militarized sovereign nation-states. If we are one human species, we no longer will accept national identities, birth certificates to prove citizenship, walls built to keep out immigrants, military preparedness against other nation-state enemies, etc. If we live within one human reality, we will want to put limits on the private accumulation of wealth that inevitably destroys human compassion, equality, and unity. As one people, we witness the living contradiction between what we are in our deeper, truer selves and the insane political-military-economic system in which the world appears mired.

Mahatma Gandhi also was a holistic thinker who understood that each unique person is an expression of the whole, an expression of "Truth." His fundamental principle of ethics and nonviolence was *satyagraha*, literally "clinging to Truth." If we respect the uniqueness of every person, rather than the differences that set us apart, then the truth of wholeness becomes evident.

Moreover, our unity, our mutual participation within larger societal groups becomes clearer. For Gandhi, the modern nation-state is "violence in a concentrated and organized form" and capitalism is a system of structural violence in which "the few ride on the backs of the millions." Unless the world rids itself of these two monsters, he believed, we will sooner or later destroy ourselves (1972, 120-132). Consequently, understanding the fundamental links between democracy, nonviolence, and holism is critical.

In the 18th century, Immanuel Kant affirmed that "rational beings all stand under the law that each of them should treat himself and all others, never merely as a means, but always at the same time as an end in himself. ... Morality consists in the relation of all action to the making of laws whereby alone a kingdom of ends is possible" (1964, 100-101). For Kant, the kingdom of ends as a command of morality means that each of us adopts moral law for ourselves with a view to a world in which every person treats everyone else as a unique and infinitely valuable "end in himself." Such a world of universal moral laws demands unreserved respect and dignity toward all and inevitably links us with all our brothers and sisters.

In addition, morality directly connects us with the holism of humanity. G. W. F. Hegel developed his notion of holism by embodying it concretely within the whole of society, showing the interrelation of part and whole at every level of society. Human value-based intentionality – epitomized by the integration of reason, love, and intuition – is not an incidental epiphenomenon within a soulless billiard-ball universe. Rather, it is an expression of the very foundational intelligence behind all life.

Out of the dozens of thinkers affirming ethical and human holism since the 20th century, I will cite just three more. Throughout his long lifetime of philosophical output, American philosopher John Dewey affirmed the inseparability of the individual and the community, ultimately viewing the human community as a whole and as the matrix for freedom and the development of individual potential. For Dewey, the concept of democracy itself simply "projects to their logical and practical limit forces inherent in human nature" (1963, 497). The democratic ideal is a logical and spiritual projection of our common human potential beyond, for example, "the secondary and provisional character of national sovereignty" (1993,120). Dewey's life work articulates the common ideal of an ever-greater actualization of our potential for free and open association with one another within the context of an emerging planetary democratic community.

Second, contemporary social scientist David Korten contrasts what he calls "the empire," a world of dominator-relationships, with the "Earth Community" that "unleashes the human potential for creative cooperation and allocates the productive surplus of society to the world of growing the generative potential of the whole." "Spiritual Consciousness," he writes, "approaches the practice of democracy as a process of collective problem-solving aimed at enhancing the well-being and potential of all." For Korten, the holism of humanity is manifested in our growth toward an "evolving Integral World, which they engage as evolutionary co-creators" (2009, 25, 53-4). Here, democracy is understood in its deeper meaning as a planetary cooperative polyarchy of communities focused on the common good of each within the collective framework of the whole. Once again, the holism of humanity is described as implicit in our creative potentialities and able to actualize itself in genuine unity in diversity.

Third, Errol E. Harris affirms that "in human self-awareness, the nisus to the whole has become conscious of itself, so the self, being apprised of its own desires and their aims, strives to organize them, in order to attain coherent wholeness, in which it can find complete self-satisfaction; that is, to make them mutually compatible, so as to remove the frustration inherent in internal conflict. It is this self-realization that determines the ultimate standard of value" (2000a, 251). The universal drive at the heart of the evolutionary process (its nisus), operates in us (as it does everywhere) to promote wholeness, holism, and the removal of internal and external conflict so that the individual person, group, or nation can live at peace within a dynamic and diverse yet ordered whole. Reason, Harris adds, eventuates in "rational love." This point can also be associated with the *agape* taught by Jesus, a love which is universal, all-embracing, and non-discriminatory. Harris writes:

> There is a sense in which love pervades the entire universe, as that universal tendency towards unity and coherent harmony previously stressed. ... While the emotional and conative aspects are not lost, love now becomes rational. It is the genuine concern for the welfare of each and every individual, it is the universal respect for persons, that treats each as an end and none merely as a means. And because every individual is a social being, it is concern for the welfare of the entire community, in which all are integrated and on which all depend. Nor can that welfare, for which this rational love is concerned, be confined to any one local or national community, because the welfare of every community depends, like that of each individual, upon the welfare of all. (1988, 162-63)

As human beings mature through the worldcentric to the cosmocentric levels of consciousness, reason and love coalesce. They unite in an orientation that embraces both subject and object, resulting in a whole world presupposed within each dimension of this relationship. This loving relationship, predicated on the harmony and integration of the whole, becomes the foundation for the biospheric consciousness called for by contemporary environmentalists.

We need to create human institutions which mirror this interdependence of parts with the whole and which reflect the ideal that "love pervades the entire universe." Our present economic and political institutions clearly defeat this harmony and coherence, giving rise to militarism, competition, absolute winners and losers, and the internecine rivalry that is leading humanity toward the brink of extinction. Uniting humanity under the *Earth Constitution* is our closest approximation to a holarchy premised on institutions that mirror universal love.

Professor Ilia Delio declares, "As this incredible cosmos unfolded love deepened through the unification of elements into more complex conscious entities ... and emerged in the human being as yearning and desire for one another" (2013, 180). If we allow this all-pervasive love to inform our lives, we could give birth to a truly new world, which on some level, we all deeply desire. Philosopher Nicolas Berdyaev says of love: "Our task in life is to radiate creative energy that brings with it light and strength and transfiguration" (160, 139). Transformative praxis must become loving transformative practice with an eye to transfiguration. We know that our institutions powerfully influence human consciousness. Law and love can be institutionalized.

Because human beings are self-aware, goal-oriented creatures, all of our ends or purposes constitute value-oriented activity. Our highest value involves the fulfillment of our potential as individual human beings – to become what we are capable of being – which is not possible when the matrix of society and civilization are viewed as inseparable parts. Holistic values, therefore, seek to actualize self-fulfillment within the empowering framework of the larger social

Humanity has the power to unite love and reason, create an orientation that embraces subject and object, and build a new world presupposed within each dimension of this relationship.

wholes that encompass us and make our self-actualization possible. Love then informs all dimensions of human existence. Our individuality and our humanity become inseparable. This inseparability is actualized not in the submergence or loss of our individuality but rather in a rational love that embraces both part and whole. Democratic world law becomes the 21st century form of love.

3.4 Social Holism: Planetary Democracy and Cooperative Economics

We have seen that ethical and social holism mirror the ecological holism revealed by scientists and economists concerned with biospheric sustainability. The volume *Valuing the Earth: Economics, Ecology, Ethics* (1993), edited by Herman E. Daly and Kenneth N. Townsend, brings together a collection of writings by experts and ethicists that reveal the deep parallel between human and ecosystem holism. Fundamental to both is the fact that relationships are internal rather than external. The notion that relationships are primarily external derives from the early modern paradigm in which the world was thought to be composed of independent "substances." This was the atomistic model in which everything was reducible to the atoms of which materials are composed.

The 20th and 21st century revolutions in all the sciences has given us an entirely different model, one in which the parts can only be understood in terms of the fields or wholes of which they are a part. As a result, relations now are viewed as internal and interdependent. The part cannot exist nor be what it is apart from its place and set of relationships within the whole. In *The Liberation of Life* (1990), Charles Birch and John B. Cobb, Jr. write:

> The ecological model proposes that on closer examination the constituent elements of the structure at each level operate in patterns of interconnectedness which are not mechanical. Each element behaves as it does because of the relations it has to other elements in the whole, and these relations are not well understood in terms of the laws of mechanics. The true character of these relations is discussed in the following section as "internal" relations .... Internal relations characterize events. For example, field theory in physics shows that the events which make up the field have their existence only as parts of the field. These events cannot exist apart from the field. They are internally related to one another. (1990, 83, 88)

What kind of institutions would reflect this holism and these internal relationships? The most basic answer is familiar yet strange to us. Again, it is democracy, properly understood and properly implemented. The uniting of persons within a democratic community means that people understand they are bound to one another by internal relationships, traditionally known as a "social contract" but now understood as a synergistic arrangement. Democracy, properly understood, envisions a world community of respect and concern for each and all. Authentic democracy implies a communicative and embracing world society of unity in diversity – the very cornerstone for coming together to establish a sustainable civilization in harmony with the holistic principles of ecology and evolutionary social science.

In the face of the immense terrors of our time, and the on-going collapse of our planetary ecology, we need to identify the fragmentation of our thought and the polarization caused by our outdated institutions. We must act to discover the holism within ourselves and how it may be reflected in holistic, nonviolent, and sustainable institutions. My contention is that uniting humanity under a *Constitution for the Federation of Earth* constitutes our best hope for creating and actualizing a human holism in harmony with our planet's ecological holism. Just as fish living in a seacoast coral reef exist within a set of internal relationships with the other fish and their environment (water temperature, health of the reef, rhythm of the tides, the seasons, protection of the ozone layer, etc.), so human beings exist within a set of internal relationships that correspond with their natural environment, one another, and human civilization.

Thinkers such as philosopher Jürgen Habermas (1998) and cognitive scientist Steven Pinker (1995) have shown the universality of language and its necessary connections with the selfhood of each of us and the communities that sustain and make possible our individual survival and flourishing. Pinker concludes that this "reveals the unity of our species" (ibid., 448). Yet nation-states are structured as if their relations to the rest of the world are external. Similarly, corporate capitalists compete and exploit both people and the environment, as though their victims and the natural world are external to their success and autonomy.

In a number of books and articles, American philosopher Nolan Pliny Jacobson has applied the holistic principles of Buddhism to contemporary social, economic, and political issues. About the system of sovereign nation-states, he writes: *"The major obstacle is the kind of selfhood in which the terrors of the modern nation are rooted. It is the archaic legacy of a self-substance, mutually independent of all others, which supports the entire superstructure of Western nations"* (1982, 41, emphasis in original). Nations think and act as "self-substances," as if they were somehow realities independent of humanity, the Earth, and all other nations. From this delusion arises the terroristic quality of today's world disorder.

Citing a profound Buddhist understanding of the human situation as "interdependent co-origination" (*pratitya-samutpāda*), Jacobson shows that the illusory character of the nation-state system is akin to the illusory character of egoistic atomism. Sadly, both illusions have been adopted by political liberalism. People believe that arbitrary borders and absurd conflicts reflect substantial realities, like the "atoms" of the Newtonian paradigm. People believe that these entities, whether nations or persons, exist largely in external relationships with one another, another ignorant illusion. In essence, the terroristic character of modern nations is rooted in the illusion of external rather than internal relationships.

The current global economic crisis is a consequence of this fragmentation and lack of holism in our supposed democratic institutions. Today, this crisis is exacerbated by a global pandemic. Patch-worked systems then attempt to find a solution, led by the inept politicians of today's nations and myopic international actors such as the United Nations and the World Bank. The current disastrous world order of poverty, misery, war, disease, and violence is a direct and predictable consequence of the lack of holism, planetary democracy, and focusing on external rather than internal relationships.

Thus, the disintegrating integrity of our environment is an accurate reflection of global economic and political disorder premised on outdated mechanism, atomism, and egoism derived from the early modern paradigm now hopelessly anachronistic. "Our ego-centeredness prevents us from enjoying the wonder of Existence," writes Panikkar (2013, 344). We must value existence for its own sake and its correlate in human dignity. Our survival on this planet, along with the future of our children and other precious living creatures, depends on our ability to establish holistic awareness, maturity, and institutions within the very near future.

Obviously, we need to become a planetary community. Errol E. Harris argues that this is impossible without democratic world law because a collection of sovereign nation-states can never become a genuine community (2014, 107-08). For centuries, this compelling argument had been made by prominent thinkers, including Emery Reves in his bestselling book *The Anatomy of Peace* (1945, 121). Renowned philosopher of law John Finnis draws the same conclusion. If there is a "common good" for humanity, he argues, then we must actualize this goal through democratic world law. Failing this, humanity will never become a global community (1980, 150-56).

The international organizations that dominate our planet are non-democratic and actively prevent us from unifying on a planetary scale. Only worldwide social democracy will mirror the holism at the heart of the human situation.

None of the international organizations that now dominate our planet – from the banking cartels, to the United Nations, to the World Trade Organization – even remotely resemble a democracy. Planetary social democracy, by contrast, makes possible the self-actualization of every person precisely because it integrates their internal, reciprocal relationships to embrace cooperative communities, from local to global scales.

As it stands now, the dominant political and economic world system *actively prevents* us from actualizing holism in human affairs. Conversely, planetary democracy provides the holistic framework needed for each human being to realize his or her potential to the maximum extent possible, by guaranteeing equal rights and equal freedom to everyone and transforming institutions at all levels, from the grassroots to the whole. In fact, planetary democracy embodies holism in three essential ways: for our ecological survival, for the continued evolution and fulfillment of the human project, and for our personal fulfillment as individuals.

As I demonstrated in *Triumph of Civilization* (2010), authentic democracy is always predicated on cognitive and moral growth (not limitless economic growth). Within an environment of open inquiry and honest information, free people will be capable of growing rapidly. Human beings governed by institutions that promote cooperation, respect, and concern will reach maturity. Humanity will become increasingly capable of choosing the common good over unhealthy and unsustainable goals.

In my books bearing on world law (*Ascent to Freedom*, *Triumph of Civilization*, and *One World Renaissance*), I have shown that the concept of law presupposes universality – the unitary dimension of reason that implies democratically legislated law for all humankind. Sound and just world law represents the holism of our common humanity actualized and concretized. It has been nascent yet inherent in the human project from the very beginning. Now that we face a possible terminus for the human project (whether from climate collapse, nuclear holocaust, pandemic, etc.), we must confront our one chance, our one real opportunity to ascend to a higher level of existence based on peace, justice, and planetary sustainability.

Many so-called progressive, holistic thinkers opine that we need cooperation, equity, and universal minimum prosperity rather than division, exploitation, domination, and extreme income disparity. But they naively fail to question the system of nation-states that divide humankind into 193 separate territories with absolute boundaries, national institutions of militarism, self-promotion, and international competition. They have not fully ascended to a holistic comprehension or consciousness, for such a perspective necessarily supports institutionalizing global democracy.

We will see in the next two chapters what the experts are saying and why their views are only partially correct, only partially practical. They see the problems and may even recognize the need for holism, but they offer *no design for our living planet*. Often, they suggest some sort of action but have no concrete plan that will lead to a paradigm shift. More to the point, they fail in various degrees to recognize our emergent human potentialities for cooperation, coordination, and compassion on a planetary scale. I believe humanity *is* ready to employ these emerging values. I also know that time is of the essence. We cannot limp along anymore. Without seriously challenging the current global system, we will not reach the root cause of our problems. Indeed, only a novel approach like planetary federation – as offered in the *Earth Constitution* – can meet the unprecedented challenges we collectively face.

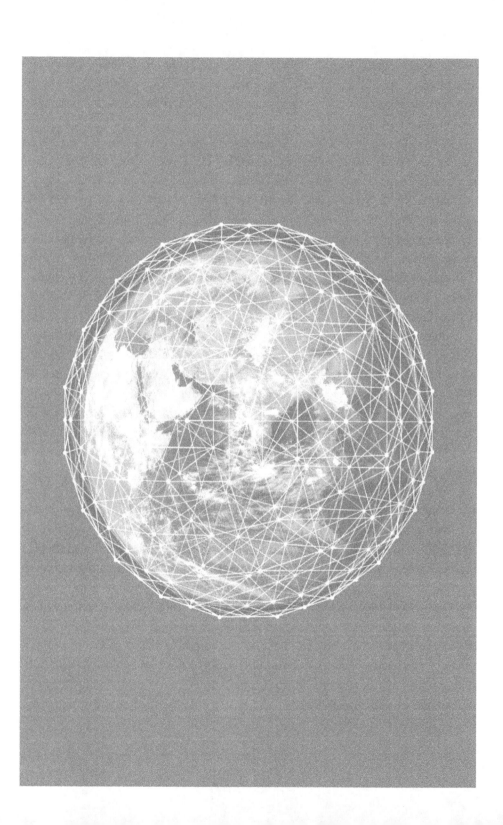

CHAPTER FOUR

Climate Collapse:
The Need for Transformational Ecology

No one can afford to assume that the world will somehow solve its problems. We must all become partners in a bold effort to change the very foundation of our civilization.

Al Gore

We live on a dying planet, and the confluence of forces leading toward massive climate change may soon become unstoppable. We already may have passed the point of no return. By most informed accounts, we most certainly will pass that point within the next ten years if radical changes are not made. What we do know for certain is that we are destroying the very foundations of the biosphere that gives us life. We are defiling the beautiful, rich, and plentiful world that nature bequeathed to us. We also know that every year that passes without fundamental change brings us closer to the abyss of no return. Al Gore is correct that we need to change the very foundation of civilization, though he has been substantially wrong in his concrete proposals.

This chapter examines the severity of the climate crisis through the lens of top environmental experts. The picture does not look good. The loudly proclaimed science-based warnings have been circulating through the global media since the 1960s. Yet in the succeeding six decades, very little has been done to mitigate the on-going destruction. Indeed, we shall see that vast corporate, financial, and national forces have aligned against effective action to protect the environment. The selfishness and greed of those dominating the global system of militarized nation-states is in lock-step with the capitalist economic system that generates unimaginable wealth for the few at the expense of the many. The name of their game is "greenwashing," coopting the language of environmentalism and sustainable development while changing as little as possible. The other name of their game is "war," though they prefer to use the sanitized term "national security." Neither game is concerned with genuine environmental integrity.

In their book *Break Through: Why We Can't Leave Saving the Planet to Environmentalists,* Nordhaus and Shellenberger correctly state that "the problem is so great that before answering *What is to be done?* we must first ask *What kind of beings are we?* and *What can we become?*" (2007, 8). These are the essential questions of this book as well, because effectively addressing our collapsing biosphere means addressing ourselves:

- ➤ Who are we as a collective global community?
- ➤ Can we become more conscious and more compassionate?
- ➤ What sacrifices are we willing to make to protect our planetary home and bequeath a resilient Earth to future generations?
- ➤ How can we redesign and transform our institutions to both craft environmental protections and enhance the process of spiritual and moral awakening?

In my opinion, humanity has immense potential to reach the level of moral and spiritual awareness required to properly steward our planet. Chapter One examined our potential for inner growth toward planetary maturity and the spirit of global brotherhood/sisterhood. Chapters Two and Three detailed the negative consequences of today's dominant world system, contrasting that system with the new holism of the 21st century. Here I will focus on climate.

With respect to the old system, Ervin Laszlo declares, "The still dominant vision has strongly negative consequences. It creates separation and self-centeredness, encourages all-out competition, and warrants the pursuit of individual, corporate and national interests without much concern with the effects on others" (2020, 85). Regarding the new system, Ken Wilber points out that many progressive, concerned citizens around the world are rightly against "hierarchies," but they do not understand the differences between superimposed hierarchies and naturally occurring holarchies.

For instance, many liberals fear "dominator" relationships and want to replace them with "partnership" relationships (Eisler 1988). Yet the entire universe is made of holarchies in which simpler elements are nested within larger embracing wholes in ever ascending levels, wholes that also make them what they are. Similarly, environmental thinkers often fail to distinguish between pathological hierarchies and healthy holarchies. The ecosystem of our planet is a natural holarchy embracing a vast number of lesser holarchies in a descending series of ecosystems within ecosystems. If we want to be in harmony with those ecosystems, we must transform ourselves, our societies, and our institutions into healthy human holarchies.

Consequently, we must question how the terrible consequences of human fragmentation and lack of holism impact our approaches to ecology. We must discern clearly what is happening if we are going to effectively address the crisis. We then must ask ourselves if the solutions proposed by environmental thinkers are themselves adequate. Have they really addressed the core questions: *What kind of beings are we? What can we become?* If the "foundation of our civilization" must be changed, as Al Gore asserts, then it will entail political, economic, moral, cognitive, and institutional transformation on a planetary scale. As we shall see, the unique design of the *Earth Constitution* brings all of human civilization into a multi-tiered living holarchy that can flourish in harmony with the organic holarchy of our biosphere.

4.1 Computer Modeling of Possible Futures

Rachel Carson published *Silent Spring* in 1962, the year I finished high school. It is often said that this best-selling book launched the modern environmental movement, as the book won many awards and went through numerous printings. People became more concerned about pollution and our impact on the environment as they began to conceptualize and envision this impact on the entire planet. During the 1970s, James Lovelock and Lynn Margulis developed the Gaia Hypothesis, the idea that Earth is a living, interconnected whole and that "the biosphere regulates the Earth's climate by acting as a thermostat" (Ellis 2018, 18). Further, the many dimensions of the planet – its geological, chemical, atmospheric, hydrological, biological, and human components – were found to function as a *system* that affects the global climate. This realization gave birth to climate science, the scientific approach to studying our entire planetary system that has grown and flourished worldwide over the past half century.

Dr. James Hansen was Director of the NASA Goddard Institute for Space Studies during the 1970s. Later, he served as Director of Columbia University's program in Climate Science, Awareness and Solutions. He also is the founder of the Citizen's Climate Lobby. While at NASA, Hansen modeled climate change on the agency's supercomputers. What he discovered was frightening, and he immediately began to sound the alarm. During the 1980s, he gave repeated testimony on climate change before Congressional committees, thereby helping to raise awareness of global warming and the climate crisis. He also helped initiate the worldwide practice of using powerful computers to explore possible human and planetary futures.

Computer modeling also was developed by Donna Meadows and her colleagues at MIT during the 1970s. Computerized models of possible planetary futures in relation to human activity have since become a fundamental feature of contemporary Earth system science (Lenton 2016). Today, many factors are included in the model such as: economic and population growth; the corresponding growth of carbon emissions, temperature, methane content, and other forms of pollution; the corresponding growth of extracting resources from Earth, often the very resources we need to support human life; and "sinks" that discharge our heat and material wastes back into the ecosystem of the planet. Variations on these models also are run, and computers are able to generate possible futures depending on the variables programed into each model. With the digital revolution making computers ever-more powerful, advanced computer modeling of all dimensions of our planetary climate is now standard scientific practice.

Consequently, climate science rapidly developed into a major field of science with many sub-disciplines. Some climate scientists specialize in the oceans, others in meteorology. Some specialize in ice fields and the polar regions, while others are expert in atmospheric gases like carbon dioxide (CO_2), hydrogen, and methane. Some take a cross-disciplinary approach, studying the relationship between geology and climate, or the nexus between ecosystems and species extinction, or how astronomy relates to climate, including cycles of the Sun, the Earth's orbit, and other external factors influencing climate. Still others are climate paleontologists who study historical climate records over millions of years. And some specialize in computer modeling and depict likely future climate conditions in relation to CO_2 concentrations and other gases in the atmosphere, the oceans, and on land.

All these experts seek to understand the planetary environments necessary to support life. They work worldwide at thousands of colleges and institutes, and some are international organizations, like the huge U.N. Intergovernmental Panel on Climate Change (IPCC: www.IPCC.ch, see below) dedicated to coordinating and synthesizing the results of the research going on in all these areas. The IPCC's *Fifth Assessment Report* (2014) summarizes our worldwide knowledge of climate change to that date, and it paints a truly frightening picture of our future if we maintain business as usual, a picture that became even more frightening in its *Sixth Assessment Report* that came out in 2019.

The fact is that we humans possess a deep understanding of climate dynamics in terms of the paleontological record, present worldwide symptoms, and the range of possible futures. We also possess credible and scientifically corroborated knowledge of the ways that human activity has contributed to climate change since the Industrial Revolution. We therefore have the knowledge, technology, and tools to establish a sustainable civilization.

The broad consensus of tens of thousands of variant climate models is that we are currently in "overshoot," which means that without major changes in the way we do things (which also can be modeled), we are rapidly heading toward planetary collapse and major civilizational disaster. William R. Catton, Jr. drew these same conclusions his book *Overshoot* (1982). Growth beyond the carrying capacity of our planet or "overshoot" will lead to system-wide "crash" and a major "die-off." Ecosystems are in fact crashing as I write this, and eventually our planet may no longer be able to support higher forms of life – like us.

Donna Meadows and her colleagues published *Limits to Growth* in 1972. The book became popular and has been reprinted numerous times. It chronicles the input and output for twelve different computer models (representing the hundreds they ran) that powerfully show the environmental options available to us and how most of our possible choices will lead to planetary disaster. In sum, the path to avoid disaster is fairly narrow and clear.

In 2002, they published their updated research entitled *Limits to Growth: The 30-Year Update*. They were joined by climate scientists from around the world, many of whom are part of the vast IPCC established in 1988 by the World Meteorological Association in coordination with the U.N. Environmental Program (UNEP). The IPCC has produced six global reports since its founding and provides fundamental scientific information required by the U.N. Framework Convention on Climate Change, signed by 194 countries, that emerged from the 1992 Conference on Climate Change in Rio de Janeiro.

Meadows and her colleagues filled their book with graphs and charts showing that population growth, extraction, production, and pollution have not abated since *Limits to Growth* was first published. Across nearly all sectors on the global scale, environmental destruction has grown and we continue in overshoot. Our stock of renewable resources is dwindling, grasslands are becoming deserts, fresh water is disappearing, agricultural lands are shrinking, while world population – and therefore demand for food, water, and resources – continues to increase dramatically. Meadows and her coauthors correctly outline what is necessary for a sustainable world system:

➤ **Extend the planning horizon for the world.** This would include plans for pollution reduction in soils, water and atmosphere, along with plans for technology use and production goals, extraction practices, and loss of nonrenewable resources – all geared toward the future health and maintenance of our entire planetary ecosystem.

➤ **Improve signals for monitoring the real impact of human activity.** All aspects and dimensions of our planetary ecosystem must be monitored, and in response to monitoring, we must be able to take action.

➤ **Speed up response times.** The goal is to keep resource extraction, production, and waste disposal to sustainable levels, and we must be able to respond effectively on local, regional, and planetary levels.

➤ **Prevent erosion of renewable resources.** Protect soils and forests and minimize the use of non-renewable resources such as minerals and fossil fuels, and practice acute conservation consistently and worldwide.

➤ **Use all resources with maximum efficiency.** This includes repairing, recycling and innovating, as well as a massive technology transfer to the third world.

➤ **Stop the exponential expansion of population and physical capital.** Limitless growth on a finite planet is impossible.

All these recommendations presuppose a united world that is capable of planning for the whole, monitoring the entire planet, responding quickly at a global scale, preventing the erosion of renewable resources, and limiting population as well as capital expansion. "Overshoots can become catastrophic when the damage they cause is irreversible," Meadows and company declare (2002, 163), and their research shows we are causing irreversible damage. Unfortunately, these authors offer no convincing solutions to these holarchical problems.

4.2 Looking into the Abyss

In 2004 when James Gustav Speth published his book, *Red Sky at Morning: America and the Crisis of the Global Environment,* he was Dean of Environmental Sciences at Yale University, founder of the World Resources Institute and co-founder of the Natural Resources Defense Council. He also advised Presidents Carter and Clinton on environmental matters and was the head of the U.N. Development Program (UNDP), having already received the prestigious Blue Planet Prize.

Speth begins this book by declaring that we cannot address our most fundamental problems unless we are willing to "look into the abyss" (2004, xiii). His penetrating expertise in climate science lays out the abyss before us, and he gives us the facts of environmental destruction over ten key sectors: (i) ozone layer depletion, (ii) climate change and global warming, (iii) desertification, (iv) deforestation, (v) biodiversity loss, (vi) population growth, (vii) freshwater losses, (viii) marine environment deterioration, (ix) toxification damage due to chemicals, and (x) acid rain. We need to face up to this abyss, Speth insists, if we are going to find credible solutions.

Four years later, Speth published *The Bridge at the End of the World: Capitalism, the Environment, and Crossing from Crisis to Sustainability.* This 2008 book investigates the vast literature on climate change and the environmental crisis. It analyzes the problems, discusses what top environmentalists have proposed, and lays out a range of options if we want to leave a habitable planet for future generations. His title is auspicious. We are facing the environmental ruin of our planet and possible human extinction if global warming increases beyond our ability to mitigate and adapt. Speth wants to build a bridge over this abyss, but he fails to offer a meaningful, full-spectrum solution to the underlying problem – lack of planetary coordination. So yes, we do need to build a bridge, but the bridge needs to lead to a new world system, a model for genuine transformation and transcendence. It cannot simply consist of a few economic and environmental adjustments, but must be total and transformative.

When he worked for the Jimmy Carter administration (1977-1981), Speth helped produce the *Global 2000 Report* predicting dire consequences if major changes in the way corporations did business were not immediately forthcoming. Yet nothing significant was done. He now writes that the report's predications are coming true (2008, 18). He admits that the capitalist world system is failing to address the fact that the global biosphere is in system failure.

Meanwhile, every year the world has been getting warmer with record temperatures increasing steadily. Severe and more prolonged droughts continue to occur and are becoming worse. The frequency of heavy precipitation events has increased everywhere with consequent flooding and major destruction. Coastal areas worldwide are under assault from rising oceans, superstorms, and tidal surges. Speth also focuses on the failure of the global economic system (capitalism) and political system (sovereign nation-states) to deal with these crises.

The required worldwide changes are deep and fundamental. Speth writes, "We must look beyond the world of practical affairs to those who are thinking difficult and unconventional thoughts and proposing transformative change" (xiv). Right on! He also seems to understand that we now live in a "full world" in which the gigantic planetary economy has grown beyond sustainable limits and is now digesting the biosphere itself – those planetary ecosystems that sustain human and all life on Earth (4-10). He writes that growth-capitalism originated in the "empty world" of several centuries ago, a world with seemingly unlimited resources and unlimited ability to absorb the wastes produced by our ever-increasing industrial output.

The present global crisis places us at the cusp of a possible tipping point, beyond which we will no longer be able to stop or significantly mitigate global warming and climate disaster. Speth writes, "The crystalizing scientific story reveals an imminent planetary emergency. We are at a planetary tipping point. We must move into a new energy direction within a decade to have a good chance to avoid setting in motion unstoppable climate change with irreversible effects" (27). The key factor, of course, is the reduction of the greenhouse gases from fossil fuel burning that multiple industries and transportations systems pour into the atmosphere of Earth. We must rapidly reduce and limit the parts per million (ppm) of CO_2 in the atmosphere as quickly as possible. "The worst impacts can still be averted, but action must be taken with swiftness and determination or a ruined planet is the likely outcome" (29).

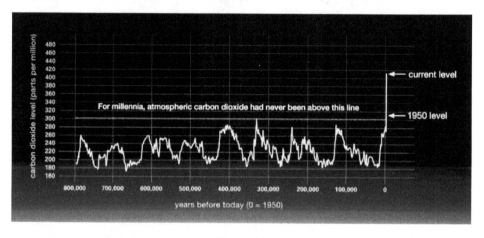

Figure 4.1
Perhaps the best-known NASA graph showing CO_2 has spiked far beyond levels in the Holocene era (never above 300 ppm). Today CO_2 is above 400 ppm and rising. The same dramatic spike shows up in charts on population growth, temperature rise, use of fossil fuels, and other indicators of our current crisis.
Source: https://climate.nasa.gov/system/content_pages/main_images/203_co2-graph-061219.jpg

Speth concludes that "most environmental deterioration is a result of systematic failures of the capitalism that we have today and that long-term solutions must seek transformative change in the key features of this contemporary capitalism" (9). He understands that we need to transform the market to work for the environment, rather than against it. He agrees that we must move to a "post-growth society" and reduce our affluent materialism and consumerism by finding life's meaning in the quality of living rather than in consumption. Our "growth fetish" disregards ecosystem limits, including our planet's ability to absorb our waste products. Hence, it ignores the "carrying capacity" of Earth and encourages "externalities," that is, shifting the costs of production to the environment or to society at large.

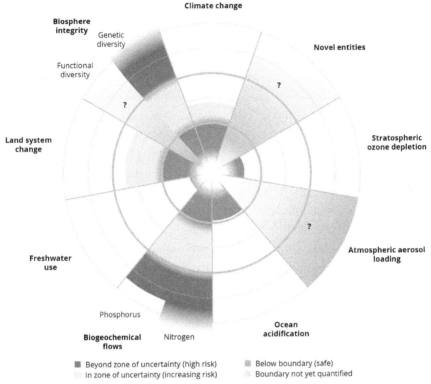

Figure 4.2

The U.N. Intergovernmental Panel on Climate Change (IPCC) developed this framework of Nine Planetary Boundaries. In the outer ring, three boundaries already have been crossed and three are near violation. All numbers (except ozone depletion) are steadily increasing.
Source: https://www.eea.europa.eu/soer/2020/soer-2020-visuals/status-of-the-nine-planetary-boundaries/view

Extensive work has been done to move beyond the commonly accepted standard of economic health called "growth in GDP" (Gross Domestic Product). Alternative indices have been developed, such as the Index of Sustainable Economic Welfare (ISEW), the Genuine Progress Indicator (GPI), and the Happy Planet Index (HPI). The new conceptions of economic health involve how well genuine human needs are satisfied (such as education, healthcare, social security, and human rights), and how meaningful and fruitful human lives can and should be lived, not simply in terms of personal prosperity, but in terms of satisfying and well-lived lives.

In sum, growth in human and planetary welfare must become the new standard, not GDP growth. Sustainability requires an economy that does not grow in input and output, but that is directed toward genuine human welfare, including the welfare of other life forms, future generations, and the environment that sustains us all. But where is this new economic and environmental paradigm to come from?

Speth states that one key to transforming the economy involves changing the legal characteristics of corporations. By law in the U.S., corporate management is required to "maximize the interests of the shareholders," which in practice means maximizing profits and growth, usually at the expense of the environment and society at large. In the United States, laws also allow corporations to have immense political influence through lobbying and campaign contributions to politicians. Yet with the process of globalization, the situation has grown even worse. Big corporations have become multinational and above the laws of any one nation. Recommendations include revoking corporate charters, expelling unwanted corporations, rolling back limited liability laws that shield managers and shareholders, eliminating corporate personhood, and getting corporations out of politics by reforming lobbying regulations (178-79).

> The multinationals have a huge impact on the global environment, generating, for example, half the gases responsible for global warming. They also control half of the world's oil, gas, and coal mining and refining. ... When unfettered by national or international laws, ecological understanding, or social responsibility, this freedom can lead to enormously destructive acts. ... They are engineering a power shift of stunning proportions, moving real economic and political power away from national, state, and local governments and communities toward unprecedented centralization of power for global corporations, bankers, and global bureaucracies. (170-172)

Indeed, we must profoundly change the nature of the corporate capitalism and place it under democratic controls. We also must evolve a new consciousness and harness new governmental goals that are strongly democratic, egalitarian, cooperative, and community oriented. What is required is a new way of thinking. Speth writes:

> Many of our deepest thinkers and many of those most familiar with the scale of the challenges we face have concluded that the transitions required can be achieved only in the context of what I will call the rise of a new consciousness. For some, it is a spiritual awakening – a transformation of the human heart. For others, it is a more intellectual process of coming to see the world anew and deeply embracing the

emerging ethics of the environment and the old ethic of love thy neighbor as thyself. (199-200)

Speth and I therefore agree that the new consciousness of holism – an awareness of the interdependent ecosystems of Earth and human life – is required. Some have called for an "intergenerational consciousness" and a renewal of our values, religion, and spirituality. Psychologists such as Erich Fromm have called for "a radical change in the human heart" as a condition for "the sheer survival of the human race" (201-202). "Today's dominant worldview," he writes, "is simply too biased toward anthropocentrism, materialism, egocentrism, contempocentrism, reductionism, rationalism, and nationalism to sustain the changes needed" (204).

Similarly, the new politics must be radically different from what has hitherto dominated governments. It must be participatory, localized, and based on human solidarity rather than competition and narrow self-interests. It also must be global, if we seek to energize and transform policies on environmental sustainability. Speth quotes with approval some thinkers who argue that global environmental protection "must be centered elsewhere than in the state system, international conferences, agencies, bureaucracies, and centers of corporate capital" (220). He also quotes political philosopher David Held who holds that we need to become "cosmopolitan citizens" and build levels of effective governance beyond the nation-state (223).

So why not ratify the *Earth Constitution*, which includes all the changes Speth and others are calling for? Without such a comprehensive democratic framework, will we ever find a bridge over the abyss? I hope to make clear in this book that the insightful recommendations of all these thinkers are only possible under the *Constitution for the Federation of Earth*.

4.3 The Role of Resistance and Struggle

Naomi Klein brings an astute social awareness to her work, including insights into the dominant capitalist ideology, unknown or misunderstood by many environmental thinkers. Her earlier books, such as *No Logo* (1999) and *The Shock Doctrine: The Rise of Disaster Capitalism* (2007), have demonstrated her credentials as a significant critical social thinker. She spent five years researching her 2014 book on the environment entitled *This Changes Everything: Capitalism vs the Climate.*

Klein travelled around the world to places where environmental struggles were taking place between the dominant economic model of "dirty extractivism" and people struggling to protect their land, water, and air from the onslaught of the capitalist exploitation model. She interviewed hundreds of people – indigenous leaders, climate activists, climate deniers, scientists, journalists, political leaders, U.N. officials, heads of major environmental groups – to give us a book brimming with relevant data, thought provoking ideas, and descriptions of local struggles and their global implications.

As such, *This Changes Everything* is a gold mine of perspectives and insights into the struggle to save our planetary environment. It lays out the broad landscape of climate controversy and responses, from business as usual, to "free-market" solutions, to geoengineering, to truly transformative visions. She argues that we must overcome the worldview associated with capitalism, with its assumptions about innate human greed and selfishness, and begin to understand our human project as a communal, democratic, planetary endeavor premised on the common good of both humanity and nature.

Klein makes clear that "dirty extractivism" is raping Mother Earth, from the jungles of Ecuador and Brazil to the Niger Delta, from the Tar Pits of Alberta to the fracking frenzy in Pennsylvania and Texas. This destructive extractivism mines fossil fuels that subsequently are burned to pollute our planet's atmosphere, oceans, forests, and agricultural lands. The planet is heating up with major destructive consequences everywhere we look. We now find ourselves, she writes, at "Decade Zero." There is no more time for delays, detours, or half-hearted compromises with the neoliberal free-market ideology that is rapidly losing credibility worldwide. She writes:

> John Kerry has likened the threat of climate change to a "weapon of mass destruction," and it's a fair analogy. But if climate change poses risks on par with nuclear war, then why are we not responding with the seriousness that the comparison implies? Why aren't we ordering companies to stop putting our future at risk, instead of bribing and cajoling them? Why are we gambling? (225)

Klein attended meetings of climate deniers, who receive major funding from the big fossil fuel companies. In Chapter One on why "The Right is Right," she describes the climate denier think tanks, such as the Heartland Institute in the U.S., as "right" because they truly understand the implications and the seriousness of the climate crisis. The climate deniers understand what is at stake – total change of their ideology and their way of life – and they want none of it.

Klein also wrote a chapter on "free trade" agreements like NAFTA, which are disastrous for the environment because they give rise to an international fossil fuel economy where everything from food to useless consumer goods are transported around the planet while burning fossil fuels. Under such trade deals, corporations have the legal power to bring lawsuits against governments where environmental or labor laws cut into their profit margins. The anti-human features of global capitalism are identical with its anti-nature features – private profit accumulated through exploitation and destruction: "The same logic that is willing to work laborers to the bone for pennies a day will burn mountains of dirty coal while spending next to nothing for pollution controls because it's the cheapest way to produce. So when factories moved to China, they also got markedly dirtier" (81).

Throughout her book, Klein develops a picture of the worldview behind the climate crisis. Coming out of Francis Bacon's 17th century predictions that the new sciences could lead to human domination over the natural world, the emerging capitalist ideology included both colonialism and slavery. Profit and power were the fundamental principles allowing dominators to extract from colonial subjects and from slaves as much profit as possible, treating their human victims as expendable.

Figure 4.3
Tar sands mining in Canada, Klein's home country,
where "dirty extractivism" still is permitted

Klein points out that Adam's Smith's *Wealth of Nations* was published in 1776, the same year that James Watt's steam engine was invented. The steam engine was touted as the invention that would free the producer from dependence on nature (i.e., the previous water-wheel technology had required dependence on flowing waterways). The steam engine could be moved anywhere and operated anytime using the fossil fuel coal. Man's dependence on nature was now nearly eliminated and his capacity for domination and exploitation of nature nearly unlimited. But this attitude spelled disaster then as now. She quotes one political scientist who said facing truths about climate change "means recognizing that the power relation between humans and the earth is the reverse of the one we have assumed for three centuries" (175). Klein writes:

> We know that we are trapped within an economic system that has it backward; it behaves as if there is no end to what is actually finite (clean water, fossil fuels, and the atmospheric space to absorb their emissions) while insisting that there are strict and immovable limits to what is actually quite flexible: the financial resources that human institutions manufacture and that, if managed differently, could build the kind of caring society we need. (347)

Then there are the geoengineers, the "mad scientists" as she calls them, who envision putting trillions of sulfur dioxide particles into the atmosphere, reflective mirrors into space, and "cloud brightening" agents into the cloud cover. Klein describes their arguments and proposals in some detail. She clearly sees this as the climax of the false relation to nature initiated by Francis Bacon and the early-modern paradigm of the 17th century. Instead of changing our worldview and our ways of relating to nature, our arrogance now leads us to envision engineering the entire planet in ways that are both untested and untestable, with unknown consequences that could be catastrophic. She points out the broad resistance to geoengineering among climate scientists and biologists, and the striking lack of "humility before nature" exhibited by the geoengineers (267).

It is not all that different with the big environmental organizations such as the Nature Conservancy, the Environmental Defense Fund, Sierra Club, the Natural Resources Defense Council and the World Resources Institute (both founded by James Gustav Speth), the World Wildlife Fund, and Conservation International. Their millions of dollars in funding comes from big oil, multinational corporations (like Walmart), or big corporate foundations (like the Ford or Rockefeller foundations). Their environmental advocacy, in return, emphasizes corporate friendly "solutions," such as carbon trading, green investment, self-regulation by corporations, and other options that never get close to addressing the roots of our environmental crisis. The Nature Conservancy even maintains its own oil well in Texas, pumps its own fossil fuel, and reaps profits through pollution (192-94). Similarly, Conservation International has partnerships with some of the worst polluters on the planet, such as Walmart, Monsanto, Shell, Chevron, McDonalds, and BP (196).

Consequently, Klein does not count on a sustainable world coming from this quarter. Like the mad scientists of the geoengineering movement, many of the big green organizations are staffed by people still operating under the capitalist paradigm that ostensibly requires nothing more than proper "market solutions" to address climate change. The most promising and inspiring encounters from her five years of travel and research come from "Blockadia." This term refers to the rapidly growing civil disobedience campaign to the extractivist projects of the big corporations. These brave activists block mining operations, pipelines, transport systems, and bring lawsuits to halt the rape of land and ecological devastation.

The Blockadia movement has brought together traditionally unlikely bedfellows. For example, ranchers in Montana have partnered with local indigenous groups that assert their treaty rights to an environment flourishing with clean water, soil, and air. The Northern Cheyenne broke legal ground by arguing that the 1977 Clean Air Act in the U.S. includes their right to breathe clean air (390). Klein chronicles the legal challenges being made by indigenous peoples around the world to the invasion of their lands and environments by the dirty extractors. This global struggle is more than a resistance movement; it now embraces a conscious worldwide struggle to save our planet.

However, the Blockadia team leaders have learned that resistance is not enough. They realize we must actively convert our local economies to sustainable, renewable, fossil fuel free community systems that work with nature's rhythms and requirements. These new systems must work "synergistically" with Earth and "require a humility that is the antithesis of damming a river, blasting bedrock for gas, or harnessing the power of the atom" (394).

Blockadia and other environmental movements serve a purpose, but activism alone is not enough to shift the paradigm. Climate planning needs to be coordinated, strategic, and planetary in scope. We must employ both a grassroots and top-down approach simultaneously.

Klein adds, "There is no more potent weapon in the battle against fossil fuels than the creation of real alternatives. Just the glimpse of another kind of economy can energize the fight against the old one. ... It must be accompanied by a power correction in which the old injustices that plague our societies are righted once and for all. That is how you build an army of solar warriors" (397-99). In other words, there are limits to resisting the system; we must build convincing alternatives.

The army of solar warriors envision another kind of economy that will require strategic planning. Klein understands that people at the grassroots level will need "the tools and the power to build a better life for themselves" (133). Planning must be decentralized as much as possible but still integrated into a global plan. She apparently envisions planetary coordination along with decentralized planning and control. However, "the failure of our political leaders to even attempt to ensure a safe future for us represents a crisis of legitimacy of almost unfathomable proportions" (364). This is all well and good, but once again I ask: *Where will this new legitimate authority for planetary coordination and vision come from?*

Klein's answer is hope arising from the fact that history is full of sudden changes of thought and attitude. It is true that needed change arises from higher values and from those who base their resistance on humanitarian goals. She cites the end of colonialism, the end of slavery, and the civil rights movement in the U.S. as evidence that change can happen unexpectedly. Yes, it can happen suddenly. One day we are a lone, morally indignant voice crying in the wilderness, then suddenly we find that everyone is speaking the same message. But the time crunch for environmental change is real and alarming. We cannot wait any longer for worldcentric values to filter up to the corporate and nation-state level.

Klein agrees that the massive planetary change needs to be "democratic," by which she seems to mean grassroots, bottom up, and concerned with the common good. She states that we need to delegitimize the dominant extractivist-capitalist worldview with its "stifling free-market ideology," and that we must create a "Marshall Plan for the Earth" (458-60). Indeed, there must be a new understanding that Earth is here for us all to live upon and for future generations to enjoy. She adds, "The climate challenge will be fruitless unless it is understood as part of a much broader battle of worldviews, a process of rebuilding and reinventing the very idea of the collective, the communal, the commons, the civil, and the civic after so many decades of attack and neglect" (460).

Klein's Marshall Plan for Earth includes a universal social safety net and a guaranteed annual income for everyone on our planet. Such a revisioning of economics must be directed toward the common good, rather than private profit, because it "opens up a space for a full-throated debate about values – about what we owe to one another based on our shared humanity, and what it is that we collectively value more than economic growth and corporate profits" (461). Let me add that if people have a guaranteed income, they no longer will be forced by economic necessity to compromise values. A planetary debate about who and what we are as human beings and the values we share could then become a living reality.

Clearly, all these fantastic ideas need to come together to provide moral clarity and new definitions of human dignity, freedom, equality, community, and effective democracy. Klein agrees that progressive forces must be prepared to seize the moment and move the world away from a "free market ideology that has been discredited by decades of deepening inequality and corruption" (465-66). I strongly concur that we need a plan which deals with the planet as a whole, while simultaneously empowering local democracy and communities worldwide.

The *Earth Constitution* meets every tenet of Klein's Marshall Plan for Earth ... and more. Indeed, none of Klein's proposals are even remotely possible unless we ratify the *Earth Constitution* and place economic and political authority in the hands of world citizens. Isn't this exactly what democracy is and should be about? Authentic democracy links the bottom-up with the top-down. It is time for grassroots environmental activists and democratic thought leaders of the Earth Federation Movement to join forces so that we may collectively solve the existential crises facing humanity.

4.4 Is Civilization Beginning to Falter?

Bill McKibben is probably the most popular and effective environmental activist on the planet. He is the founder and director of the global environmental organization 350.org, and he is a lifelong environmental scholar, writer, and thought leader. Among his many awards are the Gandhi Peace Award and the Right Livelihood Award. His many books include *The End of Nature* (1989), *The Deep Economy* (2007), *Earth* (2010), and *Falter: Has the Human Game Begun to Play Itself Out?* (2019).

In this latest book, McKibben introduces important themes not often found in other literature on the environmental crisis. These themes include the cult of Ayn Rand individualism among the superrich, the development of super-intelligent computers that might replace humans, and the breakthroughs in genetic engineering that may actually lead to designing future human beings.

McKibben divides the book into four parts that relate to his overall theme of the "human game," which now may be playing itself out. Part One is entitled "The Size of the Board," Part Two is called "Leverage," Part Three is "The Name of the Game," and Part Four is "An Outside Chance." In the first part, he describes some of the many horrific disasters that are happening around the world. While the dominant idea among economists and politicians continues to be "progress" – the notion that the size of the board on which the human game is played can be continually increased – the reality is that the board is shrinking:

> In November 2017, fifteen thousand scientists from 184 countries issued a stark "warning to humanity." [Like the proponents of progress] they had charts, but theirs depicted everything from the decline in freshwater per person to the spread of anaerobic "dead zones" in the world's seas. As a result, the scientists predicted, we face "widespread

> misery and catastrophic biodiversity loss"; soon, they added, "it will be too late to shift course away from our failing trajectory." A third of the planet's land is now severely degraded, with "persistent declining trends in productivity".... There are half as many wild animals on the planet as there were in 1970. ... The planet's oldest and largest trees are dying fast. (11-12)

The list of the calamities McKibben cites goes on and on. We are threatened not only by largescale nuclear war and destruction of the ozone layer, but also by relentless climate change which is "perhaps the greatest of all these challenges" (21). The planet's hydrology is changing fast, with some regions vastly drying up, while massive flooding and torrential rains occur in others.

McKibben adds that "as land dries out, it often burns" (24). He cites the astonishing and unprecedented fires in California from 1987 to the present and the burning of vast regions of Siberia. Since his book was published, there have been hundreds more fires in Australia and within the Amazon basin which threaten the vital "lungs of the Earth." These fires are destroying the very forests we need to survive, while unprecedented torrential rains now fall due to increased super-storms, both nearly impossible to fight. Additionally, melting ice and permafrost activate "feedback loops" that impel the climate beyond tipping points from which it is impossible to return. He describes the record temperatures being set across the globe in India, China, the U.S., Canada, Australia, Iran, the Persian Gulf, Pakistan, Egypt, Vietnam – virtually everywhere. He also discusses the rising oceans and their effects on coastal areas worldwide. As he sums up (italics his): *This is our reality right now. It will get worse, but it's already very, very bad"* (25-33).

McKibben goes on to describe the five massive geological extinctions that have occurred during the billions of years of life on Earth. The conditions under which the current "Sixth Great Extinction" is happening are worse than most of the previous ones. The difference is that we are the geological force responsible for the current destruction. The Holocene geological period of the past ten thousand years provided humans with a fairly stable climate with which to develop civilization and spread over Earth. That epoch is now over, and geologists have renamed our current period the "Anthropocene" because we are the main reason the climate is changing.

McKibben then turns to the politics of climate change, the basic principle of which is that many people "literally don't want to hear about it" (67) – the same denial James Speth and Naomi Klein also point out. The entire global economy is based on burning fossil fuels. President Obama turned to fracking in the U.S. because this increased energy independence and because gas burns cleaner than coal or oil. But scientists soon found that a certain small portion of the gas inevitably leaks from the processes of mining and transporting methane gas. And this leaked gas "traps heat in the atmosphere about eighty times more efficiently than carbon dioxide" (68). Similarly, Canada contains one of the two largest deposits of tar sands on Earth. "In the spring of 2017, [Canadian Prime Minister] Trudeau told a cheering group of Houston oilmen that 'no country would find 173 billion barrels of oil in the ground and just leave them there'" (70). Yet people don't want to hear or assimilate this frightening information. Instead, politicians continue to promote an economic system based on endless growth.

> It's not a good sign that the largest physical structures on our planet, its ice caps and barrier reefs and rain forests, are disappearing before our eyes. So: problem from hell. Governments prefer to evade it. Human psychology is not designed to cope with it. It's happening too fast. (70).

McKibben also addresses the cover-up of problems created by multinationals such as Exxon and Shell. These corporations have their own in-house scientists, and they knew early on about the effects of fossil fuels on the environment. They intentionally presented misleading data and mounted multi-million-dollar campaigns designed to cause doubt about climate science. "Thus began the most consequential lie in human history," according to McKibben (76). He also describes the massive influence these conglomerates have had on the U.S. government during the Bush, Obama, and Trump administrations.

A fascinating part of the book traces the influence of Ayn Rand novels on the political, economic, and corporate oligarchs of the United States. The basic principle of Rand's writings is simple: "Government is bad. Selfishness is good. Watch out for yourself. Solidarity is a trap. Taxes are theft." He quotes her as saying, "all codes of ethics they'll try to ram down your throat are just so much paper money put out by swindlers to fleece people" (91-92).

Specifically, Rand impacted a number of oligarchs, including Tom Perkins (one of the richest men on the planet), the Koch brothers (Charles and his late brother David led this vast economic empire), U.S. Senator Mitt Romney, Rupert Murdock (media mogul and owner of the infamous Fox News Corporation), senior officials in the Trump administration (including Trump himself), and the CEO billionaires who run the high-tech empires out of Silicon Valley in California.

Of the many books coming out on the climate crisis, McKibben's book is of particular interest because it delves into two technological breakthroughs that create deeply problematic possibilities for the future of humanity. First, the super-computer revolution portends a future where computers may be smarter than humans and might conceivably replace humans. Second, there are breakthroughs in genetic engineering that now allow scientists to genetically design new human beings.

This book raises fundamental questions such as: *What is humanness?* McKibben's exploration makes clear how hard it is to find meaning in a world hell-bent on perpetuating climate-changing economic growth, replacing humans with computer robots, and designing "better" humans by means of genetic engineering. Eventually, McKibben answers these questions himself in the last part of the book titled "An Outside Chance." He argues that we must alter our current trajectory in all three of these areas. Stuffing these lethal genies back in the bottle seems improbable, but there is still some hope.

> *The moral development of every person expands freedom for all, because higher consciousness calls us to stand with others to protect our shared home and reduce inequality and suffering.*

McKibben focuses on *resistance* and *maturity* as key strategic moves on the game board. There is an immense economic juggernaut impelling civilization toward climate collapse and human extinction. The needed resistance has grown, in part due to McKibben's own efforts and those of 350.org. Most recently, their efforts led to New York divesting all fossil fuels from the state pension fund. Nevertheless, the juggernaut of wealth, power, and institutional inertia may not be possible to stop before the human game falters for good ... and forever.

On the positive side, two other technologies might turn the tide and give us a chance at survival: "One is the solar panel and the other is the nonviolent movement," writes McKibben. He adds that to make effective use of nonviolent direct action means overcoming libertarianism – the selfish worldview focused only on individual freedom (201). We must realize that the moral enhancement of every person expands freedom for all because it calls us to stand with others to protect our shared home and to reduce inequality and suffering.

The price of solar panels is dropping, and they are being installed in places like Ghana in West Africa, making a huge difference in people's lives. Clean energy through wind and solar is an essential factor in avoiding planetary scale climate disaster. "The latest studies, from labs such as Mark Jacobson's at Stanford, make clear that every major nation on earth could be supplying 80 percent of its power from renewables by 2030, at prices far cheaper than paying the damage for climate change," writes McKibben. Of course, we also need to "eat lower on the food chain, build public transit networks, densify cities, and start farming in ways that restore carbon to soils" (211). Today, we have the technical capacity to make this mandatory transition, if we can muster the political will.

McKibben mentions the nonviolence legacies of Henry David Thoreau, Leo Tolstoy, Mahatma Gandhi, and Martin Luther King, Jr. He includes in his definition of "nonviolence" not just acts of resistance, but also "building mass movements whose goal is to change the zeitgeist and, hence, the course of history." Movements also are the method by which "the active many can overcome the ruthless few," since the fight for planetary sustainability ultimately is about "power and money and justice" (219-20).

> As Naomi Klein has said, if we can't get a serious carbon tax from a corrupted Congress, we can impose a defacto one with our bodies. And in so doing, we buy time for the renewables industry to expand – maybe even fast enough to catch up a little with the physics of global warming. ... In the early summer of 2018, Pope Francis used precisely the language we'd pioneered in that fight: most oil, gas, and coal, he said, needed to "stay underground." We'd begun to change the zeitgeist, which is the reason we'd gone to work in the first place. ... Anyone who thinks that time is therefore on the side of the oil companies is reading history wrong. This movement will win (though, as we've seen, it may not win in time). (222-24)

Thus, solar energy and nonviolence can be viewed as "technologies less of expansion than of repair, less of growth than of consolidation, less of disruption than of healing" (226). In this manner, McKibben agrees that maturity, balance, and scale are perquisites to meaningful change. We need the maturity to understand that there is force in numbers, we have the power to stop global polluters and unethical uses of science, and we must be strategic and use our power wisely. As Pope Francis said, we can keep fossil fuels underground. We also can decide to not replace humans with computers and not make designer babies.

We need to balance right-wing ideologies with progressive structures that make fraternity real. McKibben lists "labor unions, voting rights, and a social safety net" as relevant goals (230). Finally, we need proper scale to counterbalance capitalism's obsession with unchecked growth. In fact, we need to slow things down according to McKibben: "Taken all together, the results suggest that instead of dreaming about utopia, we should be fixated on keeping dystopia at bay" (235).

McKibben ends by contrasting the space program at Cape Canaveral with the surrounding beach and wildlife protection area where it is located. Instead of shooting for the stars or other planets, as if this could somehow save us, we need to be protecting our "unbearably beautiful planet." Regarding the Golden Rule, we also should be loving our brothers and sisters around the world, feeding the hungry, protecting wildlife, welcoming newborns to starship Earth, and surrounding the elderly during their final years on this precious planet. The human game is and can be "graceful and compelling," and only our love of it will allow us to save it (296).

In sum, McKibben's appeal concerning the ways we can resist and change the course of human history is moving and deeply humanistic. But will 350.org change the zeitgeist? If by "zeitgeist," we mean the spirit of radical individualism and the Ayn Rand orientation of power, domination and egoism, then perhaps his recommendations address changing this predominant worldview. But if what we really need is something much more – a total redesign of our world system, including ethics and ecology – then *Falter* falls short.

I therefore want to ask Bill McKibben: If it is true that the outdated world order must be replaced and that the emerging zeitgeist must include moral, technological, and institutional transformation, then will you support the *Earth Constitution*? Solar panels and the nonviolence movements are only two components of the fundamental paradigm shift needed to take us from fragmentation to holism. Our situation is so bad that McKibben is correct – we must indeed keep dystopia at bay. But shouldn't we do this by envisioning a practical utopia? Ratification of the *Earth Constitution* will embrace a new sense of our humanness, convert the world to solar panels and clean energy, and inspire all of humanity to fight for what is right, to flourish and not falter.

4.5   The Stark Realities of Climate Change

Dr. Joseph Romm is a trained physicist who specializes in climate change research, and who has worked, studied, and written within the domain of climate science for many years. In 2009, *Time Magazine* named him one of its "Heroes of the Environment." He also is a Senior Fellow at the Center for American Progress and was chief science advisor for the documentary series *Years of Living Dangerously,* which won an Emmy Award. His 2018 book *Climate Change: What Everyone Needs to Know* (2nd Ed.) synthesizes his substantial knowledge and systematically describes the tragedy of our current situation.

Romm worked for fifteen years at the U.S. Department of Energy advising businesses on how to become more energy efficient and reduce their carbon footprint. As of the year 2018, when he published the second edition of this book, Romm was working at the Center for American Progress which states on its website that it is a "nonpartisan policy institute" working "to improve the lives of all Americans."

The premise of his book is that human beings and most life on Earth are in great danger. We already are condemned to a drastically changed world in which our creativity and adaptability will be severely challenged over the next few decades as eco-systems continue to degrade. Every year we wait to take drastic CO_2 reduction measures (that is, stop burning fossil fuels), adds to a future significantly more horrible by magnitudes of destruction. It is not an incremental worsening; every year we delay means increasing magnitudes of suffering and horror for future generations.

Climate change is not an incremental but an exponential process. Every year we delay means increasing magnitudes of suffering and horror for our children and future generations.

The exponential rates of damage result from the nature of Earth's climate system. This system is very sensitive to a number of positive or amplifying feedback mechanisms, but there aren't any known negative or diminishing feedback loops to counteract the amplifications. For instance, a central positive feedback mechanism for warming is the amount of greenhouse gases in the atmosphere (predominately CO_2). From the preindustrial average of 280 parts per million (ppm) we now are at about 412 ppm, and the number is steadily rising at the rate of approximately 2 ppm per year. This in turn generates hotter summers, droughts, desertification, wildfires, melting polar caps and glaciers, ocean acidification and superstorms, because warm air holds more water and is a key ingredient in generating thunderstorms, hurricanes, and cyclones. Incidentally, 350.org chose its name based on 350 ppm – the safest upper limit of carbon dioxide to support life on Earth.

With regard to feedback loops, here is another example. Melting sea ice reflects nearly all of the sunlight falling on it back into space, but dark blue ocean water absorbs nearly all the sunlight hitting it, thereby becoming another irreversible loop. The more sea ice that melts, the more the planet absorbs the sun's heat. Another feedback loop involves melting permafrost in northern regions like Canada, northern Europe, and Siberia, which constitute vast tracts of land. Permafrost has remained frozen throughout recorded human history, but now it is melting at unprecedented rates, releasing not only CO_2 but great quantities of methane into the atmosphere. Methane (CO_4) is a greenhouse gas that is 34 times more potent than CO_2 (85; compare McKibben, p. 101). As with melted sea ice, permafrost melt passes an irreversible tipping point because it is impossible to refreeze the permafrost. Therefore, continued warming is unstoppable. The only question remaining: *What can we do to mitigate the coming disasters?*

Romm quotes the U.S. National Oceanic and Atmospheric Administration (NOAA), from its 2009 report: "The climate change that is taking place because of increases of CO_2 concentration is largely irreversible for 1000 years after emissions stop. ... Among illustrative irreversible impacts that should be expected if atmospheric CO_2 concentrations increase from current levels ... are irreversible dry-season rainfall reductions in several regions comparable to those of the 'dust bowl' era and inexorable sea level rise" (29). Romm describes in detail more consequences of global warming – all of which will happen. All we can do is mitigate these effects; we cannot stop them. Even if we take serious worldwide action now, we only have the ability to limit warming to 2 degrees centigrade, thereby mitigating the damage to the point of making life tolerable.

This would mean keeping CO_2 concentrations well below 450 ppm from the year 2050 through the end of the century and beyond. If we continue with business as usual and fail to limit the rise in temperature, life on Earth will feel like hell, with massive starvation, disease, and death for the majority of people and other living creatures. Here are some of the horrific consequences of climate change from Romm's book.

➤ **Dust-bowl conditions and disappearance of agricultural lands.** From southwest U.S. to sub-Saharan Africa to the breadbasket regions of China and India, global warming will end productive agriculture and turn these areas into uninhabitable deserts. Forests will dry out and be regularly ravaged by fire, adding even more carbon to the atmosphere. Already, we have seen extreme and prolonged southern droughts that have led to massive crop failures. Based on scientific studies, Romm concludes that the "coming multidecadal megadroughts will be ... worse than anything seen within the last 2000 years. ... They will be the kind of megadroughts that in the past destroyed entire civilizations" (103).

Figure 4.4
Spreading dust-bowl and desert conditions worldwide.
Source: https://www.tehrantimes.com/news/437156/Iran-becoming-more-vulnerable-to-desertification

➤ **Sea level rise and flooding of coastal lands.** Much of the increased heat is absorbed by the oceans, where heat expands water volume and produces sea-level rise. In addition, melting glaciers add vast amounts of water to the oceans. There also are massive amounts of water stored in the Greenland ice sheet (two miles thick), and even more water stored on the Antarctic continent (also two miles thick). Both the Greenland and Antarctic ice sheets are considered "unstable" because they are melting at unprecedented rates. Since 2015, scientists have been predicting a best-case scenario of a three to five feet increase in ocean levels by the year 2100. This estimate means that many major coastal cities will have to be moved or abandoned on every continent. A business as usual scenario (i.e., without drastic reduction of fossil fuels) will cause sea level rise of one foot per decade through the year 2100, which would put much of the habitable, coastal areas of Earth under water. Billions of people would be displaced and a great portion of the world's agricultural lands would be submerged or ruined because of ever-increasing salt-water intrusion (100).

➤ **Acidification of oceans and death of fisheries.** Romm writes, "The oceans are now acidifying faster than they ever have in the last 300 million years, during which time there were four major extinctions driven by natural bursts of carbon" (123). Most of the carbon we are emitting into the atmosphere gets absorbed in the oceans. Since ocean creatures evolved based on "normal" carbon concentrations, this acidification will accelerate massive extinctions. Moreover, the carbon will deplete the amount of oxygen in the ocean and suffocate many species, creating "dead zones" like the huge area in the Gulf of Mexico where few creatures now live. Today, a large percentage of the global population relies on food from the oceans.

➤ **Disappearance of agricultural lands.** Agricultural lands also are disappearing due to rising oceans, salt-water intrusion into croplands, lack of rainfall in some areas, flooding in others, and the process of desertification already discussed. As a result, millions of acres are destined to become useless as a source of food for human beings, livestock, and wildlife. All the while, Earth's population continues to soar well beyond seven billion people, and the best we can do is make radical changes in our CO_2 emissions to mitigate this disastrous future. In sum, food supplies from both agricultural lands and the oceans will greatly diminish, causing unprecedented worldwide food shortages, starvation, and death.

➤ **Uninhabitable regions of the world.** During the past half-century, we experienced a series of unprecedented heat waves causing human death and the destruction of crops. Noteworthy heat waves hit France in 2003, Moscow in 2010, and Texas in 2011 (43). The current heat record was set in 2020 in Nevada, which reached 130 degrees Fahrenheit (54.4 Centigrade) in the aptly named Death Valley. Researchers have concluded, "Absent strong climate action, we are headed toward levels of warming by 2100 that will expose as much as three fourths of the world's population to a deadly combination of temperature and humidity for at least 20 days a year" (109-10). As warming patterns settle in, regions of Earth will become uninhabitable and people will be unable to go outdoors without special protective gear.

➤ **Unprecedented spread of insects, pests, and tropical diseases.** Because of warmer, shorter winters, the U.S. and Canadian forests have lost some 70,000 square miles of trees to the Bark Beetle and the Pine Beetle, destructive insects whose spread is no longer controlled by harsh winter conditions (49). Similarly, diseases like Zika virus and Dengue fever are on the rise due in part because of warmer conditions everywhere (112). And I need say nothing about Covid-19, as that misery impacted the lives of every world citizen.

➤ **Massive extinctions.** We already are living in a period of unprecedented species extinction. The United Kingdom Royal Society Academy of Sciences wrote in 2010: "There are very strong indications that the current rate of species extinctions far exceeds anything in the fossil record" (in Romm, 126). Since life on Earth is an interconnected web, the die-off of even one species reduces the long-term chances of survival for those that remain (128). And though "new technologies and strategies make it easier for humans to protect endangered species" (128), the only way to preserve as many species as possible is to drastically reduce CO_2 emissions and the global warming process. Errol E. Harris points out the paradox of one species – supposedly the most intelligent on Earth, capable of perceiving systemic dangers and acting accordingly – causing the global extinction of life, including, potentially, its own. (Harris 2014, Chap. 1).

➤ **More frequent and devastating superstorms**. Around the world, superstorms are causing massive flooding, devastating winds, and billions of dollars in damage on a regular basis. Life has become much less predictable and secure due to warmer oceans – the breeding grounds for hurricanes, cyclones, and other extreme weather events. Warming also has induced changes in the jet stream and produced "blocking patterns" in which storms get stuck in a single location. This greatly magnifies the wind and flooding damage from the storm. As a result, coastal lands and cities – already subject to devastating storm surges due to rising ocean levels – will suffer even greater damages from frequent superstorms.

These climate conclusions present us with both an absolute moral and practical imperative to make immediate, radical changes in the way we live and use energy. Romm ends his book by citing the value-based appeal of Pope

Francis: "We must regain the conviction that we need one another, that we have a shared responsibility for ourselves and the world, and that being good and decent are worth it" (282).

Once again, I am curious how Joseph Romm thinks we can facilitate this moral and civilizational transformation without uniting humanity politically, economically, and morally. There are no "market solutions" to these problems. However, everything necessary to effectively deal with the climate crisis is built into the *Earth Constitution*. Within its framework, local communities and global democracy meet in holistic synergy. Why not recognize that this horrific crisis must be met by a truly transformative vision such as that supplied by the *Earth Constitution*?

4.6   The Coming Uninhabitable Earth

David Wallace-Wells is an environmental journalist and prolific writer about environmental issues. His 2017 book, *The Uninhabitable Earth: Life After Warming*, became a *New York Times* bestseller. He opens his book by reflecting on the theme "elements of chaos," and he lists the descending cascade of ill effects that characterize the present state of the world's climate woes. Lack of action cascades into the future. Each passing day that we do little or nothing to address the climate crisis brings another devastating impact we will suffer. We are passing one tipping point and one point of no return after another.

Wallace-Wells reviews the various forms of destruction already illuminated by other authors above. The only question worth repeating is whether we will act in time to salvage a livable planet. Regarding "heat death," he reports that the computer models are consistent and compelling. The consequences multiply with each degree of warming and the prospect of high-end temperatures (e.g., 6 degrees) means an uninhabitable Earth.

He also addresses hunger, already a world problem and inevitably getting much worse. The yield of staple cereal crops declines by 10% for every degree of warming, "which means that if the planet is five degrees warmer by the end of the century, when projections suggest we may have as many as 50% more people to feed, we may also have 50% less grain to give them" (49). In addition, he warns that without a major reduction of emissions, standard scientific predictions give us "at least four feet of sea-level rise and possibly eight by the end of the century" (59).

The chapter on wildfires chronicles the thousands of wildfires that have consumed Earth over the past few years. Drought has helped precipitate these enormous fires and the "cascading chaos that reveals the true cruelty of climate change – it can upend and turn violently against us everything we have ever thought to be stable" (77). His chapter on freshwater loss shows that huge inland lakes have been disappearing from overuse and underground aquifers are shrinking; they are losing water faster than the rate of recharge. He also points out that half the world's population depends on snow melts from glaciers in the Himalayas and elsewhere, all of which are rapidly melting. The chapter on air quality also deserves mention. Currently, medical studies show a high increase in respiratory infection and many other ailments due to air pollution. In addition, studies show an increase in diseases, some of which emerged from melting ice where they have been locked away for hundreds or thousands of years. Other diseases becoming more common due to global warming are yellow fever, malaria, and Lyme disease.

Wallace-Wells points out that some contemporary economists attribute the swift economic growth in the 19th and 20th centuries to the wonders of a "free market," while others credit the discovery of the fossil fuels. Some are predicting a great depression that will dwarf the one of 1929. The world is drowning in debt, as Ellen Brown and many others have pointed out, so with flooding, immense wildfires, droughts and water shortages, is it possible for capitalism to continue its growth mantra? According to Wallace-Wells, the enormous losses from climate disasters portend serious economic consequences that are "cascading through the world system. ... Every day we do not act, those costs accumulate, and the numbers quickly compound" (112-23).

He notes that the U.S. military is obsessed with climate change. Indeed, climate shocks around the world are causing instability, collapse of governments, major movements of refugees, and social instability. He also discusses studies done of those experiencing climate disasters. People experience PTSD, climate depression, and harbor "vengeful thoughts" (136-37). He then asks, "At what point will the climate crisis grow undeniable, uncompartmentalizable? ... How quickly will we act to save ourselves and preserve as much of the way of life we know today as possible?" (140).

Consequentially, he deals with the problems of capitalism and limitless consumption, and he explores technological solutions to the climate crisis. All three of these phenomena are major impediments to effective change, and technology is not likely to be the answer. Climate denial is not a legitimate option either. Overall, this book evades the task of identifying solutions. At best, he surveys a range of responses to climate collapse and the impending demise of the human project. Responses range from "hedonistic quietism" (withdrawing from the world into private satisfactions; a kind of Stoic renunciation of hope), to "climate nihilism" (with synonyms like "climate fatalism"), to a response he associates with the Dalai Lama – that we should be living as fully as possible with compassion, wonderment, and love. It evokes wonder just reading about the ways people respond to the prospect of the demise of civilization! Other writers like Roy Scranton declare that "civilization is already dead" (215).

We are in danger, Wallace-Wells writes, of "climate apathy" and "drawing our circles of empathy smaller and smaller, or by simply turning a blind eye" (215-16). He also mentions "new inhumanism" – a movement that rejects human self-focus and narcissism in favor of what is "not man" and apart from human egoism in its "transhuman magnificence." My translation: So what if human beings go extinct, since the magnificence of the natural world will continue without us. This despair only mounts when he recalls those physicists who have wondered why we appear to be alone in the universe. Is it because all human-like civilizations burn themselves out in climate suicide? Yet he opposes this pessimism by relying on the Goldilocks' Principle – that the initial conditions of the universe post-Big Bang were precisely such that human beings would eventually develop and self-consciously ask questions about the mystery of existence (225; see Harris 1991).

Ultimately, Wallace-Wells comes down on the side of "thinking like a planet," and he says we must think "like a people, one people, whose fate is shared by all" (226). Agreed, we must think like one people if we wish to protect our planet, since a united humanity will engender great hope and energy to confront the crisis. We also agree that our institutions influence our thinking. So if I ever meet David Wallace-Wells, I intend to ask him how he thinks we can accomplish this. How do we institutionally unite humanity to address climate collapse? Once again, the *Earth Constitution* is the solution. It allows us to really begin thinking like "one people, whose fate is shared by all." We will "think like a planet," when we are united as a planet.

4.7 A Common Consensus?

If we could paint a composite picture of what these climate experts are saying, what would emerge? Each is drawing on the same overwhelming set of scientific and empirical data. All of them fundamentally agree on the severity of the crisis, the multiple dimensions of the crisis, and the very real danger that runaway warming may soon become unstoppable. They all point out the multiple dimensions of environmental chaos already hitting our planet from every direction: heat, drought, wildfires, water shortages, floods, superstorms, hunger, social chaos, etc. With the exception of Joseph Romm, who believes growth can be decoupled from fossil fuel use (2018, 205), they all agree that economic growth, like population growth, is taking us beyond the carrying capacity of the Earth in ways that produce increasingly "cascading" consequences for all civilization and future generations.

In sum, all of them see our planet in a state of absolute crisis that must be dealt with now, for tomorrow is simply too late. Indeed, Speth, McKibben, Romm, and Wallace-Wells emphasize the very real possibility of disrupting our planetary ecosystem to the point where we bring ourselves to extinction. Meadows and her colleagues speak of the courage to discern the truth and to follow up on it. Yet, none of them discern that the world as a whole lacks a system, that the world suffers from economic and political anarchy. Speth writes about constructing a bridge to a more sustainable future, but he never challenges the militarized chaos nor the immense waste of the planet's spiritual, moral, and monetary resources.

Naomi Klein alerts us to the money spent on spreading doubt about climate collapse, and she calls for an international "Marshall Plan" as if the nations will join together to accomplish this goal in a timely manner. In essence, she brings us right back to a U.N. level of commitment (i.e., nonbinding), since she does not question the system of militarized sovereign nations. Joseph Romm, on the other hand, lays out realistic facts about the future, but he assumes the only viable solutions are "market solutions," so he never challenges the antiquated and ailing assumptions behind capitalism and nation-state sovereignty. Bill McKibben gives us an outside chance of success, symbolized by solar panels and Gandhian nonviolence (i.e., the technical and the spiritual). However, it does not seem to occur to him that we have the means to *design* our planet for sustainability.

Thus, it appears that none of these well-known environmental experts are considering world disorder as a whole. They recognize that all our problems are interconnected, but they see no solution that addresses all the problems at once, a system-level design change for the whole of humanity. They do recommend a major civilizational, coordinated effort to address the crisis before it is too late. They also advocate for major economic change, recognizing the destructive role of greed in the corporate world. And they present climate collapse as mandating a moral imperative, an ascent human beings must make to mutual recognition and solidarity.

Yet, none of them seem aware of the historic World Federalist Movement that rapidly expanded after World Wars I and II, and which has been reignited by the *Earth Constitution* – the key to future sustainability. The *Earth Constitution* is a systems theory approach to our planetary chaos. We simply must contend with the structures causing the chaos – corporatocracy and militarized nation-states. These powers currently recognize no effective laws above themselves. Surely, planetary lawlessness plays the major role in the destruction of the climate.

Even Naomi Klein, who examines the immense damage done by U.S. economic, political, and military imperialism in her book *The Shock Doctrine,* does not appear to understand the nexus between capitalism and nation-states. Corporatocracy cannot exist without the protection of the sovereign nation-state system, with its global "North-South" subsystem of exploiting cheap labor and stealing resources from the global South to benefit the global North.

How can people unite to address climate change when they are dominated by a world of imperial nations, corporate exploitation, global private banking, and all the attendant corruption of these forces? How can we unite when the world spends 1.8 trillion U.S. dollars per year on militarism and wars? Wallace-Wells describes the "climate nihilism" that has led many people to give up hope – and no wonder! How much longer can humanity hang on hoping for peace, justice, dignity, and a resilient planet?

> *How can the people of Earth unite to address climate change when they are dominated by a world system founded on militarized nations, corporate exploitation, and fear?*

All the thinkers cited in this chapter are well-informed on climate science, but they appear ill-informed regarding the vast literature that analyzes the current world system – the root cause of our woes. These exploitative relationships have clearly existed since the origins of the modern world system in the 16th century. This literature looms in the background, uncited, making the system all the more destructive for being unacknowledged. How is it possible to ignore this immense historic record of militarism, imperialism, rabid nationalism, genocides, repeated treaty violations, national-security ideologies, inter-state competition, domination, and exploitation? And in the case of some nations like mid-20th century Germany and the United States, why is the mythology of exceptionalism, superiority, exclusivism, and collective self-adoration ignored?

Therein lies our answer. We can't achieve the unity and solidarity necessary for transforming the entire world system because some of our best thinkers are not using a true system theory approach to our problems, including sustainability. In one part of his book, Speth briefly mentions Immanuel Wallerstein as a founder of "world systems theory." However, neither he nor any of the others follow through on Wallerstein's analysis of the world system as dominated since World War I by the global economic and military hegemony of the U.S., a system in which cheap labor and resources from "peripheral" nations remain fundamental to U.S. wealth and power. The most glaring example of this blindness is Joseph

Romm, who appears to affirm "American values" without question or significant qualification.

Unless we can transcend this system of imperialism, domination, and corruption that interfaces with the capitalist system worldwide, there is no hope of uniting the world to effectively address climate change. What body is going to institute global disarmament, global environmental laws, equitable global economic relationships, and global human rights? What agency is going to pursue companies that avoid environmental laws in developed countries in order to maximize profit by operating in poor countries? Who is going to educate the masses about sustainable and equitable ways of living, trading, and interacting? For that matter, who is going to educate the public on what it means to be a world citizen and foster the new consciousness that is arising on a planetary scale?

Figure 4.5
Literature on the Dangers of the Current World System and the Promise of the Earth Federation Movement

✓ Christopher Chase-Dunn: *Global Formation: Structures of the World Economy* (1998)

✓ Noam Chomsky: *Hegemony or Survival: America's Quest for Global Dominance* (2004)

✓ Chalmers Johnson: *Sorrows of Empire* (2004), and *Nemesis: The Last Days of the American Republic* (2006)

✓ Errol E. Harris: *Earth Federation Now! Tomorrow is Too Late* (2005)

✓ Petras and Veltmeyer: *Empire with Imperialism* (2006)

✓ Greg Grandin: *Empire's Workshop* (2007)

✓ Pepe Escobar: *Globalistan: How the Globalized World is Dissolving into Liquid War* (2007)

✓ Naomi Klein: *The Shock Doctrine and the Rise of Disaster Capitalism* (2008)

✓ F. William Engdahl: *Full Spectrum Dominance: Totalitarian Democracy in the New World Order* (2009)

✓ Michael Parenti: *The Face of Imperialism* (2011)

✓ Ernesto Screpanti: *Global Imperialism and the Great Crisis: The Uncertain Future of Capitalism* (2014)

✓ David Korten: *When Corporations Rule the World* (20th Anniversary Ed. 2015)

✓ William F. Blum: *Rogue State: A Guide to the World's Only Superpower* (2016)

✓ John Smith: *Imperialism in the Twenty-First Century: Globalization, Super-Exploitation, and Capitalism's Final Crisis* (2016)

✓ Douglas Valentine: *The CIA as Organized Crime* (2016)

✓ My own books: *Millennium Dawn: The Philosophy of Planetary Crisis and Human Liberation* (2005), *Triumph of Civilization: Democracy, Nonviolence, and the Piloting of Spaceship Earth* (2010), and *One World Renaissance: Holistic Planetary Transformation Through a Global Social Contract* (2016)

Einstein said we must learn to think in an entirely new way as citizens of "one world." Ervin Laszlo declared that changing the zeitgeist must include the realization of global consciousness, the realization that "our individual consciousness is an element of the one-consciousness of the world" (2020, 124). Thinkers from Karl Marx to Erich Fromm have declared that real change in human institutions will precipitate a corresponding change in human consciousness. The *Earth Constitution* holistically and synergistically activates both system change and consciousness change.

The new zeitgeist must be worldcentric, grounded in universal moral truths, free of dogma, and selfless in the sense of *agape* and *karuna* (Christian love and Buddhist compassion). *The Earth Constitution* represents these design features and closely correlates to the elevation of human consciousness we seek. With all due respect to Bill McKibben, the idea that nonviolent resistance will slow down the juggernaut enough for solar panels to cover the globe is not proactive enough. Only a template for universal solidarity under a truly humanistic global political and economic system will save us and our planet. Only a constitution that mandates an end to war and ensures planetary sustainability will suffice to perpetuate the human game.

The *Earth Constitution* is therefore both means and ends. It is the key to addressing the climate crisis as well as a potent strategy to raise humanity to a higher level of moral and spiritual awareness. Its holism brings all humans together to solve our global problems, something sovereign nation-states cannot do. It sets up the institutional mechanisms for a democratic global economy that works for everybody, and a comprehensive environmental program in which all nations and persons are on the same page. It inspires hope and vision, as well as loyalty to Earth and future generations, something national patriotism cannot possibly accomplish.

The ecological ruin of our planet can only be mitigated through a fundamental actualization of our highest human potential, in true human solidarity. The *Earth Constitution* empowers local communities to reach this potential, freeing them from the militarized and capitalized corruption that now disempowers them. With the help of the Earth Financial Administration, local communities will receive the technology, financing, and knowledge to regenerate their bioregions and make them truly sustainable. Working from the bottom up and top down, the *Earth Constitution* forms an integrated partnership and cooperative synergy that will lead to success, for it truly embodies a proper and effective design for our living planet.

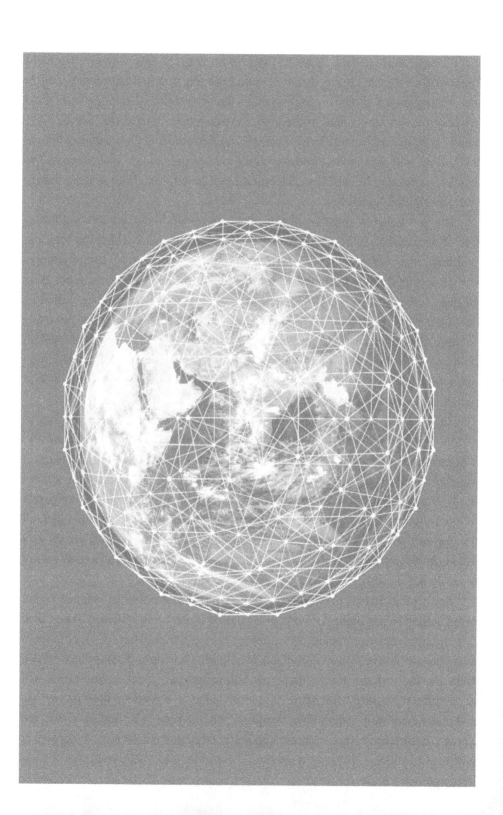

CHAPTER FIVE

Sustainable Economics:
The Need for Transformative Business Models

*Global warming is an objective reminder that it is either
the end of capitalism or the end of the world. For it is
"growth" itself, that is, the capital-driven expansion of
economic product, that effectively drives this process with
its dire and growing implications.*

Joel Kovel

Economics is about vision, but what is our vision of the world? What is the
meaning and purpose of human life? Economist Herman E. Daly believes
we begin our analysis with a "paradigm" of unquestioned assumptions, a
"preanalytic vision" of the world and how it works. Daly goes so far as to declare
that "whatever is omitted from the preanalytic vision cannot be captured by
subsequent analysis" (1996, 46).

The economists considered in this chapter have a broader vision than the
flawed theoretical premises of classical economics. They have expanded their
minds and hearts to include new and better assumptions. Daly writes that our
concrete experience is that of "persons in community" (ibid., 55). As community
members, we experience our relationships with one another, with nature, and
with the ground of Being as having aesthetic, moral, and spiritual dimensions,
and we often approach these relationships with a sense of the sacred. We may
relate to persons as having dignity, as being manifestations of the divine, or as
made in the image of God. All these aspects should be part of our preanalytic
vision, particularly when we consider economics.

These economic thinkers see the current global economic model, with its
growth imperative, as disaster for the environment. I will consider both their
critiques of the present system and their proposed solutions and then paint a
common picture of their views. In each case, I will ask: *Are their solutions really
commensurate with the immensity of the problem? Does their vision encompass
the fullness of our human situation?* These questions may bear on whether the

Covid-19 pandemic will cause the collapse of the corrupt global economic system. Then what? Could it get even worse and more corrupt? Will we experience more economic cycles of boom and bust, more privatization of drug and health-care systems (that the pandemic has exposed as utterly inadequate)? Will nations continue to invest in weapons of war? Will the super-rich exploit the masses even more? Will the "growth" mania utterly destroy the biosphere that makes life possible on this planet? Surely, it is time for a paradigm shift!

5.1   Growth Fetish *vs.* the Reality of a Finite Planet

Richard Heinberg's *The End of Growth: Adapting to Our New Economic Reality* (2011) contains a powerful argument that goes to the heart of the "growth fetish" of capitalism. Heinberg has mastered a broad series of scientific studies and literatures in economics, including analysis of the 2008 world economic bubble and collapse, along with an analysis of the global economic system of banking, money creation, growth, and debt. He has studied the economics of energy sources, extraction and use of natural resources (oil, water, food), and limits on world production and trade issues in various major countries. He also cites the extensive literature on climate change and environmental limits, including pollution, environmental decline, and natural disasters.

Heinberg describes what many thinkers have said about how to adapt to this new reality with resiliency and a minimum of suffering. He expertly puts all this data together to make a formidable argument that we are at the end of the line vis-à-vis growth. The global economic system as we have known it for at least two centuries is over. Furthermore, we must urgently adapt to a post-growth world in order to avoid the planetary collapse that will cause devastation to human well-being everywhere. As Heinberg argues, the climate crisis can only be addressed by transitioning to a truly new age of steady-state economics, politics, and culture.

The worldwide 2008 recession began with the collapse of the U.S. real estate market. Citing many studies, Heinberg shows that this collapse impacted the very structure of the globalized banking, debt, and money creation system. Capitalism famously cycles through economic bubbles and busts – including the big collapses of 1873, 1907, 1929, and 2008 – with many small recessions in between (39-40). Under this system, continued economic expansion eventually returns as an essential feature.

Today, most money is created by banks under the fractional reserve system, which requires banks to keep on reserve a certain proportion of the money they lend to borrowers (3% for smaller banks and 10% for large banks in the U.S.).

This means that a bank may lend up to 97% beyond whatever actual assets it possesses. This loan account goes on the bank's balance sheet and exists nowhere else. It is a "virtual" asset to the bank, but a much more real debt to the borrower, who promises to repay the loan with interest. Under this system, banks are leveraged far beyond their actual assets and if there are major loan defaults or a run on the bank by its customers, the bank will fail and all the money and credit are lost – unless there is a bailout by the federal government or a larger bank swallows the failed bank and its toxic assets. In short, the system as a whole is leveraged by debt far beyond its actual assets.

The assumption behind this entire system (i.e., today's capitalism) is that perpetual growth of the economy is assured. Loans are made in the fractional reserve system with the expectation that borrowers will be able to prosper so they can pay back both the principal and interest on their loans. And most businesses need loans for their projects, big or small. Capitalism also requires continual investments, not just growth, to perpetuate the cycle. Banks fail if they cannot make loans or if the loans default, since economic growth requires both lending and repayment.

Similarly, nations operate under a debt system in which they themselves borrow from the world banking system in order to invest in infrastructure and other initiatives with the expectation that growth (measured in ever-increasing GDP) will allow them to pay back the principal and interest on the loan. In reality, most countries end up paying interest in perpetuity. Under such a system, Heinberg writes, there is "a built-in expansionist imperative" (37). He adds, "The end of growth," is the ultimate credit event, as everyone gradually comes to realize there will be no surplus *later* with which to repay interest on debt that is accruing now" (103).

Heinberg reviews the two main economic theories that developed in the 20th century regarding this expansionist imperative. The great debate has been between followers of John Maynard Keynes ("Keynesians") and the "free market" thinkers like Friedrich von Hayek (often called "Neoliberals"). Keynesians advocate for regulatory intervention by the government and significant spending in order to keep the system growing and healthy. Neoliberals advocate a free, laissez-faire market with minimal government interference. In

Keynesians prefer government regulation and spending to keep the system growing, while Neoliberals favor "free markets" and minimal governmental interference. But both camps assume unlimited growth is possible.

both camps, the assumption is that a healthy economic system requires sustained growth.

Heinberg notes numerous contradictions within the system itself (40-41). The most fundamental contradiction is the assumption that unlimited growth is possible on a finite planet in which there are physical limits to natural resources and eco-system integrity. Moreover, the capacity of the planet to regenerate itself sufficiently to sustain an ever-expanding industrial economy, ever-increasing use of fossil fuels, and an ever-increasing human population is limited. He writes:

> One such error is the belief that economies can and should perpetually grow. ... This fundamental logical and philosophical mistake, embedded at the very core of modern mainstream economic philosophies, set society directly on a course toward the current era of climate change and resource depletion, and its persistence makes conventional economic theories – of both Keynesian and neoliberal varieties – utterly incapable of dealing with the economic and environmental survival threats to civilization in the 21st century. (39-40)

Using extensive sources and the latest scientific and economic studies, Heinberg shows how we have reached peak oil, peak water, peak food production, and peak extraction of certain essential minerals. He points out that "peak" does not mean the end of these things, but the point at which production "achieves its maximum rate before beginning its inevitable decline" (107).

Resources are extracted according to the "low hanging fruit" principle, which means the most accessible are taken first and then those more expensive to access and mine are exploited later. Oil companies now are undertaking mining in the deep ocean shelves and inhospitable Arctic regions, even though these operations are very expensive. Why? Because the accessible oil fields are declining in output and yielding less than global demand. According to a report of the International Energy Agency (IEA), global crude oil production will likely never surpass the 2006 level, and fossil fuels from all other sources such as natural gas will peak around 2035 (107-08).

Water is used in practically all production processes, often in great quantities for cooling and for the extraction of fossil fuels. People, of course, need fresh water for drinking, washing and cooking, and immense amounts of fresh water are used for irrigation to grow crops and feed livestock. The main sources of fresh water are the melt from snowpacks and glaciers, underground aquifers, and the world's major rivers. All these sources are rapidly and visibly shrinking: aquifers are depleting faster than they are recharging, rivers are shrinking and in some cases drying up entirely, and snowpacks and glaciers are melting at rates that will mean the end of these water sources within just a few decades (124-29). As a result, there is a growing shortage of fresh water worldwide, along with ominous signs of a severe global water shortage in the near future.

Food production also is running up against severe environmental limits that portend its inevitable decline. Food production requires not only great amounts of water but immense inputs of fossil fuel to run farm machinery, transport food to processing plants, run the plants, and then get the food to local markets. Around the world, forests are being cut down to plant more crops, and pesticides and fertilizers continue to pollute wetlands, fresh water, and oceans. Agricultural lands are being over-farmed and soil fertility diminishing accordingly. Consequently, food production has peaked and is beginning to rapidly decline.

In addition, oceans are severely over fished, and the ecological resiliency of nature seriously diminished from all these bad practices. Plants require phosphorus for growth, and worldwide phosphorus mining has reached its peak production and is declining (135). This means that a global food crisis is looming, since the current process of producing food is destroying the biological base that makes food production possible (129-138). Moreover, increasing natural disasters and accidents (such as the Deepwater Horizon oil blowout in the Gulf of Mexico) limit rejuvenation. We also depend on the very fossil fuels causing major climate change:

> The billions of tons of carbon dioxide that our species has released into the atmosphere through the combustion of fossil fuels are not only changing the global climate but also causing the oceans to acidify. Indeed, the scale of our collective impact on the planet has grown to such an extent that many scientists contend that the Earth has entered a new geologic era – the Anthropocene. Humanly generated threats to the environment's ability to support civilization are now capable of overwhelming civilization's ability to adapt and regroup. (145)

Heinberg addresses the dogma of economics that claims these crises can be surmounted because of three key economic principles: substitution, efficiency, and innovation. Mainstream economists claim that energy, mineral, and other natural resources can find endless substitutions and that under a free market, human creativity will be able to meet these challenges. Similarly, the market assumes ever increasing efficiency, such as fuel-efficient engines, machines that use less electricity per unit of power, or turbines in dams that produce more electricity per unit of water than drives the turbines.

Often, defenders of unending growth and free markets also claim that innovations are potentially limitless. For example, they anticipate inventions to remove carbon dioxide from the atmosphere or to purify ocean water into fresh water. These arguments ignore that fact that there are absolute limits built into the laws of nature. The second law of thermodynamics, the law of entropy, determines that all forms of "organized" energy will run down, leaving unorganized thermal waste (e.g., burning the "organized" energy in fossil fuels inevitably produces "unorganized" waste in the form of heated gasses such as carbon dioxide). Hence, there are hard and fast limits to innovative improvements:

> While we are never likely to reach zero in terms of time and cost, we can be certain that *the closer we get to zero time and cost, the higher the cost of the next improvement and the lower the value of the next improvement will be.* This means that, with regard to each basic human technological pursuit (communication, transportation, accounting, etc.) we will sooner or later reach a point where the cost of the next improvement will be higher than its value. ... For many consumer products this stage was reached decades ago. (176-77)

Heinberg also reviews the China growth phenomenon, geopolitics, currency wars, population stress, and the post-growth conflict between rich and poor. His informed discussions of each of these issues corroborate the central thesis of his book: Growth has ended, and while relative growth is possible here and there, the global economy, energy consumption, debt and loan development scenarios, and the entire idea of endless economic "development" is at its end, never to be revived.

He addresses how to manage the inevitable contraction and how the concept of growth will necessarily change from quantitative growth of GDP to qualitative, non-economic growth, such as improvements in the quality of life and the meaningfulness of communities. If we continue in denial, however, assuming growth can continue with the right stimulus measures in the form of interest rates, tax deductions for investors, etc., then we may well experience a catastrophic global meltdown unlike anything ever previously seen in history (233-36).

Recognizing the dynamics of our current situation should lead governments to take measures to circumvent the crisis, by at least minimizing its impact and justly equalizing the painfulness of the contraction. But such measures would require "a radical simplification of the economy," for which a general reorganization and transformation would be required.

Two options stand out. First, we could "slice a decimal place off everyone's debts," including businesses, while protecting assets below a certain level to protect the poor. Those who have little or no debt could be compensated accordingly with money added equitably to their accounts. This would be very painful, but would constitute the necessary reset in the relationship between the rich and poor in terms of real assets (238). Second, as Ellen Brown suggests in *The Web of Debt,* we could convert to *public* banking with the government creating debt-free money to address the crisis. Some combination of these two options might make our economic and financial systems "more sustainable and resilient" to face the inevitable climate limits and disasters to come. The present debt-driven system of financial speculation and borrowing based on perpetual growth would need to be given up entirely. We would have to "reinvent money" in ways that extricate its value from speculative money markets.

Heinberg offers a window into new "post-growth" economic systems proposed by various thinkers, such as Frederick Soddy, Henry George, Thorstein Veblen, E. F. Schumacher, Nicholas Georgescu-Roegen, Herman E. Daly, etc. He quotes Daly concerning a "steady-state economy," defined as "an economy with constant stocks of people and artifacts, maintained at some desired, sufficient levels by low rates of maintenance 'throughput,' that is, by the lowest feasible flows of matter and energy from the first stage of production to the last stage of consumption" (250). Success would not be measured by increased GDP. Alternative models for success include the Genuine Progress Indicator (GPI), the Gross National Happiness (GNH) measure, and the Happy Planet Index (HPI). For example, Gross National Happiness measures success across nine dimensions: (i) time use, (ii) living standards, (iii) good governance, (iv) psychological well-being, (v) community vitality, (vi) culture, (vii) health, (viii) education, and (ix) ecology (256-259). Hence, "progress" can be made in all these dimensions regardless of GDP.

Heinberg concludes by exploring a range of literature envisioning alternative community and lifestyle models that are consistent with sustainability, including transition towns, common security clubs, and community economic laboratories – all springing up around the world. People are cooperating at the community level to begin living without fossil fuels, off the electric grids of big utility companies, and out of the big commercial banking systems and into community credit unions or other people run banks. Others are forming food coops and growing much of their own food. People are forming community health clinics independent of big government or corporate health systems. They are sharing tools, developing their own alternative currencies, and engaging in labor/barter transactions. At the local level, people are creating resilient and independent communities that are much more likely to flourish as the globalized economy fails and the great contraction and transition takes place.

Heinberg ends the book by placing our current crisis within a broad perspective of the great societal transitions previously made by humanity. First, humans harnessed fire nearly two million years ago. Second was the development of language. Third was the agricultural revolution ten thousand years ago. Fourth was the industrial revolution about two hundred years ago. The fifth transition is the great contraction happening right now:

> Now we are participating in the turning from fossil fueled, debt-and growth-based industrial civilization toward a sustainable, renewable, steady-state society. While previous turning entailed overall expansion (punctuated by periodic crises, wars, and collapses), this one will be characterized by an overall contraction of society until we are living within the Earth's replenishable budget of renewable resources, while continually recycling most of the minerals and metals that we continue to use. ... The remainder of the current century will be a time of continual evolution and adaptation ... which will itself be a dynamic rather than a static condition. (284)

I wish to ask Richard Heinberg if he is aware that the *Earth Constitution* outlines the reinvention of money in ways that do not make it dependent on speculative money markets. Such a reinvention would obviate much of the fighting over how to transform the present world system that answers to no democratic authority protecting the rights of people. The experimental projects he mentions will not solve the problem of a global currency based on debt and financial speculation. If economics is about "vision," the *Earth Constitution* is the solution, since it fully addresses the complexities and inequities of our current human situation.

5.2   Can Economics Put People and Planet First?

Kate Raworth's *Doughnut Economics: Seven Ways to Think Like a 21st Century Economist* is a truly illuminating book in harmony with both human welfare and planetary sustainability. It provides a deeply informed and positive vision of how human beings can create a decent world for all and a credible future for our planet. It is a vision fleshed out by hundreds of excellent examples of what is being done worldwide to build a new world. The book summarizes both the history and dynamics of economics, plus what she calls the "seven fundamental principles" of economics.

➤ **First Principle: Change the goal from GDP to the Doughnut.** Raworth quotes Buckminster Fuller as saying "You never change things by fighting the existing reality. To change something, build a new model that makes the existing model obsolete" (4). She then proceeds to punch holes in the existing paradigm, one built on a single story, a single index, and essentially one erroneous theory of economics. This worldview became fixed in textbooks and the minds of economists worldwide. It was built around a linear model that essentially ignored our rootedness in the planetary biosphere and only focused on growth. Growth became measured worldwide in terms of GDP, which ignores human rights as well as our planet's ecology. This story became systemized worldwide in immense institutions like the World Bank and World Trade Organization. Then the system either accepted or ignored ever widening inequality, vast poverty, and growing evidence of climate change.

Raworth argues that economics can no longer be framed on this antiquated linear growth model because it is based on limitless extraction, production, distribution, consumption, and waste disposal. This model ignores nature and our planetary boundaries. She quotes a recent declaration:

> We are the first generation to know that we are undermining the ability of the Earth system to support human development. This is a profound new insight, and it is potentially very, very scary. ... It is also an enormous privilege because it means that we are the first generation to know that we now need to navigate a transformation to a globally sustainable future. (47)

Instead, we need to imagine a doughnut shape. The outer ring of the doughnut is composed of the "ecological ceiling" expressed by nine planetary boundaries that climate scientists have identified as crucial if we want conditions on our planet to remain hospitable: (i) climate change, (ii) ozone layer depletion, (iii) air pollution, (iv) biodiversity loss, (v) land conservation, (vi) freshwater withdrawals, (vii) nitrogen and phosphorus loading, (viii) chemical pollution, and (ix) ocean acidification (see *Figure 4.2* in Chapter Four). Violating these limits constitutes environmental overshoot and the "ecological ceiling" for doughnut economics.

The inner ring of the doughnut is comprised of the "social foundation" which provides "safe and just space for humanity." Just as economics fails when it goes beyond the ecological boundaries of the outer ring, it also fails when it does not provide for the well-being of human beings. Thus, the inner ring is comprised of human needs and rights, such as income and work, education, health, food, water, energy, networks, housing, equality, political voice, as well as peace and justice. In this way, Raworth expands the mission of economics from merely money supply and wealth creation to concern for all elements of human well-being. Just like all systems, economics can be *designed* to achieve favorable outcomes, like sustainability.

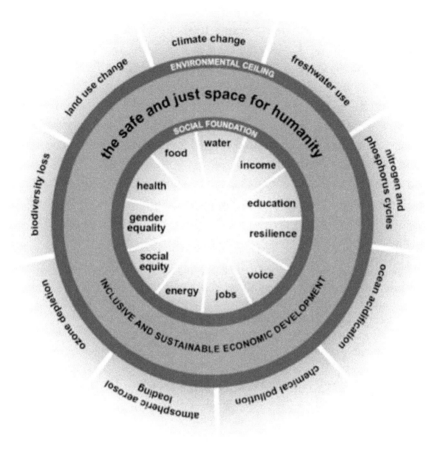

Figure 5.1
Rendering of the "Doughnut Economics" framework showing the multiple factors
that must be considered both environmentally (outer ring) and humanly (inner ring).
Sources: IPTC Photo, Högskloan | Gävle, Creator: Camilla Haglun

Raworth links this social foundation with the United Nations Sustainable Development Goals (that I will examine in Chapter Six). "Together," she writes, "the social foundation of human rights and the ecological ceiling of planetary boundaries create the inner and outer boundaries of the Doughnut" (42). She adds that we need to change the goal of economics from "the Cuckoo model" of endless GDP growth to a model based on *"human prosperity in a flourishing web of life"* (47).

➤ **Second Principle: See the Big Picture – from self-contained market to embedded economy.** Economics impacts everything, all dimensions of human life as well as the Earth that sustains us. Raworth shows the multiple ways traditional economics simply ignores most of these dimensions and operates as a self-contained theoretical model, leading to today's disastrous consequences. Conversely, the Doughnut model encapsulates the whole of our human situation, starting with the recognition that Earth has planetary boundaries which must be respected. Her big picture includes:

- ✓ Society, which is foundational, so nurture its connections;
- ✓ The Economy, which is diverse, so support all its systems;
- ✓ The Household, which is core, so embed it wisely;
- ✓ The Commons, which are creative, so unleash their potential;
- ✓ The State, which is essential, so make it serve wisely;
- ✓ Business, which is innovative, so give it purpose;
- ✓ Trade, which is double-edged, so make it fair; and
- ✓ Power, which is purposive, so check its abuse (63).

Raworth examines each of these dimensions, showing how each has been ignored or falsely conceived by GDP economics. Traditional economists never speak of "nurturing" or "wisdom" or the very purpose of business, since their model is based on early-modern ideas of mechanistic science. Raworth declares that her Doughnut Economics ends the myth of the self-contained, self-sustaining market, and replaces it with the household, the commons, society, and government – all the aspects of daily life which are embedded within our living world (79).

> **Third Principle: Nurture Human Nature – from rational economic man to social adaptable humans.** The old model of *homo economicus* states that each person (and by extension each business or company) is an atomistic, rational calculator of self-interest and advantage. Not only is this a fundamental distortion of our complex human nature, but it serves as the dominant social model for people in trade, banking, and business.

But the 21st century portrait of human nature is very different. In the new model, humans are viewed as "social and reciprocating" rather than narrowly self-interested. We have "fluid values" rather than fixed economic concerns. We are "interdependent," not isolated atoms of self-interest. We act on "approximations," not narrow calculations. And we are embedded in the web of life, which is "far from having dominion over nature" (88).

There are many ways we can overcome the destructive stereotype of "homo economicus" and empower people to economically focus on the common good and planetary sustainability.

Raworth also explores the many ways we can "nurture" and "nudge" people to do the right thing. Right behavior can be encouraged by the state, by communities, and by businesses through proper tax programs and other economic incentives, empowering cooperatives, legalizing new associations that work together for the common good, recognizing the role of households in the economy, empowering women, educating for ecological sensitivity, and promoting "generosity and public spirit" (110).

➤ **Fourth Principle: Get Savvy with Systems – from mechanistic equilibrium to dynamic complexity.** Under systems theory and complexity science, we now understand "how relationships between the many parts of a system shape the behavior of the whole" (117), and how the composition of the whole helps to shape the behavior of the parts. Natural systems – with their feedback loops and integration of many dynamic factors – operate according to complex designs of dynamic balance and mutual integration. This type of feedback loop is absent in traditional economics, which is why we suffer from boom and bust bubbles and on-going financial instability, unable to experience economic history as anything other than "a rolling cycle of dynamic disequilibrium" (126).

This dynamic disequilibrium has led to major tragedies, such as drastic inequality of wealth. Raworth adds that similar feedback loops have allowed governments to be taken over by an "oligopoly" of the rich and powerful who direct law and institutions toward their own interests. She also recalls "the damage wrought by the shock policies of privatization and market liberalization implemented in Latin America, sub-Saharan Africa, and the former Soviet Union during the 1980s and 1990s" (137).

Similarly, today's economy has ignored pollution and is bringing the world to the brink of ecological collapse. She explains why pollution is "unlike metals, minerals, and fossil fuels – it typically carries no price and so generates no direct market feedback." She concludes, "Today's economy is divisive and degenerative by default. Tomorrow's economy must be distributive and regenerative by design" (133). We also need to acknowledge "nested systems" that "serve the whole of which they are a part." In this way, we can promote diversity and create a healthy resilience to rebound from storms. We can create "open-source" design business models, find "leverage points" that help balance the entire system, and bring ethics into economics. Traditionally, economics pretended to be a "science" that abjured ethics. However, today we recognize our responsibility to act in the service of "prosperity within a flourishing web of life." Lastly, we need to "be prudential in policymaking," minimize risk, attend to the most vulnerable, and "work with humility," recognizing the limitations and shortcomings of our models (138).

➤ **Fifth Principle: Design to Distribute – from "growth will even it up again" to distributive by design.** A fundamental model of traditional economics is the "Kuznets Curve," a simple chart that shows increasing economic inequality (linked to endless GDP growth) that peaks at a certain point and then begins to descend toward ever more equality. As Raworth states, "It was a clever theory, but it was wrong" (142). Inequality was seen as an inescapable side-effect of the endless growth model, but its many negative effects were ignored, including the fact that it is destructive of democracy. It allows the few to distort not only the market but also politics. It does not help developing poor economies grow faster, and it promotes economic instability leading to recessions and depressions.

Designing to distribute, on the other hand, looks at the economic world as a network of flows, just as ecology looks at the world as a dynamic and interdependent network of flows of energy, air, water, nutrients, etc. So a systems thinker views the world as "a distributed network whose many nodes, larger and smaller ... are interconnected in a web of flows" (148).

Consequently, we must re-examine wealth, how it is owned and distributed, and find ways to establish dynamic flows that benefit everyone. A wealth tax, maximum and minimum income rules, democratizing of ownership, reexamining property and land ownership laws, low-cost loans, fair intellectual property rules, and land-value taxes have all been proposed by economist Henry George. We also must reexamine banking and money creation which, until now, have largely been monopolies of private, profit-making banks that create money as interest-bearing debt. In addition, we should ask why the lion's share of money or created value goes to "investors," who rarely produce, manage, or even participate in the enterprises in which they invest.

> But what determined each group's respective share of earnings? Economic theory says that it is their relative productivity, but in practice, it has largely turned out to be their relative power. The rise of shareholder capitalism entrenched the culture of shareholder primacy, with the belief that the company's primary obligation is to maximize returns for those who own its shares. ... Employees, who turn up for work, day-in and day-out, are essentially cast as outsiders. ... Shareholders, meanwhile, who probably never set foot on the company premises, are treated as the ultimate insiders. (160)

Raworth provides many interesting examples of communities that have devised alternative ways of dealing with income, trade, and distribution. By employing such innovations as block-chain currencies to track and distribute value, and "time-care credits" in which people accumulate credit for helping a community's more-needy or elderly residents, communities have combined personal care with income distribution and community solidarity. In addition, not for profit enterprises are spreading, along with community interest companies and cooperatives.

> **Sixth Principle: Create to Regenerate – from "growth will clean it up again" to regenerative by design.** Raworth states, "Recently compiled international data reveal that when a nation's global material footprint is taken into account – by adding up all of the biomass, fossil fuels, metal ores and construction minerals used worldwide to create the products that the country imports – then the success story seems to evaporate" (179). Instead, business needs to play a critical role in helping us achieve a "circular economy" in which we restore, repair, reuse, recycle, and regenerate value at each stage of decomposition, all the while using renewable materials and energy in harmony with ecosystem limits (188). What we need are regenerative solutions if we want to shift the paradigm.

We also need to reexamine the idea of the commons. There is no "tragedy of the commons" if we redefine and share the commons in multiple efficient and complementary ways – and that means sharing intellectual property too. The super-profitable digital companies are being reexamined with an eye to how we can create a "knowledge commons" for the world, with everyone benefitting from robot technology (164). Contemporary capitalism is still focused on attitudes that are "the opposite of generous." They focus only on creating financial value "for just one interest group: shareholders" (193). On the other hand, the Open Source Circular Economy (OSCE) movement strives to "unleash the full potential of circular manufacturing." Such movements emphasize: modularity (products with parts that are easy to assemble, disassemble, and rearrange), open standards (designing components to a common shape and size), open source (full information on the composition of materials and how to use them), and open data (documenting the location and availability of materials) (195-96). Transparency is the key to building a truly regenerative and restorative economy.

Raworth wants us to understand that the economic system itself is causing the environmental crisis. "The global financial system as we know it needs to shrink, simplify, diversify and deleverage" (199). And the state must act as a partner within the new paradigm. There is no such thing as a "free market," so various regulations and incentives are needed to give rise to desired results. The state must work with businesses, the market, the commons, and the citizens to establish a planetary network that regenerates and restores, inclusive of the common good of everyone.

➤ **Seventh Principle: Be Agnostic About Growth – from growth addicted to growth agnostic.** We must be agnostic about growth because the growth dogma is destroying the planet, creating ever greater inequality, and ruining democracy everywhere on Earth. This doesn't mean that growth in certain sectors cannot be helpful (e.g., solar panel manufacturing). It means we should redefine growth to include a wider and more progressive definition of success and well-being. It also means that we must decouple growth from "resource use," including pollution and other damaging "externalities" ignored by classic economics. Raworth calls this "sufficient absolute decoupling" in which our planetary system moves back within the nine boundaries established by climate scientists.

She also reminds us that the rapid growth of the past two centuries was largely due to the supply of cheap fossil fuels (coal, oil, natural gas). Even though we live on a planet daily flooded with clean solar, wind and water energy, the global economy is still directly dependent on fossil fuels. We must decouple the economy from this dependence. "Growth" can no longer mean using fossil fuels. She adds that at present, "GDP brings both global market power and global military power" and that we need "innovative thinkers in international relations to turn their attention to strategies that could help usher in a future of growth-agnostic global governance" (238).

In her conclusion, Raworth declares, "Ours is the first generation to deeply understand the damage we have been doing to our planetary household, and probably the last generation with the chance to do something transformative about it. ... Once we accept the economy's inherent complexity, we can shape its ever-evolving dynamics through smart stewardship" (243-44). That is, we can frame a new story. As with the Gandhian declaration, "Be the change you want to see in the world," we also need to *design* the change we want to see in the world. Changing our goals will change the paradigm. For instance, the New Economics Foundation has summarized findings that are proven to promote human well-being, such as "connecting to people around us, being active in our bodies, taking notice of the world, learning new skills, and giving to others" (240).

Raworth has done an excellent job clarifying that economics is about how we design our "global household," not about iron laws of supply and demand, extraction, production, consumption, and disposal. She says we need "a vision of a new human community." This is exactly right. And what could more adequately embody such a vision and superior design than the *Earth Constitution*? It is premised on exactly where she says we need to focus: human welfare and the quality of our lives in deep synchronicity with our planetary biosphere.

5.3   Calculating the Value of Human Life

Frank Akerman and Lisa Heinzerling wrote their book, *Priceless: On Knowing the Price of Everything and the Value of Nothing*, to examine the cost-benefit analyses undertaken by the U.S. government regarding environmental issues and the climate crisis itself. They begin by looking at the reasoning behind the creation of important environmental laws, such as the Clean Air Act (1970) and the Clean Water Act (1972). However, beginning in the 1980s, the U.S. government has been colonized by neoliberal economists who devised "cost-benefit" formulas for environmental action that routinely inflate the costs and deflate the value of the benefits of environmental laws and regulations.

These economists embrace the ideology of capitalism that declares markets to be the supreme arbiter of efficiency and value. Under their analysis, a private company only incurs the costs of acquiring resources or materials and of the labor and machinery required for production. In this way, the company's profit is the sales price of the item minus the costs of materials and production. If consumers do not want the product and refuse to buy it, then the market "efficiently" will stop producing it. Market competition, as this mainstream economic theory claims, keeps costs to a minimum while producing products people want.

Advocates of this free market doctrine detest government regulation of businesses – whether for reasons related to the environment, worker safety, decent pay, or otherwise – since regulation interferes with the "efficiency" of the market. The issue is how to price "externalities" like the cost of healthcare when a worker who deals with toxic waste gets cancer, the cost of removing lead poison in our children's schools, or the cost of cleaning up previously unspoiled wilderness areas? (3-7). How much are the lives of future generations worth? The assumption, critiqued by these authors, is that the market *can* put a price on all these things and give us quantifiable answers:

> Cost-benefit analysis sets out to do for government what the market does for business: add up the benefits of a public policy and compare them with the costs. ... In principle, one could correct for the potential sources of bias in estimating the costs of regulations and other public policies. No such correction is possible in assessing the benefits of regulation, because the benefits are, literally priceless. Herein lies the fatal flaw of cost-benefit analysis: to compare costs and benefits in its rigid framework, they must be expressed in common units. Cancer deaths avoided, wilderness and whales saved, illnesses and anxieties prevented – all these and many other benefits must be reduced to dollar

values to ensure that we are spending just enough on them, but not too much. ... Most or all of the costs are readily determined market prices, but many important benefits cannot be meaningfully quantified or priced, and are therefore implicitly given a value of zero. (37-40)

With the 1979 U.S. election of conservative President Ronald Reagan, the great moral principles that undergirded the Clean Air Act and Clean Water Act began to be attacked and eroded. These moral principles included the precept that people have the right to breath clean air and drink clean water, and that deaths from bad air and water are wrong because lives are priceless. Period. However, according to the neoliberal point of view that triumphed worldwide with Reagan in the U.S. and Margaret Thatcher in the U.K., regulation interferes with the free market, as does the public ownership of forests, resources, and many government services. The new mantra was "deregulation" and "privatization," converting everything possible to profit-making businesses with minimum government regulatory interference. This alone, they declared, could give us the cost-benefit efficiency provided by capitalist markets.

The Office of Management and Budget (OMB), founded in 1970, also played a big role, as it was colonized by advocates of cost-benefit analyses for all government functions. There are many examples of the absurd calculations made by OMB, as it attempted to show the high costs and low benefits of government regulation, including quantifying human lives. E. F. Schumacher declares that the very existence of cost-benefit analysis is an absurdity because it assumes that you can quantify everything, and its deeper assumptions are even worse: "The logical absurdity, however, is not the greatest fault of the undertaking: what is worse, and destructive of civilization, is the pretense that everything has a price or, in other words, that money is the highest of all values" (1973, 48). Neoliberalism's "free market" ideology is destructive of civilization itself.

The Environmental Protection Agency (EPA) also developed a cost-benefit analysis to measure the effectiveness of regulations. For example, the EPA analyzed whether arsenic should be removed from drinking water. Arsenic, of course, is a deadly poison for human beings that often appears in drinking water drawn from various underground sources. So what exactly does it "cost" to reduce arsenic levels to a determined "safe" zone, which the EPA defined as 10 parts per billion? Turns out that the EPA's complex formula determined that the value of a human life equals 6.1 million dollars. Thereafter, this figure became a commonly used value in many other cost-benefit calculations, as in the case of estimating wages for workers in jobs with high risk of injury or death (75-81).

To be clear, according to the EPA you can't do a cost-benefit analysis without putting a price tag on human life. Only then can you analyze whether a regulation is beneficial and efficient or whether it is inefficient and needlessly costly. I see this as an outrage and agree with the authors who assert, "Human life is the ultimate example of a value that is not a commodity, and does not have a price" (67). Yet the OMB and EPA do these analyses all the time. What is the price of health versus sickness? What is the price of dealing with bladder cancer versus dealing with a common cold? They have calculations for it all, since they believe government cannot perform efficiently unless a determinate value is ascribed to every benefit sought, whether saving human lives, regulating job safety conditions, creating good traffic laws, determining if whales are worth saving, or removing poisons from drinking water.

Akerman and Heinzerling advocate the "precautionary principle, calling for policies to protect health from potential hazards even when definitive proof and measurement of those hazards is not yet available." They quote from the Rio Declaration adopted by the U.N. in 1992 at the Conference on the Environment in Brazil: "In order to protect the environment, the precautionary approach shall be widely applied by States according to their capabilities. Where there are threats of serious or irreversible damage, lack of full scientific certainty shall not be used as a reason for postponing cost-effective measures to prevent environmental degradation" (117-118).

Not surprisingly, the government considers income, expected and lost. In these calculations, fairness is not an issue. As it turns out, the lives of high-income people are deemed more valuable than the lives of low-income people (71-74). The same applies for older, retired people, who no longer earn an income. According to these calculations, their value is considerably less per life than the value of younger people (101-02). The planetary environment also is discounted in these cost-benefit calculations. There is no such thing as "intergenerational equity" since the lives of future people are worth considerably less than consumer preferences of present generations. This book is full of interesting examples of the absurd conclusions and calculations made by the OMB and EPA over the last several decades, conclusions that defy common sense, repudiate the precautionary principle, and violate human decency, fairness, and universal moral values like the Golden Rule. As the authors explain:

> The imperatives of protecting human life, health, and the natural world around us, and ensuring the equitable treatment of the rich and poor, and of present and future generations, are not sold in markets and cannot be assigned meaningful prices. ... Our view is sharply at odds with the contemporary style of cost-benefit analysis in Washington.

The new conventional wisdom assumes that the priceless is worthless: today's decisions require calculations and bottom-line balances, and only numbers can be counted. ... An alternative method of decision-making is badly needed. (207-08)

Akerman and Heinzerling recommend an alternative "holistic approach," rather than the "atomism and reductionism" of the cost-benefit analysis. Under the holistic approach, decisions "depend on multiple quantitative and qualitative factors," and the qualitative factors include human "rights and principles, not costs and benefits" and "fairness – toward the poor and powerless today, and toward future generations" (210, 213). None of this is ever included within the neoliberal, free market capitalism that has dominated the U.S. (and much of the world) over the past four decades. They write: "Health and environmental protection ultimately involve our values about other people – those living today, and those who will live in future generations." This is "an ethical question that must be answered prior to detailed decision-making" (229).

The logic of the market deals with cost-benefit efficiency, never moral values, which in the final analysis cannot be quantified (for decisive criticisms of Utilitarianism, see Finnis 1983). Ethically, we must be concerned both with people in the future and those trapped today in poverty. We must face the tradeoffs and dilemmas honestly and pragmatically, with a "sense of moral urgency" and without compromising what is priceless (234). Indeed, these same considerations should pertain to valuing human life in all situations, they assert, since "no one ever imagines it when it comes to defending ourselves against military threats" (220). Here they betray the limitations of their own worldviews, since the example of the military as not needing this cost-benefit analysis raises the question about just how "priceless" they consider non-American lives to be. Economic holism must embrace all aspects of human life and our planetary biosphere, including the absurdity of militarism, which values non-American lives at close to zero. It is precisely this breadth of holism that is addressed by the *Earth Constitution*.

5.4 Will the Third Industrial Revolution Abolish Capitalism?

Jeremy Rifkin's books have long been a major source of creative and deeply informed thought, and *The Zero Marginal Cost Society* is no different. The book chronicles the vast transformations included in what he calls the "Third Industrial Revolution" (also the title of his earlier book). It describes the immense import of the Third Industrial Revolution for a liberated human future on Earth – a future of abundance, fulfillment, justice, and sustainability.

Rifkin begins by discussing the "Great Paradigm-Shift from Market Capitalism to Collaborative Commons." In economics, marginal cost is defined as the incremental cost of producing one additional unit of a product or service. The vaunted "efficiency" of capitalism is that competition forces companies to continually introduce new technologies and ever-leaner forms of production that lower the cost of goods and services. With automation and mechanized assembly lines, for example, automobiles can be reproduced at an ever-lower marginal cost. However, the Third Industrial Revolution – which has engendered the digital, robotic, computerized "collaborative commons" – is lowering the marginal cost of goods and services in many domains to nearly zero.

For Rifkin, the fact that computerized machines will bring the marginal cost of goods and services to nearly zero means that "the ultimate triumph of capitalism ... also marks its inescapable passage from the world stage" (11). The digital revolution or Third Industrial Revolution changes everything. During the First Industrial Revolution the steam engine was invented, and it required coal for fuel and the telegraph for communications. Rifkin says this transformed the world within thirty years, from 1860-1890. The Second Industrial Revolution involved the invention of the internal combustion engine, which needed gasoline and oil for fuel and the telephone for communications, once again transforming the world in roughly a twenty-five-year period between 1908 and 1933. Both these eras gave rise to huge "vertically integrated" systems that were operated top-down by managers and boards of directors distributing goods and services through vast centralized industrial systems.

However, the current Third Industrial Revolution has given us locally generated, low cost, green energy, often empowered by green transport systems as well as the collaborative communications commons of the worldwide web. Within the next twenty years, this new system has the potential to transform the world from centralized, vertically integrated fossil fuel systems to laterally networked, locally powered, and sustainable systems (2019, 244). Moreover, with our geometrically expanding capacity to do 3D printing, consumers are now becoming "prosumers," who can individually or collaboratively produce goods and services themselves and share them with their local community. We rapidly are moving toward "technological unemployment," as people replaced by machines discover entrepreneurial and cooperative responses that mutually empower everyone with greater and virtually free goods and services. Thus, the new system operates outside of the traditional workplace, where employees labored in vertically integrated industrial systems. Yet another reason why traditional capitalism will become obsolete.

Rifkin details at least two ways that classical capitalism failed to comprehend the reality of this new situation. First, "in classical and neoclassical economic theory, the dynamics that govern the Earth's biosphere are mere externalities to economic activity" (12). The law of entropy necessitates that all production and uses of energy necessarily become useless heat and waste. For example, the burning of fossil fuels releases CO_2 into the atmosphere. Today, "the entropic bill for the Industrial Age has arrived" in the form of "wholesale destruction of the Earth's biosphere" (2013, 13).

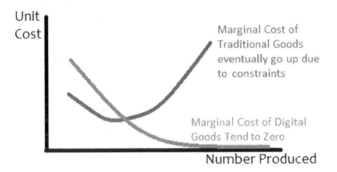

Figure 5.2
Graph of the tendency to Zero Marginal Cost
while the cost of traditional goods continues to rise
Source: IP Carrier

Second, Rifkin argues that capitalism framed both the natural world and human nature in its own image "suggesting that its workings are a reflection of the way nature itself is organized" (ibid., 70). Today, human beings no longer are framed as autonomous, competitive self-seeking creatures. The emerging science of ecology and the growing collaborative commons are "accompanied by a sweeping rethink of human nature that is fundamentally altering the way we perceive our relationship to the Earth" (80). The new view of human nature contends that we are intensely social creatures capable of shared, empathic consciousness and of "biospheric consciousness." Rifken previously wrote a book titled *The Empathic Civilization: The Race to Global Consciousness in a World of Crisis*, in which he asks, "Can we reach biosphere consciousness and global empathy in time to avert planetary disaster?" (616).

This *Zero Marginal Cost Society* chronicles the many collaborative and cooperative features of the Third Industrial Revolution that call out for a shared plenitude of knowledge, experience, and mutual service rather than private profit derived from scarcity of knowledge, experience, and services. He also describes how "extreme productivity" is both possible and happening now at near zero marginal cost. These developments point to a future world of abundance, solidarity, empathy, and freedom.

Rifkin provides an overview of the new "collaborative commons," and he addresses the oft-cited "tragedy of the commons," a phrase derived from Garrett Hardin's 1968 essay arguing that unrestricted use of the commons always leads to its overuse and destruction. Rifkin cites the growing research and literature showing that throughout history people can and have cooperated successfully in their use of the commons. Indeed, people are rapidly moving away from an ethos in which they define themselves in terms of their possessions and toward an ethos in which they define themselves in terms of their relationships and their belonging to larger, meaningful wholes.

Capitalism feeds off scarcity, but when the marginal cost of production approaches zero – through 3D printing, shared knowledge on the internet, and many forms of collaborative action – then "profits disappear because goods and services have been liberated from market pricing." The products and services still have use and value, "but no longer have exchange value" (333). In other words, people can get much of what they need without having to pay for it.

Rifkin refers to Mahatma Gandhi's intuitive understanding of sustainability. Local communities become self-sufficient and their citizens empowered through cooperative endeavors to produce enough "for every man's need but not for every man's greed" (334). Soon, locally generated green power from sun and wind will be networked across the globe providing everyone on Earth with the means for sustainable living, assuming we also make a conscious effort to limit population growth. He asks, "How many human beings can live comfortably without destroying the biosphere's ability to continually replenish the necessary ecological resources?" Depending on the formula used, it may be anywhere from 2.5 to 10 billion people (335-36). It is worth noting that access to electricity alone has been shown to lower the number of children in families within developing countries. If we can bring the fertility rate down to 2.1 children and eventually level the population off at 5 billion, Rifkin believes we will secure a decent future for humanity (349).

Rifkin reviews the now familiar list of lethal consequences caused by climate change, and he says it is extremely foolish to think we can deal with these challenges through a patchwork carbon-based regime (355). The urgency to shift the paradigm puts us in a race against time. Fortunately, recent studies show happiness comes through relationships, quality of life and meaningful existence, not through wealth and private property. The materialism of 20th century capitalism is being replaced with empathic relationships and the quest for sociability and community in the 21st century.

In my own books and articles on the need for a fundamental shift in consciousness, I speak of the transformations of human consciousness throughout history and the current mutations of consciousness taking place during our contemporary "second axial period" (e.g., Martin 2005, 2008, 2010b, 2018). Rifkin presents a similar progressive series of historical transformations, though under different names. His progression moves from primitive mythological to theological to ideological to psychological to advanced biospheric consciousness. Roughly speaking, what I call "holistic evolutionary consciousness," or what Brian Swimme and Thomas Berry call "ecozoic consciousness" (1992), he calls "biosphere consciousness."

So we agree that many people are in transition from a "psychological consciousness," with its great capacity for empathy with people and other creatures, to a biospheric consciousness, in which we understand how to live in harmony with the biosphere of the Earth that sustains us. He concludes by stating that "the transition from the capitalist era to the Collaborative Age is gaining momentum in every region in the world – hopefully, in time to heal the biosphere and create a more just, humane, and sustainable global economy for every human being on Earth in the first half of the twenty-first century" (380).

In his earlier book about "empathic civilization," Rifkin does chronicle the emergence of the modern nation-state system (292-300). He speaks about the collective national consciousness engendered by this system and its positive feature of uniting people in ways that allow for greater empathy among the nation's citizens (i.e., ethnocentric focus). But now we must go beyond national consciousness to planetary consciousness, and Rifkin appears to assume the latter development is compatible with retaining absolute sovereign-state borders.

As a result, Rifkin misses the dialectical relationship between institutions and consciousness at the global level, beyond borders. Militarized national borders inhibit and defeat the rapid growth required for planetary or biospheric consciousness. Not only that, the lawless character of the current system gives law itself a bad name and encourages the colonization of all governments by rulers who manipulate the law for national or personal interests, as in the case of Donald Trump.

The lawless character of the nation-state system gives law itself a bad name. We must create a worldwide federation of nations and pass binding world laws that are transparent, fairly applied, and guide us to become world citizens.

Rifkin seems unaware that uniting Earth as one political whole would quickly enhance the process of planetary mutual identification and empathic expansion. Instead, he returns immediately to his description of "psychological consciousness," which leads, he argues, toward empathy and a new understanding of what it means to be human (apparently regardless of the militarized nation-state system). In his concluding remarks, he brings up geopolitics once again, stating:

> Geopolitics has always been based on the assumption that the environment is a giant battleground – a war of all against all – where we fight with one another to secure resources to ensure our individual survival. Biosphere politics, by contrast, is based on the idea that the Earth is a living organism made up of interdependent relationships and that we each survive by stewarding the larger communities of which we are part. The new bottom-up continentalization and globalization allow us to complete the task of connecting the human race and opens up the possibility of extending the empathic sensibility to our species as a whole, as well as to the many other species that make up the life of the planet. (2009, 615)

Very good! Except Rifkin omits the insight that this "giant battleground" is *institutionalized* in the system of militarized sovereign nation-states, and that institutions inevitably condition human consciousness. To change the consciousness, we must change these flawed and fragmented institutions and "complete the task of connecting the human race," as Rifkin puts it. Only after we connect both politically and economically will we be able to reach holistic, planetary consciousness. This is precisely the role of the *Earth Constitution*, which is sadly missing from Rifkin's thoughts or writings.

5.5 A Green New Deal for All

Jeremy Rifkin believes in democracy and empowering workers, but his analysis of capitalism ignores that it is predicated on exploitation – exploitation either of the workers, society, other countries, nature (or all four together). In *The Zero Marginal Cost Society*, he describes the inevitable demise of capitalism due to the development of sustainable livelihoods that are near self-sufficiency and can meet their own needs at near zero marginal costs. He also recognizes that capitalism treats Earth's biosphere as "mere externalities to economic activity" (12).

However, in his latest book *The Green New Deal* (2019), Rifkin brings capitalism very much back into the picture. He argues for investment (i.e., private profit) in the green revolution in order to avoid the "stranded assets" resulting from abandonment of the fossil fuel economy. He sidesteps his earlier analysis projecting the demise of capitalism and instead posits a new hybrid economic model he calls "New Social Capitalism" (166). Apparently, he now thinks it is possible for corporations to be profitable without treating the biosphere as an economic externality.

Sadly, he no longer thinks that "the ultimate triumph of capitalism ... also marks its inescapable passage from the world stage." Instead, he asserts the "buyer beware" slogan of fossil fuel capitalism during the Second Industrial Revolution has been changed by the Third Industrial Revolution to a new slogan: "doing well by doing good" (167, 205). Indeed, companies around the world are now turning to "Energy Service Companies" (ESCOs), "a pragmatic business model that can speed the transition into a near-zero emission era in the short time horizon before us" (205).

Yet Rifkin entirely ignores numerous factors that undermine the supposed superiority of the new business model. First, he ignores the deleterious reality of global private banking, a powerful cabal that has condemned billions of poor people in developing nations to perpetual poverty as their governments are forced to pay back loans for falsely framed "development" projects. These poor nations are then coerced into "austerity" measures in order to ensure the continual flow of money from the poor to the rich (Chossudovsky 1999; Screpanti 2014). He totally ignores the fact that most of the world's money is created as debt to these private banking cartels, a system that creates top-down domination, perpetual exploitation, and continuing corruption everywhere on Earth (see Brown 2007).

For example, he makes no mention of the European Union recently condemning Greece (one of its member nations) to poverty and perdition in order to force the repayment of billions in loans to the gigantic E.U. banks, a move advocated by Germany – a leader, according to Rifkin, in the conversion to a green economy. In so doing, Germany effectively trampled green economic development in Greece in order to secure its own green prosperity.

Second, Rifkin ignores the reality of existing inequality of wealth worldwide. In fact, Rifkin (who regularly consults with very rich and powerful people) never claims "social capitalism" should *limit* their amount of "doing well" (i.e., the profit accumulated in the process of "doing good"). Nor does he claim social capitalists should share dividends equally with their employees. He therefore seems to bless unlimited profitmaking, whether or not the enterprise is engaged in green business. It makes me wonder what Rifkin thinks about big pharma making billions from the global pandemic – simply "doing well by doing good" – while millions lose their homes, their jobs, and their lives. Without holistic integration on a global scale, his hybrid system remains a criminal enterprise.

There is vast research showing that excessive accumulation of wealth distorts democracy, empathic consciousness, and civilizational well-being. Recently, economist Thomas Piketty demonstrated how "capitalism automatically generates arbitrary and unsustainable inequalities that radically undermine the meritocratic values on which democratic societies are based" (Piketty 2014, 1; see also Parenti 1995; Chomsky 1996). All Rifkin recommends is a graduated income tax to promote the transition to a green economy.

Most of *The Green New Deal* focuses on what is happening with green business in the U.S., E.U., and China. With regard to the proposed U.S. Green New Deal, he sets out twenty-three points outlining what he believes is necessary. Nearly every one of his points brings federal, state, and local governments into the equation (223-30). Consequently, just as capitalism is revived and renamed "social capitalism" by Rifkin, so the role of government becomes paramount in his view ... which brings us to the third reality ignored by Rifkin: the system of sovereign nation-states.

He admits that we need "binding legal standards" to ensure universal conformance to green non-carbon markets. Then he calls the U.S., E.U., and China the "three elephants in the room" whose leadership is key to a global green revolution (215 ff.). These three elephants are indeed crucial to a successful conversion, however, even if these large beasts manage to establish binding standards for their societies, they remain lawless in their refusal to recognize any effective laws over themselves.

Consequently, the real "elephant in the room" is the system of sovereign nation-states. Under this system, there can be no binding legal standards because there is no entity with the powers of legislation, adjudication, or enforcement. Moreover, their lawless use of military force breeds skepticism everywhere concerning the need for the worldwide rule of law. Militarism and terrorism are a fourth reality ignored by Rifkin. The U.S., for example, remains unchecked by the U.N. when its remote-controlled drones or CIA assassination teams lawlessly execute people around the world whom it designates (according to secret criteria) as its enemies (Gehring, ed. 2003). One again, herein lies the real elephant – the lawless system of sovereign states that promotes and protects an exploitive global economic system.

The fifth reality ignored by Rifkin is the vulnerability of the commons. Fundamentally our planet is a *global commons* that belongs to all the people of Earth, who possess the collective authority to protect and restore it. Moreover, Earth's resources should not be viewed as belonging to certain countries, as in oil somehow "belongs" to Saudi Arabia, the lungs of Earth somehow "belong" to Brazil, giant natural gas reserves somehow "belong" to Russia, and the right to exit the Paris Climate Accords somehow "belongs" to the United States. This is the "Tragedy of Our Planetary Commons" (Martin 2020). And it structurally prevents a planetary consciousness from forming.

We have seen economist Kate Raworth underscore a fundamental truth: Success in overcoming our planetary crisis will depend on how we design our systems. She writes, "Economics, it turns out, is not a matter of discovering laws. It is essentially a question of design. ... [T]he last two hundred years of industrial activity have been based upon a linear industrial system whose design is inherently degenerative. ... From a systems-thinking perspective ... far greater leverage comes from changing the paradigm that gives rise to the system's goals" (2017, 180, 182). Doing good by doing well, helpful as that may sound, does not fundamentally change the system's goals.

"Social capitalism" is nothing new since critical thinkers have been talking about "market socialism" or "cooperative capitalism" for decades (e.g., Harrington 1989; Smith 2013). Democratic "socialism" fundamentally means that law regulates markets in ways that serve the common good rather than the unlimited accumulation of private wealth (Martin 2018). It also means production by cooperating associations of workers rather than remote investors.

In his earlier book, *Entropy: Into the Greenhouse World* (1989), Rifkin seems to agree with a radical rethinking of the paradigm of private property:

> In a low-entropy culture the concept of private property is retained for consumer goods and services and family real estate but not for large tracts of land and other renewable and nonrenewable resources. The long-accepted practice of private exploitation of "natural" property is replaced with the notion of public guardianship. The orthodox economic view that each person's individual self-interest when added together with the self-interest of everyone else always serves the common good of the community is regarded with suspicion or, more appropriately, with outright derision. (1989, 245)

Nevertheless, it appears when Rifkin wrote *The Green New Deal*, he had forgotten this important principle. As we will examine in Chapter Six, this is exactly what the United Nations supports and enables: "We reaffirm that every State has, and shall freely exercise, full permanent sovereignty over all its wealth, natural resources and economic activity" (U.N. Sustainable Development Goals, Introductory Resolution, item 18).

The sixth and final reality ignored by Rifkin is how and why consciousness expands. Rifkin does devote a section of his book titled "Thinking Like a Species" in which he writes, "The great paradigm changes in human history are infrastructure revolutions that change our forms of governance, our cognition, and our very worldview" (211). He then describes how human consciousness has expanded historically, acknowledging that the empathic impulse arises within ever-larger frameworks that include both the great world religions and national loyalties to "nation-state identity" (213). Here, he recognizes that government at the nation-state level creates "figurative families" and larger collectivities for human empathy. Yet, when it comes to "thinking like a species" and establishing a planetary civilization with a "biosphere consciousness," he implies that the technical infrastructure of the Third Industrial Revolution alone is sufficient to do this. He does not state the obvious: If national governments created empathic families, then democratic world government would clearly expand consciousness to the species level for a truly adequate and global green new deal.

It is difficult to see how human beings will attain empathic consciousness and planetary solidarity without a radical rethinking of global economics. Just as the nation-state system divides people and defeats biospheric consciousness, so does the global monetary system, which fosters competition, inequality, and desperation everywhere. The *Earth Constitution* would rapidly foster this transformation of consciousness because it is *designed* to do so. It is designed to assign public ownership of our planetary commons (oceans, forests, air, and water) to the people of Earth. It also is designed to holistically and effectively deal with the technical, logistical, and local empowerment actions that Rifkin recognizes as globally necessary to transform the entire planet to sustainability.

Virtually every one of Rifkin's twenty-three recommendations for the U.S. Green New Deal is contained in the *Earth Constitution* and the World Legislation enacted to date by the Provisional World Parliament. In addition, the *Earth Constitution* makes democratically legislated world laws binding, and it provides worldwide agencies for monitoring the health of the planet and for deploying Third Industrial Revolution technology equitably around the planet to ensure real sustainability. It also institutionalizes global public banking.

Rifkin's recommendations share the fundamental ambiguity of many climate activists and economists – recommendations that are directed to the nation-state level (where they might be implemented by law) and by inference to the world level (where there is no effective law and no agent to implement them). Rifkin is talking about a U.S. Green New Deal but implicitly suggesting a world Green New Deal. Similarly, Charles Eisenstein, in his book *Climate: A New Story* (2018), presents many excellent recommendations about what mankind should do to address the population problem, the economic "growth" problem, the debt-trap issue, financing of poor farmers, etc. But he cites no agency or authority representing the people of Earth and the planetary biosphere with the power to effectively implement these recommendations.

The *Constitution for the Federation of Earth* is designed precisely to address all the crises of our fragmented world system – our economics, politics, the military-industrial complex, global poverty, and climate change. We need a true paradigm-shift if we are to achieve sustainability, and ratification of the *Earth Constitution* is our best bet for a "Planetary Green New Deal."

5.6 Capitalism *vs.* the Environment

Under the capitalist system as we have known it for several hundred years, the law of growth has predominated. Its vision of growth has ignored the limits of our finite biosphere, our finite non-renewable resources, and the finite nature of the sinks into which we dump our ever-increasing wastes. Unlike capitalism of the past, when the non-renewable resources of our planet seemed limitless and the sinks seemed unlimited in their capacity to absorb our wastes, today we live in a "full world."

The global economy now reaches nearly every region of our planet. We no longer live in an "empty world" with seemingly endless resources to exploit and endless spaces for the wastes generated by production, transportation, consumption, and disposal. The world appeared vast and empty centuries ago when capitalism was gearing up its immense productive powers. Today, the finite world has been filled by both an exploding planetary population and endlessly expanding production, consumption, and corresponding wastes. In the current over-full world, we are destroying the very natural processes and functions necessary for life (Daly 1996).

With 7.9 billion people, all of whom use resources and produce waste, the carrying capacity of Earth is rapidly disintegrating and losing its ability to sustain higher forms of life. We are far beyond the carrying capacity of Earth, both for human and many other forms of life. As a result, we are experiencing the great "Sixth Extinction" (Kolbert 2014), a die off that ultimately will include ourselves if fundamental changes are not made quickly.

Holism means that human beings have internal relationships with the ecosystem of the planet. What we do affects the ecosystem, and changes in the ecosystem affect us. Classical economic theory ignored all of this, giving us a very distorted picture, as if economics were independent of the laws of entropy, the finite nature of the ecosystem, and the strict limits to growth on our planet (Daly and Townsend 1993, Chap. 3). Economist E. F. Schumacher states, "An attitude to life which seeks fulfillment in the single-minded pursuit of wealth – in short, materialism – does not fit into this world, because it contains within itself no limiting principle, while the environment in which it is placed is strictly limited" (1973, 30).

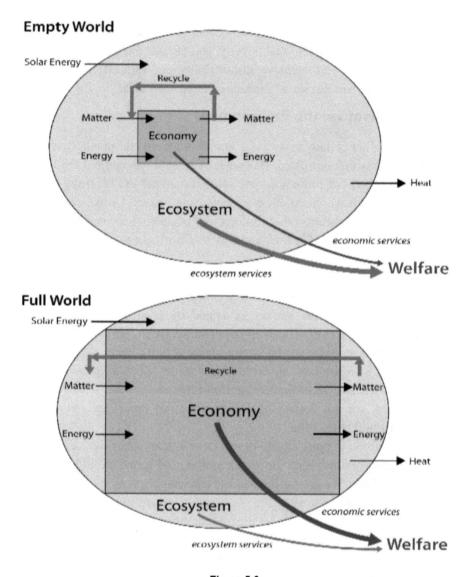

Figure 5.3
Herman E. Daly's contrast between a "Full World" and an "Empty World."
In a Full World, growth in the aggregate system can simply no longer take place.
Source: https://greattransition.org/publication/economics-for-a-full-world

Traditional economic theory commodifies everything, including human beings, their bodies, their health, and the value of life itself. Capitalism has no choice but to do this because it is based on the commodification of nature and human beings. It cannot deal with intrinsic value, those dimensions of life that Akerman and

Heinzerling call "priceless." Thankfully, many thinkers are realizing that this economic reductionism is a fundamental distortion of our human situation.

The meaning of the word "economics" comes from the Greek *Oikonomia* meaning "household management," used in this way by Aristotle and others. Our household is now planet Earth. Economics is supposed to manage the household for the benefit of all its inhabitants, since living creatures are, on various levels, *ends in themselves* and therefore priceless. But a doctrine based on commodification will eventually put a price on everything – the entire precious world of nature, animals, and human beings – which is exactly what has happened. We need to move beyond economic reductionism to a mode of economic management that serves human dignity and quality of life.

Can capitalism be transformed to accept "development without growth" and accommodate dignity and reverence for nature? Or is the essence of capitalism wedded to constant growth and respect only for what can be commodified? In *Capital*, Karl Marx saw limitless growth as its essence:

> At the end of the movement, money emerges once again as its starting point. Therefore, the final result of each separate cycle, in which a purchase and consequent sale are completed, forms of itself the starting point for a new cycle. ... The circulation of money as capital is an end in itself, for the valorization of value takes place only within this constantly renewed movement. The movement of capital is therefore limitless. (1990, 253)

Most economists agree with Marx that growth is the goal of capitalism. Not only is the system debt-based, but for the most part the legal mandate for businesses is to maximize returns for their investors, something that is measured in purely quantitative terms as increased wealth. There is no democratically chosen group managing our planetary household. The rich and powerful undemocratically manage the multinationals, national banks, World Bank, IMF, and WTO for the benefit of ... *you guessed it* ... the rich and powerful. The ideology of "free enterprise" means "every man for himself."

The resulting impact on the whole of humanity depends on some "invisible hand." Adam Smith lived in an "empty world," and his mythological invisible hand may have made some sense in 1776 when the *Wealth of Nations* first appeared, but today it is madness. This so-called invisible hand has created massive poverty for the majority of humans and obscene riches for the few. It has destroyed democracy, equality, and freedom everywhere on the planet, and it has ravaged the natural environment that sustains higher life forms on Earth.

If we wish to transform economic systems, then we need a democratically elected world government to manage our planetary economic household. We also need global public banking predicated on a currency with real value, not created as debt requiring endless growth. In truth, the invisible hand is just sleight of hand – a system that pulls the wool over the eyes of mankind by the few who run its corrupt institutions. Ellen Brown quotes Sir Josiah Stamp, director of the Bank of England and the second richest man in Britain during the 1920s:

> The modern banking system manufactures money out of nothing. The process is perhaps the most astounding piece of sleight of hand that was ever invented. Banking was conceived in inequity and born in sin. ... Bankers own the earth. Take it away from them but leave them the power to create money, and, with the flick of a pen, they will create enough money to buy it back again. ... Take this great power away from them and all great fortunes like mine will disappear, for then this would be a better and happier world to live in. ... But, if you want to continue to be slaves of bankers and pay the cost of your own slavery, then let bankers continue to create money and control credit. (2008, 2)

As we have seen in Chapters Two and Three, the logic of the sovereign nation-state system is intertwined with the logic of capitalism. These systems derive from the discredited early-modern paradigm, and they were not designed to benefit humanity. Rather, they evolved contingently along with other social phenomena, including monarchy, aristocracy, banking, and the emergence of science and scientifically-grounded technology. Today, the wealth of power brokers is dependent on the global exploitation of the masses for their prosperity and predominance. The global economic system aspires to what economist Ernesto Screpanti calls "sovereignless global governance" (2014, 205). The global debt-based banking system and capital-rich multinational corporations operate as a system that diminishes national sovereignty and makes even wealthy nations subject to its demands, thereby defeating democracy and citizen governance everywhere on the planet:

> Contemporary neoliberal capitalism seems to be activating the conditions of relative impoverishment observed by Marx in nineteenth-century laissez-faire capitalism. On the one side is a narrow class of capitalists and speculators getting progressively richer, while on the other side is a growing mass of proletarians whose income tends to approach subsistence levels, often falling below them. This tendency condemns the reformist practices of labor and social democratic parties to ineffectiveness and generates a process of cultural and political

disintegration of traditional national worker's movements, a sort of "clean sweep" of the old organizational and ideological defenses that aimed to support the working class by integrating it into national power systems. (Ibid., 207)

Those who argue that nations need to take-back their sovereignty are merely seeking to substitute slavery in a militarized world of national fragments for the economic slavery wrought by multinational globalization and corporatocracy. The only progressive and enlightened alternative is to recognize the sovereignty of the people of Earth and the need for a planetary democratic design that puts people first, not militarized nations or global capital.

Under Article 8.7.1.6 of the *Earth Constitution*, the planetary monetary system will create money "for the financing of the activities and projects of the World Government, and for all other financial purposes approved by the World Parliament, without requiring payment of interest on bond, investments or other claims to financial ownership or debt." The key to liberating humanity from the growth fetish and establishing a thriving quality of life is debt-free money creation. Money can then represent "use-value" rather than mere "exchange-value." The end of capitalism does not mean the end of fair trade; it means the end of trade for the purposes of growth and capital accumulation. Once the wealth and growth incentive behind the debt system is ended, it will be possible to conform economics to the actual limits of our planetary ecosystem.

Such a transformation also will end the imperative to commodify human beings by subordinating our needs to capital accumulation. Under the *Earth Constitution,* human dignity is no longer subordinate to either the system of militarized sovereign nation-states or capitalist debt-based growth. The *Earth Constitution* requires all government officials to take a "pledge of service to humanity." There is a new vision and new understanding of government and economics, an acknowledgement of the human community as a whole and the dignity of each member.

5.7   What Should Be Included in a Cost-Benefit Analysis?

Prior to the rise of environmental awareness, the preanalytic vision of economics was largely of the "empty world" depicted in the diagram above (*Figure 5.3*). Hence, as Raworth points out, it was built around a linear model that essentially ignored our rootedness in the planetary biosphere. This is the fundamental principle of the new economics, as Daly puts it, "the economy is a subset of the environment" (1996, 6). We realize now that economics must mirror how we manage our extraction, production, consumption, and disposal to conform with ecosystem health, viability, and regenerative capacity.

How do we do this? Daly believes the answer is "painfully simple: by population control, by redistribution of wealth and income, and by technical improvements in resource productivity. In sum, not by growth, but by development" (ibid., 7). Here, development means qualitative improvement of human lives and planetary health, not quantitative increase. It also means "development without growth – that is without throughput growth beyond the regeneration and absorption capacities of the environment" (ibid., 13).

Any ecosystem – whether local, regional, or global – has fixed limits of integrity and viability in relation to its carrying capacity for living creatures, including humans. In addition, the law of entropy puts absolute limits on human activities in relation to the biosphere. All these limits require us to deeply understand and adapt to our interdependence with nature and one another. These limitations demand that we attend to the issue of "scale" – the scope of human activities on our planet that will ensure we abide within its carrying capacity and not destroy its ability to sustain future generations.

Daly points out that markets alone cannot accomplish these objectives. Markets can help "solve the allocation problem by providing the necessary information and incentive" (ibid., 50). However, they cannot deal with optimal scale, which means conforming the economy to the carrying capacity of Earth, what Kate Raworth calls the "outer rim of the doughnut." Proper scale requires cooperation, planning, a worldcentric community spirit, and careful design that is open-ended for adaptation to change and qualitative development. Markets, he says, also cannot deal effectively with just distribution, which will fairly include the needs of all, including future generations.

Cost-benefit analyses are inevitable, and the *Earth Constitution* calls for studies performed by the agencies within the Integrative Complex to assist the World Parliament in making informed decisions. Using the arsenic example from Akerman and Heinzerling above: What are the costs of arsenic removal to a certain "safe" low level (e.g., 10 parts per billion), as opposed to the benefits of saving so many lives? At a safe level, some lives still may be lost on rare occasion, and further scrubbing of water supplies to prevent every conceivable death would be extremely expensive. Could this money be better used to preserve lives in other contexts, such as dealing with the massive displacement of people from climate disasters around the planet? But these sorts of cost-benefit decisions will always benefit humanity, not increase corporate profit.

Recall that the OMB came up with the value of a human life at 6.1 million dollars. The absurdity of this result stems from an early-modern preanalytic vision held by EPA economists. They assumed economics can and must quantify everything so that economic comparisons can be made. But this view is terribly wrong because it actually devalues human life. When economists consider values impinging on human life, equality, dignity and the well-being of unborn generations, Daly suggests the following:

> We need a new central organizing principle – a fundamental ethic that will guide our action in a way more in harmony both with basic religious insight and scientifically verifiable limits of the natural world. This ethic is suggested by the terms "sustainability," "sufficiency," "equity," "efficiency." Growth has become unsustainable. It has never been equitable in the sense that some live far above sufficiency, while others live far below. And no system that uses resources at a rate that destroys the natural life-support systems without meeting these basic needs of all can possibly be considered efficient. (1996, 219)

Daly also lists a number of priorities to be followed when making tough economic decisions. Below, I have shortened the list and adapted these priorities into humanistic economic priorities for our planetary community:

Figure 5.4 **Humanistic Economic Priorities**
✓ Natural ecosystems, soils, forests, and water supplies must be restored, regenerated, and protected.
✓ The market and capital must be guided by well-designed laws based on patterns of interaction that protect and restore the environment.
✓ The population of Earth must be stabilized to match the carrying capacity of our planetary biosphere.
✓ Extreme economic inequality must be rendered impossible and extreme poverty must be eliminated.
✓ Economic and environmental policies should always consider the well-being of future generations.
✓ When there is insufficient factual grounding for decisions, all cost-benefit calculations must use the "Precautionary Principle" (i.e., policies that protect human and planetary health from potential hazards).
✓ Freedom and creativity must flourish, and markets may operate under the banner of "fair trade" so long as there exists a common codified vision that embraces planetary community values dedicated to the integrity of each within the welfare of all.

The recognition of human dignity and the discovery of holism serve as a counter-movement, indeed a complete reversal, from the early-modern paradigm of atomism and reductionism. Holism means we are all interdependent and interrelated with one another worldwide, and dignity means that our most fundamental relationship to one another and to future generations is moral and not quantifiable. These are universal, planetary values requiring universal, planetary solutions.

A decent, healthy economics must conform to these realities, as well as to the laws and limits associated with our interdependency within the planetary biosphere. Economic decisions must be thoughtfully integrated into the holism of moral principles, human dignity and equality, as well as the holism of our planetary biosphere that supports all life on Earth. Holism acknowledges that we are the same *homo sapiens* species everywhere and one civilization, participating in one planetary destiny.

The intrinsic value of persons often is expressed by the concept of *dignity*. Kant affirms that every person is an "end in themselves" and should not be used "merely as a means." The U.N. Universal Declaration of Human Rights declares: "Recognition of the inherent dignity and of the equal and inalienable rights of all members of the human family is the foundation of freedom, justice and peace in the world." Dignity and rights cannot be commodified. They do not fit into the traditional economist's theoretical model. No wonder capitalism has not given us "freedom, justice and peace in the world."

In his classic work *Natural Law and Natural Rights* (1980), philosopher of law John Finnis argues that the purpose of law is to make possible the participation of all persons participating in seven basic *objective goods* of living. These seven goods are: (i) life, (ii) knowledge, (iii) practical reason (the capacity for making wise moral choices), (iv) friendship, (v) aesthetic experience, (vi) play, and (viii) religion (or a sense of meaning in life). Which of these goods can be commodified?

Of course, traditional economists have found ways to turn these precious aspects of life into superficial versions of each of them – all to make profit. But each is most fundamentally about the internal fullness of a human life. What is authentically internal cannot be commodified, only what is empirical and external. Law and economics, he argues, should make possible the internally motivated pursuit of such goods for all people with reasonable equity. How do we do this for our planet as a whole?

We live as persons in community dependent on our relationship with one another, with nature, and with future generations. This reality is not only economic in nature, but also moral. It includes the human spirit, which demands respect, concern, compassion, justice, and equality. None of these can be commodified, and all are connected with the intrinsic dignity (not economic value) of human beings.

Consequently, economic theory must ask how it can serve intrinsic value and sustainability, in other words, the common good rather than the selfish goal of wealth accumulation. It must become a theory of service to what is truly valuable – people and planet. It must repudiate its Promethean past, the absurd idea that everything can be commodified into external and purely instrumental relationships, suffering under what Ken Wilber calls its "monological rationality" (2017, 45).

Recall that James Gustav Speth calls for "a new economy beyond capitalism that is non-growth and dedicated to human welfare, to a new consciousness of reverence for nature and intergenerational human dignity, to a new, localized participatory democracy in solidarity with global institutions for coordinating the planetary effort." And Daly argues for a "steady-state economy" focused on qualitative improvement rather than quantitative growth. But how do we measure qualitative objectives? At a minimum, we would need to include the basic necessities required to sustain more than one billion people on our planet who currently lack basic resources, but beyond this quantitative minimum where does quality lie? In any reasonable version of a cost-benefit analysis, all these things would need to be addressed. The answer to this question, however, brings us back to the transformational psychologists and spiritual thinkers discussed in Chapter One.

5.8 Economics and Conscious Evolution

A broad consensus has developed among psychologists and ethical thinkers, including integral wisdom scholars such as Ken Wilber. The consensus is that human beings are evolving to higher levels of moral, emotional, interpersonal and intellectual maturity, both as a species and as individuals. First, we grow out of the egoism of childhood to the ethnocentrism of our immediate social environment. Second, we grow up to the worldcentric level and see ourselves as a collective of human beings with differences that become inconsequential. Third, proper growth moves us into cosmocentric levels, in which we experience a harmony with the fundamental principles of the cosmos itself and with the holistic ground of the entire universe, sometimes called God, Tao, Dharmakaya, or Brahman.

True human maturity begins at the third stage of growth – what Ken Wilber (2007) calls the "worldcentric" stage of development and psychologist Lawrence

Kohlberg (1984) calls the stage of "moral autonomy," which he explicitly identifies with Kant's ethics. At this stage, one no longer follows the values of the surrounding ethnocentric culture, but rather recognizes universal values autonomously and judges their coherence, consistency, and universality. My book *Global Democracy and Human Self-Transcendence* argues that mature worldcentric and morally autonomous human beings have progressively understood the concept of human rights within ever-larger frameworks since the 18th century, as summarized in the following chart.

Figure 5.5 **Development of Planetary Consciousness**	
18th Century: **Political Rights**	The recognition that all humans have reasoning ability and therefore the right to political freedoms that allow them to participate in government as well as independent lives.
19th Century: **Economic & Social Rights**	The recognition that all humans pursue goals in their lives, and therefore possess rights to economic and social well-being that reasonably ensure the pursuit of their goals.
20th Century: **Human Rights**	The recognition that neither political nor economic rights have meaning apart from a world system that includes guaranteed human rights, peace, and environmental protection.
21st Century **Planetary Rights**	The realization that humanity must adopt democratic world government to achieve collective human rights, universal peace, and planetary sustainability.

Human beings have been growing over the centuries toward moral maturity, progressively recognizing universal human rights and dignity. If we want a future on this planet, we must organize our economy and politics globally and democratically. We need to have our planetary rights legally protected on the global level. Once again, the *Earth Constitution* accomplishes these sacred goals by codifying our planetary rights.

Kate Raworth says that economists must "nurture human nature," since human nature is capable of growth and transformation. Therefore, we are not locked into the outdated capitalist image of *homo economicus*. We are free to envision what we can and must become. Jeremy Rifkin says we can grow to "biospheric consciousness." James Gustav Speth reviews the new criteria for economic development without growth such as the "Happy Planet Index" and "Genuine Progress Indicator" in which the quality of life is emphasized. Bill McKibben declares that we need a new human maturity characterized by intelligence and love rather than hate, greed, and fear. Herman E. Daly affirms that "we have obligations to our Creator to care for Creation ... which implies that sustainability, not growth, should become the ruling ethic for a Creation-centered economy" (1996, 224). Thus, there is a broad consensus among environmentalists and sustainable economists that we need to consciously evolve.

Conscious evolution means that human beings realize what life is truly about. On the most immediate level, one sees life characterized by objective values, as articulated by John Finnis or Eric Fromm. For Finnis, life is about developing "practical wisdom" through which we can participate in the range of objective goods that embrace our lives and make them worthwhile. For Fromm, it is about cultivating our love and reason to the point where life becomes a creative journey into ever wider and deeper relationships. Perhaps the meaning of this journey can be seen as the actualization of cosmic consciousness, with its corresponding universal love and deep transpersonal self-fulfillment.

Or, consistent with these perspectives, we might adopt a more Zen Buddhist approach and discern that life is about actualizing the astonishing fullness and freedom intrinsic to the present moment. If so, then life is about the immense value and joy of existence that appears when we stop striving and begin living for the sake of living. Reaching this level of consciousness requires that we take the necessary steps, such as meditation and mindfulness, to enhance our growth.

Indeed, we ultimately may realize there is nowhere to go because the fullness of life was there all the time, immanent within each moment. We may discover there is nothing to achieve, other than simply living the life we have been gifted. Currently, a bill is being prepared under the *Earth Constitution* for the Provisional World Parliament that will require all members of the world Parliament to undergo spiritual training in meditation and mindfulness. This, of course, would not guarantee wise

Life is about the immense beauty and joy of existence that appears when we stop striving and begin living. Ultimately, we realize that the fullness and ecstasy in everyday living is the supreme value.

parliamentarians, but it clearly helps point people in the right direction.

Abraham Maslow's famous "Hierarchy of Needs" concludes that people first seek to satisfy subsistence needs, then security needs, then relationship needs, then esteem needs. At the highest levels, people address their need for self-actualization and self-transcendence – levels few people reach under our current world system. Maslow called these highest levels "being-needs" leading to "being-values," such as beauty, goodness, truth, justice, wholeness, completion, and perfection/enlightenment (214, 75-76). At the deepest and highest levels, we find only those values having to do with the quality of life. There is nothing quantitative here, only what has intrinsic value. Economics must awaken sufficiently to address all these levels. "Transformative business" means that people are themselves transformed in the course of doing business and that business itself includes a transformative dimension.

As people ascend from egocentric love, to love of community, to worldcentric love, then to cosmocentric love, they move from the quantitative level of satisfying bodily needs and procuring physical necessities to levels populated with joy, wonder, and ecstasy in everyday life. These blissful experiences constitute qualitative journeying into an ever-greater fullness of life. Here is where the meaning and destiny of our human project lies, not in perpetual quantitative expansion. How we design our institutions cannot guarantee such growth, of course, but it can help make it possible. However, our present institutions actively interfere with and block the actualization of such possibilities.

Ken Wilber declares, "Beauty, Truth, and Goodness are very real dimensions of reality to which language has adapted" (2007, 19). Our human project is to actualize these realities in our lives, ever increasing the qualitative awareness and fulfillment throughout our life-paths. Raimon Panikkar calls this fullness the awareness of "Being" and explains, "Being, in short, is that symbol that embraces the whole of reality in all its possible aspects we are able to detect, and in whose Destiny we are involved as co-spectators, actors, and co-authors" (2010, 94). Our journeying into ever-greater fullness of life is simultaneously our partnership with Being as its "co-spectators, actors, and co-authors." The universe became conscious of itself in us for a purpose, so we may actualize these highest values. Whether we call it socialist economics or democratic capitalism is not really important. Perhaps we can call it "doughnut economics" (Raworth's term) or "steady-state economics" (Daly's term). What is important is our wise management of our planetary household for the mutual benefit of everyone, all living creatures, and future generations. This requires reframing economics to the contemporary holistic paradigm.

Ervin Laszlo summarizes the implications of contemporary science that has entirely overthrown the early-modern paradigm from which both capitalism and the sovereign nation-state system originated. He calls it the "Akasha Paradigm" after the classical Vedic concept that all existence arises from a primal energy field or God:

> We are all connected, intrinsically and permanently connected. That is the new paradigm, the Akasha paradigm, emerging at the cutting edge of the sciences. We only disregard this new insight at our own risk. If we can open up our mind and our heart to our oneness with the world we will come up with a solution. The precondition for this is to allow the wisdom that is in us to become operative. This wisdom has guided people through the ages. ... The only way we can do this is by acting together at a deep level. By sensing our oneness, by cooperating, by becoming coherent. We are no longer coherent either with each other or with the world around us. (2014: 77-78)

The *Earth Constitution* does not cite such philosophical considerations directly, of course. Nor will federating Earth automatically make us "coherent" with the holarchy of the biosphere. However, it is an elegantly drafted document that sets forth a practical plan for organizing humanity to govern our chaotic world. In this sense, it is truly a "bridge at the end of the world," to quote Speth. But behind it, and expressed in its every article, is an awareness of the equality, dignity, and relationship of every human being within our planetary community as a whole.

The *Earth Constitution* also is a ready model for meeting the physical needs and spiritual yearning of every human community on the planet. It embodies the paradigm shift from fragmentation to unity in diversity that is necessary for human survival and flourishing into the fullness of life. It establishes a coherence with the natural holarchy of humankind, because it understands that politics and economics must be predicated on human dignity and well-being. In addition, the *Earth Constitution* provides freedom for people to develop the highest quality of life they can imagine. Economist E. F. Schumacher writes:

> We need the freedom and lots and lots of small autonomous units, and, at the same time, the orderliness of large-scale, possibly global, unity and coordination. When it comes to action, we obviously need small units. ... But when it comes to the world of ideas, to principles or to ethics, to the indivisibility of peace and also of ecology, we need to recognize the unity of mankind and base our actions upon this recognition. (1973, 69)

Similarly, philosopher Joel Kovel in his powerful book on "ecosocialism" titled *The Enemy of Nature: The End of Capitalism or the End of the World?* (2007) affirms that a sustainable civilization cannot be a matter of thousands of localized communities around the planet each attempting to live sustainably in harmony with Earth. Rather, "Unless the notion of community is advanced in a universalizing way, it loses transformative power and, despite good intentions, drifts toward ethnocentricity" (ibid., 265). These economic thinkers are correct. The proper scope of community is the whole of humanity, in the sense of people participating in local communities nested within and responsible for the holarchy of civilization.

To open our minds and hearts to world-oneness means becoming coherent with one another and cooperating as a species. This is the proper domain of economics. This understanding of unity in diversity has been expressed not only within the *Earth Constitution*, but also in the work of the Provisional World Parliament, which operates under the authority of Article 19 (see Chapter Seven). If one studies the dozens of World Legislative Acts that have been passed during the fourteen sessions of the Parliament to date, one will see the principles of equality and dignity at work. Take, for example, World Legislative Act 22, the "Equity Act." This act makes it mandatory for every person to have the basic sufficiencies required for a healthy human life, and it defines limits on the private wealth any person can accumulate relative to those basic sufficiencies.

In the *Earth Constitution*, economics is understood to support the sufficiency of life for each and every person within the solidarity of the whole human community. Legislative acts of the Provisional World Parliament are not binding now or after the *Earth Constitution* is ratified, but these exemplary acts reflect the spirit of world federalization. This humanitarian spirit will guide the decisions of the world agencies as well, including the actual World Parliament, the World Financial Administration, World Judiciary, and World Ombudsmus – all of which coordinate and empower global sustainable economics.

Article 13 of the *Earth Constitution* guarantees "free and adequate" public education, public health care, social security, and "sufficient lifetime income for living under conditions of human dignity during older age." Unlike weaker stipulations in the U.N. Universal Declaration of Human Rights, these rights are not merely symbolic ideals. They are legally enforceable and guaranteed rights. As Mahatma Gandhi reminds us, "The world has enough for everyone's need but not for everyone's greed."

Sustainable economics, therefore, is first and foremost about conforming to the delicate patterns and finite limitations of our planetary ecosystem so that our human needs are supplied without detracting from the ability of the ecosystem to meet the needs of future generations. Second, it is about universal justice, peace, and well-being for Earth and her creatures. Third, it is about creating the social environment for the continued development and progressive fulfillment of the human spirit in relation to the whole of Being.

Before we explore the *Earth Constitution* in more detail, it is important to understand why the widely touted U.N. Sustainable Development Goals (SDGs) cannot address either the economic crisis or the impending collapse of our planetary environment. In the next chapter, we will see that while the SDGs describe mostly appropriate goals for a regenerative and sustainable planet, they are based on the current and unquestioned system that supports antiquated economic theory and sovereign nations-states – both of which defeat the realization of holistic, planetary goals because they were not designed to handle global problems. As such, planetary sustainability under the SDGs is little more than a pipe dream that likely will result in planetary suicide.

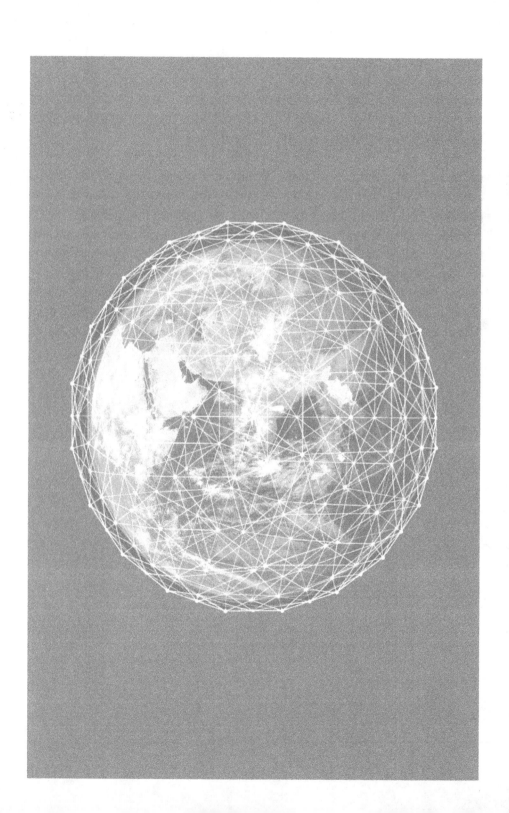

CHAPTER SIX
The U.N. Sustainable Development Goals: Trapped in a Modernist Ideology

The phenomena of the territorial state, the nation, and a popular economy constituted within national borders formed a historical constellation in which the democratic process assumed a more or less convincing form. ... Today, developments summarized under the term "globalization" have put the entire constellation into question.

Jürgen Habermas

As we have seen, the 20th century brought a dominant global world system that institutionalized "modernity," accepted militarized sovereign nation-states, and permitted the ascendency of corporate capitalism. These forces deeply colonized the consciousness of those who developed the United Nations Sustainable Development Goals (SDGs). The result is a false idealism coupled with unconscious blindness, which in the 21st century places our entire human project in serious and imminent danger. When we consider each of the seventeen SDGs in light of this background, it becomes evident that a sustainable civilization and viable future for humanity can only be achieved by transcending these "modernist" assumptions and ascending to a global perspective, as envisioned by the *Constitution for the Federation of Earth*.

6.1 The Colonizing of Our Consciousness

Many scholars have traced the evolutionary history of the capitalist system from its emergence during the Renaissance in the 15th century to the early 20th century. The evolution of the militarized sovereign nation-state system also has been widely traced from its 17th century origins to the early 20th century. Also well chronicled is the development of technology from early inventions like the steam engine in 1698 to its systematic development and pervasive infiltration into nearly all aspects of life on our planet. What is not generally understood, however, is the degree to which this triumvirate of forces – capitalism, national sovereignty, and technology – has colonized the consciousness of political leaders and academics worldwide, including those who participated in the development of the U.N. Sustainable Development Goals.

Philosopher Jürgen Habermas (1989), who built on the work of social scientist Max Weber and other thinkers, presents a penetrating analysis on why human beings seem unable to successfully address our most fundamental problems. He calls this the "colonizing of our lifeworld by the system." The economic and the bureaucratic state systems became so rooted in our consciousness that they infiltrated our "lifeworld" with erroneous presuppositions and assumptions. Even many thoughtful, educated human beings seem incapable of deeply examining their theories, thoughts, and reasoning patterns. Reason, as Habermas puts it, becomes mere "functionalist reason" operating within colonized assumptions and treating our most serious problems as if they are technical, functional problems, rather than symptoms of a much deeper pathology of modern civilization.

Systems that are so pervasively integrated into our cultural, scientific, institutional and subjective worlds tend to become invisible to thought. They become unrecognized presuppositions and the starting place from which we attempt to address problems of conflict, poverty, health pandemics, human rights violations, or sustainability. Because these presuppositions are the foundations for thought, it is not clear to most people that these "foundations" may be questioned and should be examined. For it is precisely these assumptions – capitalism, sovereign nation-states, and technological rationality – that are fast destroying the planetary biosphere that sustains human life.

Many scholars informed by neo-Marxist critical social theory have pointed out the inseparability of capitalism and the competitive system of sovereign nation-states (e.g., Chase-Dunn 1998, 61). Other scholars, like Jeremy Rifkin, chronicle the three industrial revolutions that transformed the world technologically, one after the other, to the point where technology becomes a defining feature of the collective consciousness everywhere on Earth (2011). Moreover, for Rifkin, the third industrial revolution shows us a way beyond capitalism's obsession with growth.

Many scholars and historians also have documented the endless wars that have occurred within the system of militarized sovereign states from the early conquest, theft, and enslavement of peoples by the Spanish and Portuguese to the era of slavery and colonial exploitation of subjugated peoples dominated by the Western European powers (Britain, France, and the United States), for whom slavery was fundamental to economic success for centuries. Today, wars continue over oil, pipelines, control of strategic resources, and geopolitical control of world markets. Chinese economic investments in Africa engender the creation by the U.S. of "AFRICOM" – the Africa military command of the U.S. (Engelhardt, 2014). War and economic rivalry remain endemic to the entire world system due to the early-modern economic and political assumptions built into the modern world and the 21st century.

United Nations economic ideology asserts that the job of each government is to grow its gross domestic product (GDP). In his study of GDP, John Smith shows that value production in a globalized economy can only be measured in terms of *global factors* that collectively contribute to any particular local production. Therefore, GDP as an economic measure is an "illusion." The U.N.'s concept of GDP not only ignores the immense quantity of labor required for survival (subsistence agriculture and non-export-oriented work, especially for people in poor countries), it also forces poor governments into economic policies that increase financial and foreign domination. At the same time, GDP "becomes a veil concealing not just the extent but the very existence of North-South exploitation" (2016, 266).

Capitalism has always been about expansion and economic growth, and the measure of economic success to this day has been tied to nation-states. David Harvey, in *The New Imperialism,* argues that there has always been "a dialectical tension" between nation-states territorial power and the power of global financial capital. The two are not identical but require one another because of their "internal relations":

> Imperialism of the capitalist sort arises out of a dialectical relation between territorial and capitalistic logics of power. The two logics are distinctive and in no way reducible to each other, but they are tightly interwoven. They may be construed as internal relations of each other. (2005, 183)

Can we imagine a global system of capitalist exploitation not protected and reinforced by imperialist state power? Noam Chomsky (1996) declares that the U.S. blockades and invasions of numerous countries attempting to extricate themselves from the global imperial system through some form of socialism (e.g., Vietnam, Chile, Cuba, Nicaragua, Venezuela, etc.) pose the "threat of a good example." If people cooperatively manage to thrive on their own shared work and resources, then they challenge the entire ideology of capitalism and neo-liberal globalization. Thus, military power is needed to enforce capitalist dogma.

Numerous thinkers have pointed out that incremental growth (say at 2% per year) will lead to a doubling of the initial quantity in just 35 years (e.g., Heinberg 2011, 14). Like the population of Earth which has doubled four times since the beginning of the 20th century, so the economy of the planet keeps doubling in size every few decades. More economic growth requires more energy. Globally, the trillions of dollars animating the economy are dependent upon immense, ever-increasing resources of energy. Most of this energy is from heat and pollution producing fossil fuels. The planetary biosphere is collapsing under this massive assault (Wallace-Wells 2019).

Under the globalized, neo-liberal capitalism that has dominated the world in the early 21st century, the Northern developed countries do not compete with Southern countries. As John Smith points out in *Imperialism in the Twenty-First Century* (2016): "A most striking feature of the imperialist world economy is that ... Northern firms do not compete with Southern firms, they compete with other Northern firms, including to see who can most rapidly and effectively outsource production to low-wage countries. Meanwhile, Southern nations fiercely compete with one another to pimp their cheap labor to Northern 'lead firms'" (2016, 84).

Here again, we should ask if it is possible to imagine such exploitative and imperialist capitalism without militarized sovereign nation-states? The two institutions are flip sides of the same coin. The U.N. through its General Agreement on Tariffs and Trade (GATT), the World Trade Organization (WTO), the World Bank, and the International Monetary Fund (IMF) are thoroughly compromised by this world system with its technocratic and functionalist imperatives. Both the World Bank and IMF are lending institutions comprised of consortiums of profit-making first world banks headquartered in Washington, D.C. Some employees of these U.N. agencies may feel compassion for the suffering bottom 50% of humanity, but they are extremely limited in their ability to address the carnage, not only because of insufficient funding of their agencies but because their own assumptions hamper their ability to offer sane, rational, and humanized programs.

Again, it is impossible to imagine exploitative capitalism taking advantage of poverty-stricken third world nations without a system of militarized borders preventing human beings from freely moving about the world. Currently, it is perfectly acceptable for companies to canvas territories in search of the best place to maximize profits, with money freely coming in to exploit cheap labor and resources, but with people trapped in perpetual poverty, unable to leave and search for a better life.

I have lectured at the United Nations University for Peace in central Costa Rica a number of times. Professors tell me afterward that what I say violates the basic assumptions on which the University for Peace is founded, which include, of course, the U.N. Charter and the globalized capitalist system. The fact is, even though graduates of the University for Peace find positions within U.N. Programs in countries around the world and do good work at local levels, neither peace, nor viable prosperity for the poor, nor sustainability can ever be achieved under the U.N. system as it has functioned to date.

We have seen many sustainability economists such as Herman E. Daly, Richard Heinberg, and Kate Raworth point out what is perhaps the most fundamental contradiction of capitalism: Infinite growth in a finite world is impossible. Such growth is like a cancer within the biosphere, rapidly destroying the life of the entire organism. Heinberg writes: "At the landmark international Copenhagen climate conference in December 2009, the priorities of the most fuel-dependent nations were clear: carbon emissions should be cut and fossil fuel dependency reduced, but only if doing so does not threaten economic growth" (2011, 19).

Even the U.N. Development Program (UNDP) agrees that the concept of "growth" must include quality of life development, as noted in its declarations concerning "sustainable development" (see Martin 2018, 128-31). However, the UNDP never questions the sovereign nation-state system that is inextricably linked to global capitalism. Consequently, the UNDP promotes local sustainable development within countries wracked by global economic imperialism – a situation from which there is no escape under the present world system.

At the 2009 Copenhagen Climate Change Conference, representatives of 115 sovereign nation-states met about the global climate crisis. Each of them represented their own economic activities (and hence their capitalist ruling classes), and each was deeply rooted in faulty functionalist reasoning and the technological imperative. This imperative constitutes the third element in the triumvirate of assumptions colonizing the consciousness of decision makers with the power to affect the future of humanity. Critical social thinker Jacques Ellul called this third element "the technological system." Let us note his explanation of how this system works within human consciousness:

> First of all, man, achieving consciousness, finds technology already here. For him, technology constitutes a milieu which he enters and in which he integrates. ... He is instantly within this universe of machines and products. ... Now without realizing it, this environment shapes us in the necessary forms of behavior, the ideological outlooks. ... Of course, he does not see clearly what it is all about, he does not discern the "technological system," the "laws" of technology. ... Being situated

in this technological universe and not detecting the system is the best condition for being integrated into it, being part of it as a matter of course, without even realizing it. (1980, 311-12)

This paragraph could be written verbatim substituting the phrase "sovereign nation-state" or the word "capitalism" for the word "technology." Substitute, for example, the "sovereign nation-state": *First of all, man, achieving consciousness, finds the sovereign nation-state already here. For him the sovereign nation-state constitutes the milieu which he enters and in which he integrates.* Or the capitalist system: *First of all, man, achieving consciousness, finds the capitalist system already here. For him the capitalist system constitutes the milieu which he enters and into which he integrates.* Clearly, the mindset of most 21st century persons is colonized by sovereign states, capitalism, and the technological imperative. Our consciousness has nearly completely absorbed these systems as inherently real and unalterable. Most people only "see" from this set of assumptions and, therefore, fail to see the assumptions themselves.

It is important to realize how deep this colonization of consciousness goes. Karl Marx correctly says that under capitalism we become alienated from other people, the work process, the product of our work (which is taken from us by its owners), and from our true selves or "species-being" (1972, 66-125). The militarized sovereign state acculturates us to paying war taxes, seeing military personnel and recruiters everywhere, believing immigrants are the "problem," and accepting state secrecy as necessary for our protection. Much of the world's population lives within societies where these pathological phenomena appear as "normal." The military as a way of life and fear of the other is normal. Designated "enemy" countries and suspicion of the motives of other nations is normal. And most importantly for the perpetuation of the entire system, competition rather than cooperation is normal.

Nevertheless, advanced thinkers such as Habermas, James L. Marsh (1995), and Joel Kovel (2007) point out a deeper, critical element largely ignored by the dominant culture and its ideological spokesmen. Habermas calls it the "communicative" element, the possibility of mutual understanding, growth, and solidarity with others. I submit that this broad concept includes three prerequisites: First, it is based on the discovery of the scientific method during the 17th century which, when properly understood, is capable of pursuing truth even to the point of upending and transforming our most foundational beliefs about the world. Second, it includes the power of human reason interlinked with human dignity, universal moral principles, and the resulting enlightenment that emerged in the 18th century (Martin 2018, Chap. 2). Third, it includes the dawn of systematic critical social theory through Marxism that was born in the 19th century, the realization that injustice and evil can be structural, capable of corrupting our institutions just as they can infect persons. Each of these developments provided positive, liberating elements that remain with us and are ready to blossom in the 21st century.

Together these dimensions of rationality give human beings access to a deeper moral code and advanced cognition, a worldview beyond functionalism that allows us to access our higher human potential for communicative understanding, solidarity, cooperation, love, and creative transformation. We see these higher principles in the universal teachings of all cultures, as in the thought of Rabindranath Tagore, Sri Aurobindo, Mahatma Gandhi, and Martin Luther King, Jr. Moreover, science gives us an open-ended form of rationality that can be self-critical, potentially paradigm-shifting, and liberating. Indeed, beginning with Max Planck and Albert Einstein in the first decade of the 20th century, science has revolutionized our understanding of the world. It has transformed the Newtonian assumptions that nature (including human beings) was atomistic, materialistic, deterministic, and mechanistic into a holistic worldview in which nature is integrated, relational, organic, and open to emergent freedom. Today, it has resulted in the work of such major transformative thinkers as Ervin Laszlo (2017) and Henry Stapp (2011).

Yet, this new paradigm – which loudly calls out and compels us to transform our thinking in line with it – is being ignored by most political and economic thinkers, including by those who stand by the now antiquated U.N. system and its naïve SDGs. As philosophical cosmologist Errol E. Harris points out "If the implications of this scientific revolution and the new paradigm it introduces are taken seriously, holism should be the dominating concept in all our thinking" (2000, 90).

Incorporating holism into our civilization requires a spiritual awakening at both the personal and institutional levels. Beyond functionalism, we encounter the unity in diversity of our common human reality, the oneness proclaimed in the best writings of all the great religious and spiritual traditions. Deep reason, love, and justice break down barriers, including economic and political barriers, joining human beings in a common quest for living together on Earth in joy and friendship. As Swami Agnivesh (2015) affirms, the Vedas proclaim *vasudhaiva kutumbakam,* all men and women are brothers and sisters.

Thus, empiricism, positivism, and reductionism have long since been transcended by thought leaders who see emergent human freedom as connected with the very foundations of the cosmos (Teilhard de Chardin 1961; Stapp 2011). Yet the implications of this revolution have yet to deeply reach the consciousness of those who run the nation-states or who promote a global capitalist ideology. Sadly, this elevated mindset does not even operate at the U.N., where the interdependent structure of human life, both with one another and with nature, is largely ignored. Why? Because the U.N. accepts the fragmentation of dividing the world into nearly two hundred militarized territorial fragments that compete economically and politically. Thus, the U.N. directly violates the holism and unity of spirit and matter now evident in the 21st century.

The Enlightenment gave us not only a revival of the ancient Greek concept of a universal rationality (fundamental to our collective human project), but also the concept of human dignity. As a dimension of our existence, dignity demands liberation from all that degrades, dehumanizes, dominates, and manipulates (Martin 2018, Chap. 2). Human dignity will allow us to transcend and transform our broken institutions. None of the three broken systems – capitalism, sovereign nations, or functional rationality – is capable of recognizing human dignity as constitutive and fundamental. In the 18th century, Immanuel Kant properly articulated the ancient meaning of human dignity:

> Humanity itself is a dignity, for a man cannot be used as a means by any man ... but must always be used at the same time as an end. It is just in this that his dignity (personality) consists ... so neither can he act contrary to the equally necessary self-esteem of others ... he is under obligation to acknowledge, in a practical way, the dignity of humanity in every man. (In Glover 2000, 23)

Where does human dignity arise in the nexus of systems that colonize contemporary consciousness? As human rights scholar Jack Donnelly (2003) points out, ever since the U.N. Universal Declaration was formulated in 1947, human rights have routinely taken a back seat to economic and military imperatives, in both corporate and nation-state decision making. Even nations whose constitutions include the idea of human rights are inhibited by the current economic system from actualizing these ideals, since state sovereignty precludes addressing human rights violations in other states.

In the 19th century, Marx gave a systematic analysis of capitalism and revealed its structures of domination, exploitation, and dehumanization. He properly linked capitalism with nation-state imperialism, which together seek to gain and secure international markets and cheap resources. Although communism failed to deliver the promised paradigm shift, a deep understanding of how the functionalist world system works is liberating, exposing its inherent violation of universal human dignity (see Miranda 1986). Today, our very survival depends on our ability to draw on these liberating elements and to reach true planetary maturity. We must awaken and free ourselves from this colonized consciousness.

Militarized sovereign nation-states have long been understood to operate out of power politics and unmitigated self-interest, ignoring human dignity, especially the dignity of those considered competitors or enemies. These nation-states are substantially controlled by their capitalist ruling classes. In his 1916 essay, "Imperialism: The Highest Stage of Capitalism," V. I. Lenin pointed out the connection between the immense concentrations of finance capital within national

> *Our very survival depends on our ability to draw on all these liberating elements and to reach true planetary maturity. We must awaken morally and spiritually, and free ourselves from the dominant colonized consciousness.*

ruling classes and their need to use military power to colonize and protect ever-more markets and regions of Earth. Today, the exploited Southern nations are under the domination of global finance capital, the World Bank, and the IMF – all instruments of a planetary system dominated by the Northern nations, with their international conglomerates and banking cartels (cf. Smith 2016, 266).

In addition, corporate capitalism has long assumed the private accumulation of wealth as an unlimited right, with the owners of that wealth possessing rights to dominate, exploit, and degrade both employees and the public. Similarly, technology involves a one-dimensional system that enhances the power of the state and capital to dominate and exploit both nature and persons in their employ. The U.N. dare not question the nation-state any more than it may dare question the capitalist system.

These interlocking systems form the presuppositions of the dominant ideology. In the light of this analysis, we need to examine the U.N. Sustainable Development Goals that guide economic and political behavior of nations and corporations. Before doing so, however, it is important to concretize and specify this global system of domination and exploitation as it has existed since at least 1945.

6.2　Everyone Pretends to See the Emperor's New Clothes

Since the United Nations founding in San Francisco in 1945 and its establishment in New York City, the United States has dominated what goes on at the U.N. The U.N. professes to pursue the ideals of peace, justice, and environmental sustainability but only within the conceptual limits provided by its main financial contributor. As long as the organization is supported by voluntary contributions from its members, the actual framework for thought and most decision-making will be "one dollar one vote," which is why the U.S. always has the lion's-share of the votes.

Not that the U.N. is otherwise structured democratically. It is not. The Security Council, largely under the domination of the U.S., controls everything that goes on. In *Rogue State: A Guide to the World's Only Superpower,* William Blum constructs an entire chapter on the hundreds of votes issued by the U.N. General Assembly over a period of twenty-six consecutive years (Chapter 20). On issues that would increase peace, justice, or environmental protections in the world, most votes have pitted the vast majority of nations against the United States and (often) Israel (i.e., the latter, when not abstaining, always votes with the U.S.)

For example, the proposal to expand U.N. approaches that "emphasize the development of nations and individuals as a human right" resulted in a vote of 120 nations *for*, and the U.S. *against*. On a "declaration of non-use of nuclear weapons against non-nuclear states," the votes were 110 nations *for*, and the US and Albania *against*. On holding "negotiations on disarmament and cessation of the nuclear arms race," 111 nations voted *for*, and the U.S. and Israel voted *against*. On affirming a "world charter for the protection of ecology," the vote was 111 nations *for*, and the U.S. *against*. The list goes on and on ... and the votes mean little. The General Assembly has no real power, since any resolutions that require action come to the Security Council and are routinely vetoed by the U.S.

Former U.S. Attorney General Ramsey Clark detailed the American manipulation of members of the Security Council to get the go-ahead for the military attack against Iraq after Iraq's invasion of Kuwait in January 1991. In order to secure a "Yes" vote from the then members of the Security Council, the U.S. doled out these bribes and threats: Ethiopia and Zaire were given new aid packages; China was awarded $114 million in deferred aid from the World Bank; Saudi Arabia provided $4 billion to the then disintegrating Soviet Union; Malaysia eventually caved to "enormous pressure"; Cuba was subjected to severe punishment for its "No" vote; and Yemen's ambassador was told this would be "the most expensive 'no' vote you ever cast," after which a $70 million U.S. aid package to Yemen was cancelled (1994, 153-55). Of course, the invasion went ahead in spite of the few "No" votes.

Today, despite the wanton and reckless Trump presidency, the U.S. maintains a global empire and the U.N. functions as an integral part of that empire. The richest nation not only uses the leverage of its monetary contribution to the U.N. to control that organization; it also buys, bullies, and blackmails its way to domination within the U.N. Top secret U.S. documents such as the 1948 "Policy Planning Study 23," written by George Kennan for the State Department when planning for the post-war world, declared very clearly that the rhetoric about human rights and democracy was intended only for public consumption. Kennan said that in the future "we will have to dispense with all sentimentality and day-dreaming; and our attention will have to be concentrated on our immediate national objectives. ... The day is not far off when we are going to have to deal in straight power concepts. The less we are hampered by idealistic slogans, the better" (in Chomsky 1996, 9-10).

The U.S. knew it was the sole remaining superpower after World War II and affirmatively planned to consolidate a global empire. When Kennan wrote, the U.N. had only been in existence for three years, and he clearly did not view the U.N. as an impediment to American plans to employ "straight power concepts." John Perkins, in *Confessions of an Economic Hitman,* details the way U.S. embassies are tasked to promote economic colonization by American-run multinational corporations. Similarly, Douglas Valentine, in *The CIA as Organized Crime: How Illegal Operations Corrupt America and the World,* details how the CIA operates out of U.S. embassies and supports bribery, extortion, torture, and murder in the service of supposed U.S. interests.

For the majority of the world's small and weak nations, it is bad enough to have a U.S. embassy watching every political and economic decision of their government and promoting neo-colonial domination in their midst. But that tragedy compounds if they are unlucky enough to also be hosting a U.S. military base. Chalmers Johnson, in *Sorrows of Empire: Militarism, Secrecy, and the End of the Republic*, details the 725 known military bases that exist outside the U.S., all of which give the U.S. another place to monitor local movements that advocate for change or social justice, and a local site from which to launch assassinations, coups, or threats against governments that fail to follow American "free market" ideology (i.e., countries must open their resources to penetration by U.S. foreign capital).

William Blum, in *Killing Hope: U.S. Military and CIA Interventions Since World War II,* describes the major interventions. Leaving out "minor" U.S. bullying – like foreign election interference, economic sanctions, and other forms of intimidation and control – the book still contains fifty-five chapters detailing major military interventions by the U.S. since 1945. In sum, the U.S. has invaded or overthrown close to fifty countries (several countries more than once). Clandestine and open warfare has killed millions of people, mostly civilians, in such diverse places as Vietnam, Cambodia, Laos, Nicaragua, Angola, the former Yugoslavia, Iraq, Afghanistan, Libya, and Syria.

The Provisional World Parliament (PWP) under the *Earth Constitution* issued a study of U.S. interventions, which showed a strong correlation between military imperialism and U.N. Peace Keeper operations (World Legislative Act 64; see full description of the PWP in the next chapter). After the imperialists destroy a society and create chaos among the people, the U.N. Peace Keepers are sent in to clean up the mess and restore order. Naomi Klein in *The Shock Doctrine and the Rise of Disaster Capitalism* details the carnage created in country after country in the service of imposing "free market" policies on victim populations. In addition, within the U.N., no person is "elected" to the position of Secretary General without the approval of the U.S. In short, the U.N.'s job is to create a façade, superficially claiming dedication to peace while serving as a cover for the machinations of empire. As sociologist James Petras writes in *Empire with Imperialism: The Globalizing Dynamics of Neo-Liberal Capitalism*:

> The economic interests represented by these capitalist corporations converge with the national interests advanced and protected by the nation-states that make up what can be termed the "imperial state system," a system currently dominated by the US state. ... The US imperial state, both directly (via the departments of state and defense) and indirectly (via control over financial institutions such as the World Bank and International Monetary Fund), constitutes a directorate to manage the global system. (2005, 25-26)

Michael Parenti details how the U.N. General Agreement on Tariffs and Trade (GATT), signed by over 120 nations, produced its brainchild the World Trade Organization (WTO). The WTO has the power to override the labor laws or environmental laws of member nations in favor of maintaining the profit margins of multinational corporations (1995, 32). Local democracy does not matter. Corporations, including bankers, rule the world (Korton 1996; Brown 2007). What matters is profit for multinational corporations, banking cartels, and for the imperial center itself. John Perkins writes:

> In the final analysis, the global empire depends to a large extent on the fact that the dollar acts as the standard world currency, and that the United States Mint has the right to print those dollars. Thus, we make loans to countries like Ecuador with the full knowledge that they will never repay them; in fact, we do not want them to honor their debts, since the nonpayment is what gives us our leverage, our pound of flesh. (2004, 212)

There is a large body of scholarly literature describing the U.S. empire and how it works; what I have cited here simply scratches the surface. My main point is that the U.N. was created within the scope of and as an adjunct to this global empire. Do not think for a minute that the U.N. is an independent entity. Major as well as minor players at the U.N. are required to pretend they do not see the heavy hand of empire.

Of course, just like any government, corporation or organized social enterprise, the U.N. has its own subculture, its own criteria for hiring and firing, and its own employees who don't want to lose their jobs. All activity is carefully watched by the imperial powers, primarily the United States, to ensure conformity within ideological parameters. Critical thinking is constrained and never allowed to go very far. Rarely, if ever, is capitalism or the system of sovereign nation-states questioned. As Mark Twain quipped: *You cannot get a man to understand some idea if his job depends on his not understanding it.*

Currently, the fate of humanity hangs in the balance, while the old guard struggles to preserve the fragmented modern ideological system. Trillions of dollars have been invested in this dominant and undemocratic world economic and military system, to which the U.N. must conform, rendering any internal critique of the U.N. practically impossible. Yet in this book, we are free to study the U.N. Sustainable Development Goals in terms of their unspoken presuppositions. As we proceed, I will examine what is explicitly stated in the SDGs, as well as what is missing and should have been included.

6.3　Context for the U.N. Sustainable Development Goals

We have seen that the first major U.N. Conference on the environmental crisis took place in Stockholm in 1972. Thereafter, meetings took place and agreements were formed in the Montreal Protocol of 1987 and the Kyoto Protocol of 1997, which committed state parties to reduce greenhouse gas emissions. Another big meeting was held in Rio de Janeiro in 1992, during which the participants formulated the famous "Agenda 21" document that demanded significant reductions in CO_2 emissions by the year 2000. Then in March 1994, the U.N. Framework Convention on Climate Change (UNFCCC) went into effect with 197 member countries – nearly all the world. UNFCCC committed nations to act even in the face of scientific uncertainty concerning climate change.

However, at the 2002 U.N. environmental conference in Johannesburg, South Africa, it was clear the Agenda 21 goals had not been reached. This admitted failure of nations and corporations to address the crisis in meaningful ways led to a new and more comprehensive set of goals called the Millennium Development Goals (MDGs). The MDGs were declared in effect between 2000 and 2015.

The historic Paris Climate Agreement took place in 2015 and included 196 counties. This agreement modified development priorities so the planet collectively would not continue warming more than 1.5 to 2 degrees centigrade above pre-industrial levels. The Intergovernmental Panel on Climate Change (IPCC) agreed on this limit as absolutely imperative for our human future, stating that each nation-state must "prepare, communicate and maintain successive nationally determined contributions" to achieve these objectives.

However, the 2015 Paris Agreement also stated: "Each climate plan reflects the country's ambition for reducing emissions, taking into account its domestic circumstances and capabilities. Guidance on NDCs [nationally determined contributions to CO_2 emissions reductions] are currently being negotiated under the Ad Hoc Working Group on the Paris Agreement (APA), agenda item 3."[1] Based on this nonbinding and inadequate standard, participants concluded that the Millennium Development Goals were themselves not sufficiently encompassing. So a more thorough set of 17 Sustainable Development Goals (SDGs) was drafted, including 169 specified "targets" with more concrete details. In the end, the elaborated SDGs were approved by the General Assembly, but the framework assumptions of our world system were never examined.

To summarize, U.N. and national representatives have been meeting annually since 1972 to focus on climate change. Though agreements have initiated changes in the behavior of some nations and corporations, there is consensus among climate scientists and environmental experts that the modest improvements have been wholly inadequate to address on-going climate collapse (Maslin 2013; Lenton 2016). The most recent *Sixth Assessment Report* of the Intergovernmental Panel on Climate Science describes our planetary situation as "dire."[2] Today, there is common recognition that humans have failed to deal with the problem and the crisis is worsening year by year.

The U.N. document embodying the SDGs is entitled "Transforming Our World: The 2030 Agenda for Sustainable Development." This document resounds with high-minded ideals agreed to by all the signatory nations. Of the 91 articles, Article 3 is a prime example of the U.N.'s beautiful yet nonbinding language:

> We resolve, between now and 2030, to end poverty and hunger everywhere; to combat inequalities within and among countries; to build peaceful, just and inclusive societies; to protect human rights and

---

[1] https://unfccc.int/process-and-meetings/the-paris-agreement/the-paris-agreement/nationally-determined-contributions-ndcs.

[2] https://www.ipcc.ch/report/sixth-assessment-report-cycle/.

promote gender equality and the empowerment of women and girls; and to ensure the lasting protection of the planet and its natural resources. We resolve also to create conditions for sustainable, inclusive and sustained economic growth, shared prosperity and decent work for all, taking into account different levels of national development and capacities.

This statement is excellent and it expresses fundamental and noble ideals (with the exception of the false ideal of "growth"), which are replete throughout the "Transforming Our World" document. Nevertheless, the first thing to note about the SDGs is that there is no mention of "global economic exploitation," no mention of "imperialism," and no mention of "militarism" anywhere in this document of nearly 15,000 words. The word "exploitation" does appears three times, but only in relation to child trafficking or the exploitation of women. Apparently, the framers of the SDGs appear unaware of other forms of exploitation, such as the abuse of impoverished workers. Apparently, militarism and imperialism are not major world problems affecting the environment. The horrific existence of nuclear weapons or WMDs appears nowhere in this document, despite their obvious relevance to the attempt to create a sustainable and just world system.

The document makes a pervasive distinction between "developed" and "developing" countries, and it treats nations as individual units, each pursuing increased GDP but with no recognition of the exploitative relationship between the wealth of developed countries (a.k.a. the Global North) and the poverty of developing countries (a.k.a. the Global South). Thus, the U.N. and its member states pretend the emperor *is* wearing clothes, that the obfuscating rhetoric of imperial countries embodies a good-faith description of the world system. Indeed, the U.N. is required to pretend that capitalism and endless growth – controlled and managed by first world wealth, loans, and economic management – is the sole legitimate path to sustainability. Contrary to the economic dependency of the Global South on financing provided by the Global North, each country is responsible for increasing its own GDP, just as each is responsible for attaining the targets set by the SDGs.

U.N. personnel pretend that the immense militarism of empire nations – which conduct assassinations, blockades, drone strikes, invasions, sanctions, and political manipulation of weaker nations – is used only to keep worldwide peace, often assisted by U.N. Peacekeepers or complicit NATO military forces. They also pretend the militarism of competing nations that fear and/or resist the empire (such as China, Russia, and Iran) is not a significant problem with respect to climate destruction and wasted resources that could be used for climate protection and regeneration. Approximately $1.8 trillion in worldwide expenditures pour down the toilet of militarism annually, but this monstrous misuse of wealth is off the table for discussion when the U.N. addresses climate change.

Within this willful blindness, the SDGs were formulated. Otherwise the U.S. and other global forces would never have allowed them to see the light of day. As one American insider to the U.N. recently declared, "Nothing happens at the United Nations without U.S. approval." This may be an overstatement, but it contains substantial truth. The U.N. has been colonized by the economic and political ideology of the global imperial center.

It is important to note the fragmented approach to protecting our unitary global biosphere. We live on one planet with an integrated ecology, yet according to the U.N., climate collapse can only be addressed through each country's separate plan "taking into account its domestic circumstances and capabilities." A climate plan that is both fragmented and voluntary is doomed from the start, even more so when we consider that nations are competing economically and many are desperate for economic growth. Truly, it is absurd to think our planetary climate crisis can be solved by individual countries tasked with "domestic" climate responsibility. The chart below highlights this obvious conclusion.

Figure 6.1 Regulation of Planetary Resources		
World Resource	Nation-State System	Earth Constitution Solution
Atmosphere & hydrosphere	Freely exploited by each sovereign nation	Protected by World Environmental Authority
Oceans	Freely exploited by each sovereign nation	Protected by World Environmental Authority
Glaciers, lakes, rivers & fresh water supplies	Under control of nation where resource is located	Protected by World Environmental Authority
Forests and CO_2 production	Under control of nation where resource is located	Protected by World Environmental Authority
Corporate pollution and toxic waste	Under control of each sovereign nation	Prohibited globally
Sustainability goals	Discretionary	Mandatory
Militarization and war	Under control of each sovereign nation	Supervised disarmament and an end to war
Universal human rights	Freely exploited by each sovereign nation	Protected by World Court & World Ombudsmus
Third World exploitation and cheap labor	Under control of for-profit banks & multinational corps	Guaranteed living wage and global fair trade
Whether to finance transformative projects	Under control of for-profit banks & sovereign nations	Interest free financing and technology transfer

The WTO, World Bank, and IMF place conditions on countries, taking initiative out of their hands and putting control in the hands of multinational corporations and global banks. The unfair economic and structural requirements – including transnational WTO regulations that can defeat domestic climate laws – render the U.N. climate change policies fragmented and incoherent. Even though environmental collapse crosses all borders and requires massive bioregional, continental and global cooperation to effectively address it, each country is essentially mandated to fixate on increasing its GDP while simultaneously reducing carbon emissions, eliminating poverty, and creating social justice and gender equity. In this way, the system ignores and covers up both the exploitation and the interdependence of the real environmental situation.

Half a century has passed since the 1972 Stockholm Conference, and in response to repeated failures since then, the U.N. still has not questioned the presuppositions of the world system itself. We have seen how these presuppositions colonized the collective consciousness. The U.N. may have substituted the SDGs for the less detailed MDGs, but because the SDGs are based on the same faulty assumptions, the SDGs will fail as well. For example, under "Means of Implementation," the SDG document describes how this imagined and wishful transformation will take place. Item 41 states:

> **We recognize that each country has primary responsibility for its own economic and social development.** The new Agenda deals with the means required for implementation of the Goals and targets. **We recognize that these will include the mobilization of financial resources** as well as capacity-building and the transfer of environmentally sound technologies **to developing countries on favourable terms,** including on concessional and preferential terms, as mutually agreed. Public finance, both domestic and international, will play a vital role in providing essential services and public goods and in catalyzing other sources of finance. **We acknowledge the role of the diverse private sector, ranging from micro-enterprises to cooperatives to multinationals,** and that of civil society organizations and philanthropic organizations **in the implementation of the new Agenda.** [emphasis added]

Thus, financing is available on "favorable terms," but that translates into interest-bearing notes. Multinationals are given a nod of approval, since they apparently are not considered monopolistic or exploitive of cheap labor and resources. And implementation is funneled through the U.N. because: "Seventy years ago, an earlier generation of world leaders came together to create the United Nations. From the ashes of war and division they fashioned this Organization and the values of peace, dialogue and international cooperation which underpin it. The supreme embodiment of those values is the Charter of the United Nations" (Item 49). This statement is simply untrue. Those values are not the main thrust of the U.N. Charter.

Since the founding of the U.N. in 1945, most scholars count some 150 armed conflicts and wars, proving that international "dialogue" under the auspices of the U.N. routinely fails. The Security Council veto allows that body to govern nearly all decision-making, resulting in little if any genuine "international cooperation." Consequently, the five victorious nations in World War II – United States, Great Britain, France, Russia, and China – possess unchecked power over all other nations. Moreover, under Article 42, the Security Council is empowered to keep the peace by going to war!

How can peace ever be established by going to war? Real peace is only established and protected by the democratic rule of law requiring an unbiased judiciary, fair civilian enforcement, and a planetary constitution guaranteeing equal due process to all. This is precisely what the U.N. Charter forbids due to its subordination of many nations to the so-called "Security Council," which has never provided humanity with the slightest hint of real security. Instead, the U.N. blesses the current system of sovereign nation-states that recognize no enforceable law above themselves, thereby sanctioning and maintaining an international war-system. It is an ideological lie that the U.N. keeps the peace or has the power to enforce the SDGs which otherwise might help save planet Earth. Truly, the emperor is not wearing any clothes.

6.4   The *Earth Constitution* Solution

The *Constitution for the Federation of Earth* was written by hundreds of world citizens working together through a process of four Constituent Assemblies over a period of 23 years, from 1968 to 1991. At the fourth assembly in Troia, Portugal in 1991, the document was declared finished and ready for ratification under the democratic requirements set forth in its Article 17. The *Earth Constitution* establishes global democracy and creates a legal order for all nations of Earth that eventually demilitarizes the nations, ends poverty, ensures global social justice, and institutionalizes environmental protection and sustainability. In other words, it establishes not a "Security Council" that keeps the peace by going to war, but a world peace-system that includes an effective judiciary, civilian enforcement, and a true constitution that specifies the inalienable rights of all persons everywhere.

In addition, the *Earth Constitution* is designed to handle the climate crisis in a systematic and comprehensive way. It makes food, fresh water, clean air, and freedom from poverty fundamental human rights that are legally redeemable under the Earth Federation government. It also recognizes that human beings have the *right to peace*, along with *the right to a healthy planetary environment*. As we have seen, nations acting alone cannot handle global, planetary problems. The U.N. dream of delegating *voluntary* responsibility to some two hundred mostly militarized sovereign nation-states is just that – a dream. Nor can we deal with this planetary crisis without changing the economic and political systems that plainly caused the crisis.

It is important to understand that the *Earth Constitution* does not abolish the U.N. Rather, it incorporates U.N. agencies into its framework. For instance: The U.N. High Commission on Human Rights integrates with the *Earth Constitution's* World Ombudsmus, the U.N. International Court of Justice and the International Criminal Court merge into a new World Court System, the U.N. General Assembly becomes the House of Nations, etc. The only substantive change is that the undemocratic and unworkable U.N. Charter is replaced by a genuine and excellently designed, democratic *Constitution for the Federation of Earth*.

If the people of Earth really have the human rights specified in the U.N. Universal Declaration of Rights, then they have the rights declared in its Article 28: "Everyone is entitled to a social and international order in which the rights and freedoms set forth in this Declaration can be fully realized." Sadly, however, human rights are violated just as routinely around the world today as they were in 1947 when this Declaration was signed. The U.N. system clearly is not the "international order" we urgently needed. In 75 years of existence, the U.N. has failed to give us peace, provide environmental protection, end poverty, or to protect the rights of persons around the world who are routinely victimized by countries, corporations, and terror groups.

To fully illustrate why the 17 Sustainable Development Goals are ineffectual under the current system, it is worth examining each of them.

➤ **GOAL 1: End poverty in all its forms everywhere.** This noble ideal provides the gloss of the entire SDG document. Yet the document contains no analysis of the global production of poverty by the world-system. Astonishingly, the document never even mentions the planetary population explosion. Indeed, the word "population" only appears five times and even then in innocuous contexts. Yet the addition of millions of new people every year to Earth is surely a major contributor to global poverty and misery.

We have a planet whose population increases by about 80 million new persons per year. Population experts have been pointing out the calamity of global population explosion since at least the 1960s (Cohen 1995). The SDG document asserts that poverty everywhere can be eliminated by 2030 even though the planet will likely have about a billion new mouths to feed by that date. This goal not only ignores the population explosion but the fact that arable land is dramatically decreasing annually due to overuse and desertification. Similarly, the global fish supply has been steadily dwindling since the 1980s. The *Earth Constitution,* by contrast, takes a holistic approach by embracing all these factors. It encourages voluntary population control, and it provides for reproductive education and worldwide supplies of birth control technologies – the first obvious steps toward ending poverty on Earth.

➤ **GOAL 2: End Hunger, achieve food security and improved nutrition and promote sustainable agriculture**. This goal states: "By 2030, double the agricultural productivity and incomes of small-scale food producers, in particular women, indigenous peoples, family farmers, pastoralists and fishers, including through secure and equal access to land, other productive resources and inputs, knowledge, financial services, markets and opportunities for value addition and non-farm employment." How this is to be achieved in the light of the current world system remains an inscrutable mystery. And again, because these goals are strictly voluntary, nations are free to ignore them. In contrast, the *Earth Constitution* makes food security a planetary right and creates the institutions for global planning necessary to actualizing this right.

➤ **GOAL 3: Ensure healthy lives and promote well-being for all ages.** Can you imagine this being achieved in Indonesia, a global mecca for cheap labor, resources, and massive poverty? Can you imagine this being done in the U.S., a global mecca for class domination where healthcare is not a right and 40% of its population has no health insurance or resources to visit a physician? The *Earth Constitution* makes reasonable equality a binding, legally redeemable right for everyone. It makes universal health care and the other necessities for human well-being into redeemable legal rights.

➤ **GOAL 4: Ensure inclusive and equitable education and promote lifelong learning opportunities for all.** According to the Bonded Labor Liberation Front of India, there are between 20 and 65 million bonded laborers in India alone, and millions of these are children enslaved within the silk industry and other industries requiring free labor and tiny fingers. Girls in Afghanistan and other countries are often forbidden to get an education.

Without a transformed world system away from capitalism, militarized nation-states and the U.N. protection of these institutions, lofty goals by the year 2030 are mere fantasy. By contrast, the *Earth Constitution* puts democratically elected representatives of the people of Earth in charge of transforming our current exploitative economic system and allows them to actualize human potential through education for all, real protection of children, as well as providing other fundamental rights. In this manner, the *Earth Constitution* activates the grassroots participation of all people, by holding Global Peoples' Assemblies and other means of popular mobilization.

➤ **GOAL 5: Achieve gender equality and empower all women and girls.** In 1972, the Equal Rights Amendment was introduced in the United States to give equal rights to all citizens and abolish legal distinctions between women and men in divorce, property, employment, etc. Presently, this amendment remains far from passing due to immense structural and cultural forces that impede such equality in the U.S. and around the world. Nevertheless, this U.N. goal says: "Adopt and strengthen sound policies and enforceable legislation for the promotion of gender equality and the empowerment of all women and girls at all levels."

If this is next to impossible in the U.S., where fundamentalist Christians claim males must rule in the family, how will it be possible in many Muslim countries around the world, or in the many places where "genital mutilation" is still forced upon girls? It is deeply ironic that the SDGs recognize "enforceable legislation" is necessary to achieve this goal. Yet the U.N. system sanctioning national sovereignty adamantly opposes "enforceable legislation" at the world level – exactly where it is needed most. Under the *Earth Constitution*, the equal rights of women are a matter of enforceable world law.

➤ **GOAL 6: Ensure availability and sustainable management of water for all.** With global warming increasing at the same time the population of Earth continues to explode, freshwater is becoming progressively diminished. Water tables everywhere on the planet are dropping, while the demand for water continues to increase. For example, hundreds of millions of people in the countries surrounding the Himalayas depend on annual snow and ice melt from this mountain range, a melt which is disappearing as the glaciers disappear. Thus, how can this goal be achieved while ignoring a population explosion, shrinking arable land, and likely future "water wars" between India, China, Nepal, Bangladesh, and other countries whose water supplies are rapidly diminishing?

Activist Vandana Shiva in India documents the ways in which multinational corporations – with the help of the World Bank and WTO – are sucking up immense quantities of water in countries worldwide, causing the water tables to drop and then selling the extracted water back to the people inside of environmentally damaging plastic containers. Everywhere private forces are working to get control of formerly public water supplies, thereby converting resources that should belong to the people into corporate-owned monopolies (2002, Chap. 4). To solve this problem, the *Earth Constitution* designs a world system in which all people have an inalienable right to sufficient quantities of free, clean, life-giving water.

➤ **GOAL 7: Ensure access to affordable, reliable, sustainable and modern energy for all.** In *This Changes Everything: Capitalism vs the Climate,* we have seen Naomi Klein detail how big oil companies spend hundreds of millions of dollars to publicly question the conclusions of climate scientists about the effects of fossil fuel on planetary warming. Immense economic forces are allied against the goal of sustainable energy. Even today, six years into the jurisdiction of the SDGs, the nations of the world struggle for oil resources, build new pipelines for oil and gas, and engage in destructive environmental fracking for natural gas.

As early as 1981, Buckminster Fuller outlined a very doable plan for a clean global energy grid in which solar power could be brought to the entire Earth through an interlinked and international solar energy system. Such a truly planetary grid would not need extraordinary quantities of batteries because some portions of Earth are nearly always experiencing sunlight. This would, of course, require energy cooperation rather than competition. It also would require a united Earth such as envisioned by the *Earth Constitution.* Our problems can be solved, but only if we are truly united.

➤ **GOAL 8: Promote sustained, inclusive and sustainable economic growth, full and productive employment, and decent work for all.** Even elementary theories of capitalism realize that this is impossible. Capitalist profit margins require unemployment, a desperate work force willing to compete for low wages. As we have seen, economists such as Herman E. Daly, Richard Heinberg, and Kate Raworth proclaim that we are at the end of growth. Daly affirms that development must be *qualitative,* and should no longer be measured in quantitative terms.

Moreover, environmental experts like James Gustav Speth in *The Bridge at the End of the World: Capitalism, the Environment, and Crossing the Crisis to Sustainability* show that capitalist growth dogma is at the heart of the problem. Even though the Earth Federation government will employ millions of now unemployed people in environmental restoration and regeneration projects, the new global economic system will need to accommodate automation, as Jeremy Rifkin (2011) points out.

Rifkin envisions self-sufficient economic communities in which much of production is done digitally and human life is oriented to more meaningful and creative values rather than simply working to survive. Daly (1996) details how *everything* must be designed for maximum durability, for reparability, and for eventual recycling. Extracting materials from Earth must be reduced to an absolute minimum, and waste returned to Earth must also be minimal. The *Earth Constitution* makes this new world possible because it places democratically organized human beings in authority over global corporations, over private banking, and over territorial nation-states. It gives the World Parliament the means to create a world system that works for everyone in ways the U.N. system is manifestly incapable of doing.

➤ **GOAL 9: Build resilient infrastructure, promote inclusive and sustainable industrialization, and foster innovation.** This is precisely what the *Earth Constitution* is designed to do and what the present world disorder is incapable of doing. World Trade Organization (WTO) regulations contain dozens of pages dedicated to "intellectual property rights" in order to ensure the profiteering of multinational corporations in ways that defeat technology transfer, pharmaceutical transfer, and open-source information on sustainable industrialization.

The *Earth Constitution* establishes global public banking dedicated to making sustainable development available to poor people everywhere. The private banking cartels that now dominate the world make achieving Goal 9 impossible. The SDG document speaks of giving people greater "access to banking, insurance and financial services for all," but it never critically recognizes that that the lion's share of these services operate for private profit and for the exploitation of those who receive them.

Without global monetary policy and public banking for the people of Earth, life will never become sustainable or in harmony with the carrying capacity of our planetary biosphere, wars will never end, and poverty will never be eliminated (Brown 2007). As long as money creation and banking are in the hands of private banking cartels, the means of economic freedom are denied to the people of Earth. Just as clean water and air are the responsibility of sound democratic government, so money as a universal medium of exchange must be democratically owned by the people of Earth. Money creation and banking are just as much a public service as are roads. It is absurd to keep these services in private hands.

Philosopher Alan Gewirth (1996) argued correctly that government is responsible for both the freedom *and* well-being of people. Freedom is inexorably tied to well-being, and both are meaningless without an infrastructure that places democracy in the hands of the people of Earth. That is the role of democracy, and under the *Earth Constitution,* money creation and banking are predicated on the common good of current and future generations. Anyone approaching true planetary maturity should surely comprehend this principle.

➤ **GOAL 10: Reduce inequality within and among countries.** According to the Pew Research Center, 96 out of 167 countries with populations over half a million claim to be democratic. Yet most of these are ruled by a tiny super-rich class, and the other 71 countries are ruled by self-interested oligarchies. As of 2019 in the U.S., the bottom 50% of households had $1.67 trillion or 1.6% of the net worth, while the top 10% has 94.4% of all wealth. With that kind of power in the hands of so few – a power that is institutionalized toward continuing to increase their wealth at the rate of billions of dollars per day – how can Goal 10 possibly be achieved? Obviously, only global governance, such as provided by the *Earth Constitution*, would have the power to supersede the corporations and sovereign nations on issues of inequality. Consequently, only planetary democracy can bring reasonable economic equity.

➤ **GOAL 11: Make cities and human settlements inclusive, safe, resilient and sustainable.** Perhaps the authors of the SDGs all live in relatively clean, comfortable European cities like Brussels, Paris, Copenhagen, Munich, or Milan. Perhaps they have never walked through the world's dozens of nightmare megacities like Mexico City, Lagos, Mumbai, Kolkata, Dhaka, or Manila. My travels have shown me miles and miles of unmitigated slums, uncounted millions of impoverished people packed tightly into filthy cities, many of whom cannot even find a hovel or tent in the slums and who live their entire lives on the streets without shelter. If you walk through these cities early in the morning before sunrise, bodies lay everywhere, asleep on the pavement, with their entire set of worldly possessions jammed into a plastic bag, used as a pillow for their heads.

In short, there is no way these centers of chaos and misery will be converted to "inclusive, safe, resilient, sustainable" habitats by the SDG target year 2030. Even with ratification of the *Earth Constitution*, we would be hard-pressed to meet that deadline. However, we'd at least have a fighting chance, since the *Earth Constitution* was designed using an integrated and holistic approach, such that all of humanity's challenges are seen in their true interdependent and interrelated context. In essence, we either solve all the problems together on a planetary scale, or not at all. To fragment the task among some two hundred sovereign entities is madness.

➤ **GOAL 12: Ensure sustainable consumption and production patterns.** This goal is absolutely fundamental if Earth is to have a sustainable civilization. One item under this goal declares: "By 2030, substantially reduce waste gener-ation through prevention, reduction, recycling and reuse. Encourage companies, especially large and transnational companies, to adopt sustainable practices and to integrate sustainability information into their reporting cycle."

As previously mentioned, there is no critique of the profit motive in this document and no critique of capitalism. Yet even an elementary analysis of capitalism must admit its tendency to externalize costs in order to maximize profits. Waste, like unemployment, is essential to capitalism, and externalization of costs is essential to profit maximization. To "encourage" companies is not enough to make it happen. Some multinational companies have more assets and more leverage than the nations who host them. And we have seen that WTO rules prohibit these countries from even making laws that cut into profit margins.

As Annie Leonard points out in the on-line "Story of Stuff," many companies are famous for calculating planned obsolescence (i.e., how rapidly their design can make things break without losing consumer loyalty). If products break and customers repeatedly buy new ones, profit margins are vastly increased. All the sustainable economists declare that sustainable production and consumption will require an entirely different lifestyle, especially for the top ten nations of the world that today consume 80% of the planet's resources (https://internationalbusinessguide.org/hungry-planet/).

With the U.S. ruling class owning 94.4% of the wealth generated by this unsustainable system, who is going to convince them to give up the system – the U.S. government that is run by them? Both the Democratic Party and the Republican Party are run by wealthy persons and corporations. Are protestors in the streets who are beaten up by the police going to make a real difference? Only planetary federation can ensure that sustainable production and consumption patterns become effective and equitable without undue suffering or injustice with regard to any of the parties involved. Right now, the top 10% who control everything, including the U.N., have zero interest in achieving such equality.

➤ **GOAL 13: Take urgent action to combat climate change and its impacts.*** This goal hits the nail on the head in that urgent action is clearly needed. But this SDG directs nations to "Integrate climate change measures into national policies, strategies and planning." And the asterisk indicates that any urgent action means "acknowledging that the United Nations Framework Convention on Climate Change is the primary international, intergovernmental forum for negotiating the global response to climate change."

Affirming this U.N. Framework Convention does not constitute the "urgent action" we need. Only addressing the root causes of the entire nexus of global problems will produce success. The world is suffering not only from climate collapse, but also from a global pandemic, immense poverty, population explosion, endless wars, worldwide militarism, mindless competition among nations, and social, moral, and spiritual chaos. The U.N. Convention of Climate Change is in truth a prescription for human extinction because it ignores the root causes and deep interconnections of our dire planetary problems.

➤ **GOAL 14: Conserve and sustainably use the oceans, seas and marine resources for sustainable development.** This goal is critical because the oceans are dying. If the oceans die, the planet dies, and we die. Environmental experts produce volume after volume of data confirming this fact – the implicit premise of Goal 14. We have seen environmental leader Bill McKibben in his 2019 book *Falter: Has the Human Game Begun to Play Itself Out?* detail the acidification of the oceans, their carbonization, the growing multiple dead zones, the diminishing oxygen production of the oceans (already suffocating some species), their rapid warming, and the immense quantities of plastic waste polluting them. James Gustav Speth in his 2004 book *Red Sky at Morning* writes, "In 1960, 5 percent of marine fisheries were either fished to capacity or overfished; today 75 percent of marine fisheries are in this condition. ... Data reveal that the global fish catch has shown a strong and consistent downturn every year since 1988" (2004, 15, 33). The oceans are indeed dying.

Nations and corporations have developed the technology to mine the natural resources of continental shelves to some two hundred miles offshore, and a barrage of mining has erupted in the 21st century, with some immense accidents like Deepwater Horizon in the Gulf of Mexico. The U.N. Convention on the Law of the Sea (UNCLOS), which went into effect in 1994 after sixty countries endorsed it, has clauses that attempt to protect the sea. Of course, the U.N. is hamstrung by claims of national sovereignty (i.e., lawlessness), rendering the Law of the Sea Convention weak and practically unenforceable. Moreover. the U.S., which is the nation most abusive of the seas, has refused to ratify this treaty, claiming it infringes U.S. sovereignty.

Goal 14 urges nations to "enhance the conservation and sustainable use of oceans and their resources by implementing international law as reflected in UNCLOS, which provides the legal framework for the conservation and sustainable use of oceans and their resources." But is an unenforceable treaty really a "legal framework"? Because ratification is voluntary for each nation, the U.N. treaty system cannot possibly save the environment. James Gustav Speth (former Dean of Environmental Studies at Yale University) writes:

> The bottom line is that today's treaties and their associated agreements and protocols cannot drive the changes needed. ... Typically, these agreements are easy for governments to slight because the treaties' impressive – but nonbinding – goals are not followed by clear

requirements, targets, and timetables. And even when there are targets and timetables, the targets are often inadequate and means of enforcement are lacking. As a result, the climate convention is not protecting the climate, the biodiversity convention is not protecting biodiversity, the desertification convention is not preventing desertification, and even the older and stronger Convention on the Law of the Sea is not protecting fisheries. (2008, 71-72)

Clearly, the SDG goals will not prevent climate collapse nor give us effective sustainable development. Ratifying the UNCLOS will not prevent the oceans from dying either. Under the *Earth Constitution*, by contrast, the oceans of Earth belong to the people of Earth and the oceans are made a protected global commons. No longer may sovereign nations freely exploit the seas for their own interests. Similarly, the atmosphere and forests of Earth are essential to the biosphere and human life on Earth, so all these would belong to the people of Earth as well. (See also my 2019 article "The Tragedy of the Global Commons" at www.Academia.edu).

➤ **GOAL 15: Protect, restore and promote sustainable use of terrestrial ecosystems, sustainably manage forests, combat desertification, and halt and reverse land degradation and halt biodiversity loss.** Reaching this goal by the year 2030 is absolutely essential to the future of life on Earth. Yet the U.N. tells the nations of the world to "mobilize and significantly increase financial resources from all sources to conserve and sustainably use biodiversity and ecosystems." One supposes third world countries can do this while continuing to pay back their immense international indebtedness to first world banking cartels, while maintaining military preparedness by buying expensive weapons from first world arms dealing countries, and while dealing with their own internal social poverty and chaos. One supposes they are to do this while carrying out "structural adjustment" programs imposed by the World Bank and the IMF that require them to sell off their infrastructure and social programs to profit-making first world corporations. However, poor nations are not going to be able to "finance" the protection of their environment nor the natural resources that cross national boundaries, making this demand quixotic at best.

Under the *Earth Constitution*, finance for ecological protection is debt-free and non-exploitative because it arises from the global public banking that is at the heart of the Earth Federation government. Action need not be fragmented by nation-state boundaries or countries going into debt to protect their national ecological integrity. The Earth Federation government is designed precisely to address global problems beyond the scope of nation-states.

➤ **GOAL 16: Promote peaceful and inclusive societies for sustainable development, provide access to justice for all and build effective, accountable and inclusive institutions at all levels.** Here, the incomplete ideology of the SDGs becomes most glaring. Countries are expected to be internally peaceful and inclusive. There is no mention of international wars, terrorists financed by international actors, massive worldwide weapons sales by first-world countries, or even internal civil strife. And there definitely is no mention of the world pouring $1.8 trillion down the drain annually through wars and military expenditures.

Societies around the world are being torn apart by the U.S. empire attempting to maintain its global economic and political domination and by the resistance of competing powerful actors (such as China, Russia, and Iran). Major portions of the world are in chaos because of these struggles, from Afghanistan to Iraq, Syria, Libya, Yemen, Lebanon, and Palestine. Internal conflicts are raging in dozens of more countries financed by the U.S.-Israeli coalition. The U.N. is required to ignore all this and pretend that we have a world order ready to cooperate and meet the SDG goals by the year 2030. Since 9/11, when the U.S. government declared its endless global war on terror, this process has intensified rather than diminished. As historian Tom Engelhardt writes:

> Since 9/11, the result has been a religion of perpetual conflict whose doctrines tend to grow ever more extreme. In our time, for instance, the NSS has moved from Dick Cheney's "1 percent doctrine" (if there is even 1 percent chance that some country might someday attack us, we should act "as if it is a certainty") to something like a "zero percent doctrine." Whether in its drone wars with presidential "kill lists" or the cyberwar – probably the first in history – that it launched against Iran, it no longer cares to argue most of the time that such strikes need even a 1 percent justification. Its ongoing, self-proclaimed Global War on Terror, whether on the ground or in the air, in person or by drone, in space or cyberspace (where the newest military command is already in action) is justification enough for just about any act, however aggressive. (2014: 7)

The key to achieving true sustainability is a world system of governance *designed* to accomplish this goal of world peace. By and large, the SDGs contain an accurate list of laudable and necessary goals, but the U.N. lacks the power and a coherent means for achieving them. The goals simply are not achievable within the present world system. They also are incomplete, as we have seen, since they exclude the absolute need for planetary population reduction through voluntary education and global dissemination of birth-control technologies. Ultimately, the SDGs are a pipedream, so long as they ignore the need for world peace through demilitarization.

Thus, laudable goals have been thrust into the Procrustean bed of an antiquated world system that is *the* primary cause of war, poverty, and environmental destruction – a major impediment to accomplishing these goals. An article on the SDGs from the point of view of "developing countries" agrees, stating that the "promotion of Justice at the national level, which stems from the spirit of the SDGs, particularly for goals 3, 4, 5 and 10, is not pursued" (Jabbari, et al. 2020). Meanwhile, protestors worldwide cry out: "No justice, no peace!" Consequently, Goal 16 is mere empty verbiage that ignores and covers up a planetary war system and pretends that we can have peace within nations without also having justice and peace between nations.

In sum, capitalism and sovereign nation-states are centuries-old inventions, products of deeply discredited early-modern assumptions about the world. They simply are not designed for planetary, biospheric health and protection. The *Earth Constitution*, on the other hand, presents a precisely designed system that abolishes neither free markets nor nation-states. But it does convert markets to democratic, non-exploitative forms of trade. It also establishes common good forms of finance, eliminates militarism from our planet, and stops the absurd claims of nations that recognize no binding laws above themselves. Such world system changes are necessary if there is to be real sustainability and an end to the ever-growing environmental chaos and possible extinction.

➤ **GOAL 17: Strengthen the means of implementation and revitalize the global partnership for sustainable development finance.** This last goal reaffirms the present global economic system of Global North domination and exploitation. It reasserts the Addis Ababa Action Agenda which, as summarized here, relates "to domestic public resources, domestic and international private business and finance, international development cooperation, international trade as an engine for development, debt and debt sustainability, addressing systemic issues and science, technology, innovation and capacity-building, and data, monitoring and follow-up." These are all features of the U.N. Economic and Social Council (ECOSOC), the World Bank, WTO, and IMF – the "international private business and finance" organizations that have kept third world nations in poverty for generations. Short of a miracle, the system will never reverse itself nor help poor countries achieve "debt sustainability."

As a result, Goal 17 makes clear that the current system of debt enslavement will not be abandoned. At best, it will be modified so poor countries can pay down their debt "sustainably" (i.e., forever). As Richard Heinberg points out so clearly in *The End of Growth* (2011), the debt-financing system requires growth so that the surplus provided can be used to service debt. Without growth, no borrower under the present system can pay the interest, let alone the principle on their loans. Growth is an assumption of the SDGs, in direct contradiction to the fundamental principle that you cannot have endless growth on a finite planet.

Goal 17 openly states: "We recognize that domestic resources are first and foremost generated by economic growth, supported by an enabling environment at all levels. ... Private business activity, investment and innovation are major drivers of productivity, inclusive economic growth and job creation." Privatization, corporatocracy, and the private banking cartels of the Global North continue to hold all the cards while demanding that the Global South tighten its belt and strive for environmental sustainability.

By contrast, the *Earth Constitution* begins the process of rapid transition to a sustainable world by having the Earth Federation government *assume the international debts* of the poor nations, thereby freeing them and providing them with a clean slate. It does not abolish loans made by the Global North, but it mandates repayment agreements with the world government utilizing Earth Federation currency. In this way, the transition to an equitable global economic system is guaranteed. By creating global public banking and taking money creation out of the hands of private banking cartels, the *Earth Constitution* also commences debt-free money creation to provide ample resources for a truly sustainable world.

6.5   Getting Serious About Sustainability

Unfortunately, the SDGs align with the capitalist system, which is designed for private profit at the expense of the common good. The SDGs also bow to the sovereignty of nation-states, which are designed for war, power politics, and a system of dominating weaker states. By refusing to examine the world system itself, what we end up with in the SDGs is naïve ideology rather than true social science.

Articles critical of the SDGs are not uncommon. For example, a 2015 article by Thomas Pogge calls them "brilliant propaganda." Like the majority of such articles, however, Pogge fails to examine the world system behind the SDGs and only demands that the U.N. assign "clear responsibilities for achieving the goals" and set up an independent "measurement operation" to track progress. Yet such minimal, functionalist solutions will not get at the root of the problem.

If one examines the assumptions behind the SDGs, it becomes clear that the 17 Goals cannot possibly be achieved within the current system. The SDGs incorporate the right to the private accumulation of wealth, along with the equivalent dogma of nation-state sovereignty: "We reaffirm that every State has, and shall freely exercise, full permanent sovereignty over all its wealth, natural resources and economic activity." With this statement we have the U.N.'s bottom line – that the global commons and the resources all human beings need to live are somehow the private property of sovereign nations.

In sum, the SDG document exhibits the following features: (i) the population explosion is ignored and not considered a major problem, (ii) global militarism and wars are ignored and not considered a major problem, (iii) the economic system in which a tiny handful of people own more than 50% of the world's wealth while the bottom half of humanity live in poverty and without access to education or other means to develop their human potential is not considered a problem, (iv) the dogma of sovereignty, which makes all treaties voluntary and easily breakable is not considered a problem, and (v) the fetish of national sovereignty over all a nation's "wealth, natural resources, and economic activity" is not considered a problem.

Consider again the implications of this dogma of sovereignty, as in the case of Brazil, which happens to host the Amazon forest – the "lungs of Earth" – that produce nearly 50% of the world's oxygen and moderate the global climate. Under the current system, the government of Brazil has the *legal right* to timber, develop, and destroy the Amazon forest for private gain. Similarly, the United States has the *legal right* not to sign the Law of the Sea Convention and to withdraw from the 2015 Paris Climate Accord, even though withdrawal by the world's largest polluter means harming the global climate for all of humanity. And China has the *legal right* to produce all the CO_2 it wants, send it into the global atmosphere, and increase the greenhouse effect that is overheating our entire planet. All nations have these same ridiculous rights, including the *legal right* to militarize themselves to the teeth, wasting badly needed resources to protect and restore our planetary environment. To understand the breadth of sovereignty is to discern how impotent and absurd the U.N. system really is.

Will our collective consciousness awaken in time to a truly world-centric and planet-centered morality? Many humans are already there, but our institutions have not caught up. It therefore is unlikely that human beings will be able to draw upon the liberating dynamics that remain largely untapped at the heart of world civilization in time to prevent planetary ecocide. It also is unlikely

The Earth Constitution creates the social, political, and economic conditions that facilitate personal growth and civic maturity, allowing humans to reach their highest potential and practice universal ethical principles, such as cooperation, solidarity, justice, and compassion.

that the virus of nationalism will lift or that the xenophobic forces pitting the U.S. and its lackeys against China, Russia, and Iran will learn to embrace human and planetary holism. And it is unlikely that the empire, whether under an Obama, Trump, or Biden, will grow to planetary maturity and adopt unity in diversity. None of these are likely because the present world system itself blocks human growth toward a wider and fuller consciousness. Without swift transformation, ecocide appears the most likely outcome – death to our planet and her human inhabitants.

Therefore, what is desperately needed is a global public authority with the power to implement comprehensive change. What we need is a federated Earth and global democratic governance. Nothing short of a spiritual revolution is required, starting with true human liberation. If we seek a just and peaceful world, then it is important to understand this sacred principle: *Socio-political-economic conditions can facilitate personal growth and civic maturity, allowing humans to reach their highest potential and practice universal ethical principles, such as cooperation, solidarity, and justice. Or, these same factors, if badly designed, have the ability to retard growth and sever people from their innate capacities and civic duties.*

The *Earth Constitution* provides the conditions for this transformative growth, while the present world system inhibits and blocks human liberation. Under the *Earth Constitution*, nations are not abolished, they are *united*. All nations become part of the World Parliament that represents the common good of everyone on Earth, within the context of an Earth Federation government designed to solve the global problems that the U.N. system is not designed to effectively address.

More specifically, the *Earth Constitution* sets in place four fundamental issues ignored by the SDGs. First, it establishes voluntary educational programs for family planning designed to reduce our planetary population to a sustainable level. Family planning aside, simply educating women has been shown to reduce the number of children. These programs guide the people of Earth toward planetary maturity. Second, it sets up programs to carefully reduce, then abolish militarism by establishing enforceable democratic law over all nations so that conflicts are handled through courts and mediation. Third, the *Earth Constitution* is designed to convert exploitive capital markets into democratic markets serving the common good, while at the same time reducing social disparities and eliminating severe poverty everywhere on Earth. Global public banking will empower humanity and planetary sustainability.

Fourth, and most importantly given the climate crisis, the *Earth Constitution* is purposely and carefully designed to bring the world into a healthy balance with the biosphere of our planet. It places the ecosystems, the oceans, the atmosphere, and every natural resource vital to the health of our planet under the enforceable authority of the people of Earth. No longer will planetary resources be considered the private property of militarized sovereign nation-states. The *Earth Constitution* will truly unite humanity and address our collapsing biosphere along with the other lethal global problems we face.

The United Nations SDG document accurately points out that all these problems are interrelated and must be effectively addressed simultaneously. Yet the fragmented world system it presupposes is hopelessly inadequate to achieve this. The *Earth Constitution* literally is *designed* to simultaneously end war, disarm nations, protect universal human rights, diminish economic disparity, clean up our planet, and promote solidarity among the people of Earth to finally establish a sustainable civilization. The Earth belongs to all of us, and it will take all of us – democratically and globally united – to solve our planetary problems. The SDGs and the world system that they presuppose cannot possibly save us. Our immediate and most pressing goal must be to ratify the *Constitution for the Federation of Earth.*

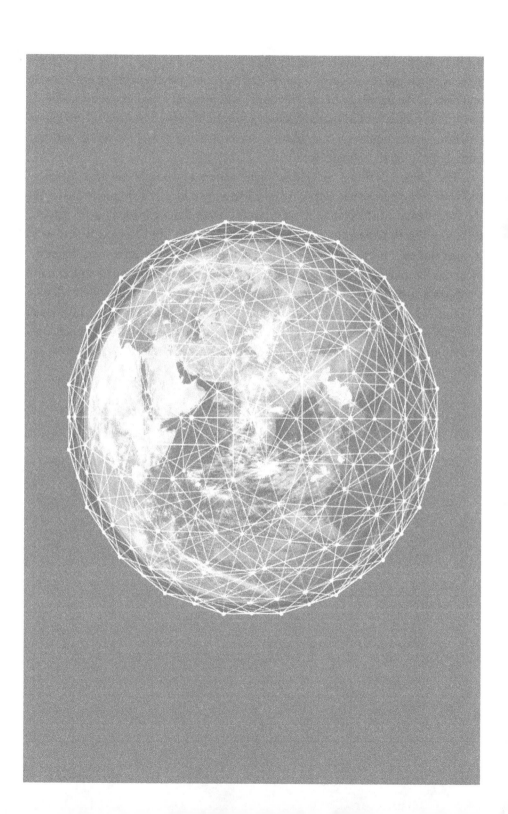

CHAPTER SEVEN
Design Features of the *Earth Constitution*

If the nations of the world could be appraised of the mortal dangers now threatening the one known habitat of life, intelligent or other, and the extent to which national sovereignty prevents the adoption of the needed antidotes, they might be prepared willingly to accept federation under a global administration, as the only course of action that can save them and their peoples, along with possibly most other living species, from extinction.

Errol E. Harris

7.1   The *Earth Constitution* as Sustainable Planetary Design

The *Earth Constitution* is a masterful document for converting our planetary noosphere into harmony with the biosphere, and a master plan for actualizing conscious evolution for humanity. This is because the *Earth Constitution* is specifically *designed* for planetary transformation to a holarchical, cooperative world system. Many decades of systems theory development have made us aware that the ways we design our institutions will determine the outcome and whether we achieve critical goals.

Systems theory has made very clear that different consequences flow from different design structures. All designers know this basic principle. All systems theorists know it as well (Laszlo 2002; Meadows 2008). Computer modeling of climate change also flows from this understanding (see Meadows, et al. 2004). Not only do empirical consequences follow from system design, the design of a human-based system also transforms human consciousness. Uniting the world into one political and economic system should be a fundamental aim of conscious evolution.

Because consequences follow design, if we change the world economic and political system to ban dumping greenhouse gases into the atmosphere, then the result of this design change will necessarily be a much more stable and life-friendly planetary environment. We can identify a long series of "if-then" propositions that link our present destructive actions with disastrous consequences. "If we stop spending 1.8 trillion U.S. dollars a year on war and weapons, then we would have vast resources to devote to climate protection and restoration."

Our present institutions are designed in accordance with the outdated and fragmented assumptions of the early-modern era, an old paradigm that permits nations and corporations to assault the biospheric, life-supporting harmonies of our planetary ecosystem. We understand very clearly what needs to be done. If we stop producing immense quantities of toxic waste and depositing these in ruined landfills and super-toxic wastelands around the planet, then future generations will have less poison in their water, air, soil and foods. Pretty simple in theory.

Ditto with human consciousness. If we give up the fragmented political system of militarized sovereign nation-states in lethal competition with one another and federate all countries under the framework of unity in diversity, then human consciousness will move to a higher evolutionary level characterized by political and economic solidarity. As early as the 15th century, Nicolas of Cusa declared "in all the parts shines the whole" (in Panikkar 2013, 265). If humanity unites politically and economically, a worldcentric consciousness will become second nature for people, and the whole will shine through each person. Our common human dignity will become self-evident, and this leap is now absolutely necessary for human survival.

The planet's life-supporting biospheric patterns evolved over billions of years. Yet, if we continue operating under antiquated institutions, within as few as fifty years, the outdated design of our institutions will wreck Earth's biosphere to the point where our planet will become incapable of supporting higher forms of life (Romm 2018; Wallace-Wells 2019). The biosphere evolved as an integrated system that stabilized during the Holocene geological period of the past 12,000 years (Ellis 2018). It supplies the resources, absorbs the wastes, and recycles the water and air necessary for life on Earth. The challenge now is for our planetary noosphere to become an integrated, balanced whole in harmony with the biosphere. It needs to become a holarchy with a brain, a holosphere of unity in diversity. The *Earth Constitution* provides a template for this conversion process.

Together, we must convert to renewable and clean energy from sun and wind power, while recycling and reusing all products. We must bring to an absolute minimum the toxic waste and heat we discharge into nature. And we must only extract a minimal amount of non-renewable resources from Earth, leaving her resilient and healthy in perpetuity for further generations.

Conversely, a patchwork of loosely connected planetary projects with no central administration will never achieve the integration of whole and part that is necessary for our future. A collection of some 193 mostly militarized and competing nation-states will never mount a cohesive plan of action in time. Our only credible option is to truly unite humanity under universal principles of good will and solidarity – the very values at the heart of the *Earth Constitution*. As philosopher Joel Kovel correctly states:

> A pure community, or even "bioregional" economy is a fantasy. Strict localism belongs to the aboriginal stages of society. It cannot be reproduced today, and even if it could, would be an ecological nightmare at present population levels. Imagine the heat loses from a multitude of dispersed sites, the squandering of scarce resources, the needless reproduction of effort, and the cultural impoverishment. This is by no means to be interpreted as a denial of the great value of small-scale and local endeavors: any flourishing ecosystem, after all, functions by differentiated, which is to say, particular activity. It is, rather, an insistence that the local and particular exists in and through the global whole; that there needs to be, in any economy, an interdependence whose walls are not confinable to any township, or bioregion, and that, fundamentally, the issue is the relationship of the parts to the whole. (2008, 182-83)

The *Earth Constitution* overcomes the fragmented governing of Earth by nations in rivalry with one another and profit-making corporations all competing with one another. This chaotic global governing model, going back some three to four hundred years, is entirely inadequate to allow the planetary noosphere to actualize its potential to become a holarchy living in peace and prosperity in harmony with the biosphere. The *Earth Constitution* does not abolish nations, nor envision them as an anarchy of local sustainable communities (a common fallacy among environmentalists, as Kovel recognizes). Rather, it establishes a proper relationship between the parts and the whole. The parts fractally pattern the whole in a process of reciprocal harmony, synergy, and empowerment. And in all the parts shines the whole.

Similarly, the fragmented economic model of corporations competing for scarce resources and private profits for their investors is utterly inadequate to allow the noosphere to emerge in harmony with the biosphere. If profit is the bottom line for the survival of your business, the temptation always will be to externalize costs of doing business onto nature or society. Not only that, but the *design* of both these systems, political and economic, is precisely what has been destroying the biosphere. Everywhere, scientists are registering the consequences of these fragmented designs in the form of environmental destruction and climate collapse.

In Chapter Four we saw Donna Meadows and colleagues sketch out some actions necessary to have a sustainable world system: (i) extend the planning horizon of the world, (ii) improve the signals for monitoring the real impact of human activity everywhere on Earth, (iii) speed up response times to keep resources and fresh water at sustainable levels, (iv) prevent the erosion of renewable resources such as soils and forests and minimize the use of fossil fuels and minerals, (v) use all resources at maximum efficiency including repair, recycling and innovating, and (vi) stop the expansion of both population and physical capital (2008, 250-60). I would add to this list: (vii) a massive restoration effort directed to rebuilding soil fertility, (viii) cleaning up poisons in soil, water, and atmosphere, (ix) replanting forests, and (x) revitalizing ecosystems.

Yet, how can we possibly do all these things on an efficient, planetary scale without a democratic holarchical world system with the resources, authority, and global infrastructure to make it happen? Under the present system of sovereign nation-states, Brazil somehow "owns" the lungs of Earth and its national government has the legal right to cut down and "develop" the Amazon basin for private and national profit. U.S. President Trump has gutted the Clean Air Act of 1970 (amended 1990). He has opened up national lands to genetically modified organisms (GMO) planting and eased the regulations on oil companies regarding the production of plastic bags. Does the U.S. have the legal right to endanger the future of everyone on the planet by doing these things? Absolutely. Clearly, this institutional framework, including the U.N. acquiescence in such suicidal, sovereign "legal rights" is absurd.

Geography professor Mark Maslin concludes that "if implemented now, much of the cost and damage that could be caused by changing climate could be mitigated. However, this requires nations and regions to plan for the next fifty years, something that most societies are unable to do due to the very short-term perspective of political institutions" (2013, 140). So how do we effectively extend the planning horizon of the world, except by establishing a government for the world? Strange as it may sound, sovereign nations today have the "legal right" to ignore the future of humankind due to their internal political considerations.

Kate Raworth says rightly that economics is about *design*, but without a planetary infrastructure that gives us economics in the service of a holistic, common good, and sustainable design, there will be no planetary transformation. Only the Earth Federation government, using debt-free, global public currency, could employ the hundreds of millions or trillions in currency necessary to restore and salvage our planetary ecosystem. Only a global government could put everyone on the same page in planning for the future. We absolutely need a design that unites the entire planet within a sustainable model, a design premised on the interdependence of the parts with the whole of humanity.

Paul Hawken has edited a book entitled *Drawdown: The Most Comprehensive Plan Ever Proposed to Reverse Global Warming* (2017). The book explains all the latest possibilities for sustainable civilization in terms of energy, food, land use, buildings and cities, transport, materials, etc. Hawken's book shows what we could and should be doing on all these fronts. However, it recognizes no planetary institutions that might adopt these brilliant ideas and effectively promote them, and/or enact legislation regarding them, worldwide. This would be the precise role of the Earth Federation government.

Goener, Dyck, and Lageroos point out that the patterns of intelligence we see in nature involve fractal designs, root-like structures and connected filaments ascending from the smallest levels and branching into larger structures similarly networked, all moving upward together in a holarchy. They write:

> To laymen, fractals appear to be simple branching structures that repeat at all levels. Fractals are useful scientifically, however, because this repetition fits a specific mathematical pattern called a *power law* structure. In power-law structures, the number of components found at each successive level has a specific *ratio* across a *hierarchy of scales*, from the extremely small to the very large. In energy flow terms, the structures of each scale represent a particular *carrying capacity* (conduit size). Fractals are important in living systems because this arrangement of big and little conduits allows processing effectively throughout the whole. (2008, 199)

Therefore, the design of any global government must unite the whole of humanity in networks of national, regional, and local governance. It also would need to network the world's communications and commerce, from the smallest businesses and municipalities to global market feedback system. As these authors affirm, the structures of each scale must conform to the "power law" – where the roots and branches of the fractal pattern move upward in the holarchy, with the larger network-trunks processing proportionally more economic, political, and communicative flows. The health of all the parts becomes reciprocally interdependent with the health of the whole, a holarchy that governs and empowers the parts within a coordinated network.

All these features are designed into the *Earth Constitution*. Its holarchical structure utilizes the "power law" by extending the planning horizon, monitoring the whole, quickening response time, recycling and restoring the planet's precious resources, and humanely reducing population and growth. The Earth Federation under the *Earth Constitution* establishes a holarchy uniting humanity under a common system for biosphere monitoring, balancing, correcting, and harmonizing.

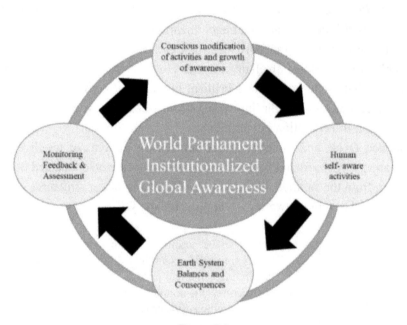

Figure 7.1

Under the Earth Constitution, institutions are designed for just these purposes. By uniting humanity under this system, human activities can be adjusted to the continuously monitored health of our planet and its future.

Ratification of the *Earth Constitution* will provide the holarchical structure needed to organize, guide, and implement the sustainable organization of human civilization in harmony with our planetary biosphere. My critique of climate and economic experts in Chapters Four and Five revealed a gaping omission from the prospective visions of these thinkers. First, nearly all of them failed to realize that our planet needs a holarchically organized democratic World Parliament if we wish to catalyze both the concentrated intelligence and authority needed to organize our planet for sustainability. Second, they ignored the domination and imperialism emanating from the anachronistic system of militarized sovereign nation-states, which recognize little or no authority above themselves, states that so far have refused to unite under a global democratic system.

We have seen that the *Earth Constitution* provides humanity with the "brain" and the effective means for dealing with climate crisis. It also ends war, militarism, national security madness, and interstate competition and rivalry – all of which must happen for sustainability. Without transforming our broken and anachronistic world system in the direction of a caring, coherent human community, there can be no effective dealing with climate change. It should now appear obvious that we cannot create a sustainability system for Earth without an Earth Federation government.

7.2   Vision Behind the *Earth Constitution* as a Whole

The *Earth Constitution* presents an integrated and holistic vision for all aspects of global governance. The nations are joined under the principle of unity in diversity, as are the people of Earth, who assume responsibility for the self-governance of our planet. Article 1 of the *Earth Constitution* states six broad functions:

Article 1: Functions of World Democracy	
1.1	To prevent war, secure disarmament, and resolve territorial and other disputes which endanger peace and human rights.
1.2	To protect universal human rights, including life, liberty, security, democracy, and equal opportunities in life.
1.3	To obtain for all people on earth the conditions required for equitable economic and social development and for diminishing social differences.
1.4	To regulate world trade, communications, transportation, currency, standards, use of world resources, and other global and international processes.
1.5	To protect the environment and the ecological fabric of life from all sources of damage, and to control technological innovations whose effects transcend national boundaries, for the purpose of keeping Earth a safe, healthy and happy home for humanity.
1.6	To devise and implement solutions to all problems which are beyond the capacity of national governments, or which are now or may become of global or international concern or consequence.

In response to the first function of ending war, one sometimes hears criticism of the stages of demilitarization outlined in Article 17 of the *Earth Constitution*. At the first operative stage when twenty-five nations join the Earth Federation, they are not required to disarm, only to turn over any weapons of mass destruction to the World Disarmament Agency. Also during this first stage, federated nations agree to abandon using any military weapons against one another. At the second operative stage when 50% of the nations comprising 50% of the world's population have joined the federation, federated nations are then required to disarm. People have asked if this is practical and whether disarmament might put the federation in a vulnerable position in relation to those nations that have not yet joined he federation?

To properly address this question, one must consider the entire vision behind the *Earth Constitution,* which is designed to ensure a positive future for humanity. Perhaps the most important question asked by people throughout history is *how do we break the chain of violence*? It is well recognized that violence breeds more violence. Therefore, some who have envisioned a world government claim that it needs a military to maintain order, but state militarism is nothing less than institutionalized violence – the very problem we need to solve.

Consequently, the drafters of the *Earth Constitution* sought a mature and practical way to break the chain of violence. But to end militarism means taking a risk ... and great courage, as world federalist proponent Mahatma Gandhi pointed out (see Martin 2017). The reason ending war is the very first goal in the first article of the *Earth Constitution* is that no other goals can proceed without planetary peace. Without this first broad function of the Earth Federation government, the other five broad functions will not succeed. It is that simple if we want a future for ourselves and our children.

Therefore, when reading the *Earth Constitution* for the first time – and it is included in its entirety in the Appendix – one should read it in this light. The entire system is designed to steadily reduce violence. In fact, it is the job of the World Police to reduce violence as well (Article 10.4.4). The entire framework is designed to first provide the essential right of peace, then to provide other human rights: education, healthcare, a healthy environment, freedom, housing, equitable rule of law, etc. Moreover, the design of the World Financial Administration makes these goals practicable, goals which are impossible under the current system of sovereign nation-states and private banking cartels.

Article 2 of the *Earth Constitution* states correctly that the people of Earth are sovereign – a realization dawning today on people all around the world. Nations are not legitimately sovereign. They are in fact illegitimate because they lack the capacity to protect the common good of their citizens, which is dependent on both world peace and a healthy planetary biosphere. Only a democratic world government can achieve these objectives. By the second operative stage, when 50% of the world has united, people everywhere on Earth will have realized these freedoms. Indeed, the remaining fragmented nations will regain legitimacy by joining the rest of humanity under the *Earth Constitution.*

Now consider Article 4, which gives the people of Earth broad specific powers to end war, protect the environment, and deal with our numerous global challenges. Article 4, taken as a whole, illustrates that the *Earth Constitution* gives us an integrated package. If we really want a decent world order, then the first four of the forty-one powers listed in Article 4 provide us with the power to reach this goal:

Article 4: Powers that Prevent War
4.1 Prevent wars and armed conflicts among the nations, regions, districts, parts and peoples of Earth.
4.2 Supervise disarmament and prevent re-armament; prohibit and eliminate the design, testing, manufacture, sale, purchase, use and possession of weapons of mass destruction; and prohibit or regulate all lethal weapons which the World Parliament may decide.
4.3 Prohibit incitement to war, and discrimination against or defamation of conscientious objectors.
4.4 Provide the means for peaceful and just solutions of disputes and conflicts among or between nations, peoples, and/or other components within the Federation of Earth.

By the time 50% of humanity has federated, people will surely understand: (i) their national governments can only be fully legitimate as part of the federation, (ii) there will be no success in protecting the environment or human rights without ending the war-system, and (iii) the *Earth Constitution* comes as an integrated whole designed to accomplish all these objectives. The people of Earth will want to disarm their nations and join the Federation.

During the initial stages of Earth Federation, when the first twenty-five nations join and ratify the *Earth Constitution*, the World Parliament already will be working with the rest of the world to plan for comprehensive, measured, and gradual disarmament. The World Disarmament Agency (Article 17.3.8 and 17.3.8) is precisely what the name implies. Therefore, by the second operative stage, the process of world disarmament will be well underway, and the focus will be how to safely and smoothly bring the rest of the nations into the Earth Federation.

The initial Earth Federation will not be another fragmented power block, like the European Union. Rather, it will be intrinsically open because it is intrinsically universal. Once the process of federation has begun, the world will rapidly unite around the *Earth Constitution's* holistic design for peace, justice, and sustainability. For the first time in human history, we will have a mature world civilization based on genuine human values and mature respect for human dignity and freedom. What Paulo Freire calls our "ontological vocation" to become more fully human will have significantly advanced (1974, 40-41).

Our planetary maturity rests on our development of reason, love, and a worldcentric perspective that transcends violence, fear, hatred, and greed. We have seen throughout this book that the moral and spiritual dimensions of human life reciprocally interpenetrate with the proper design of our institutions. We cannot elevate the collective consciousness without a global philosophy. The vision and functionality of the *Earth Constitution* applies to sound planetary governance as well as environmental resilience.

7.3   Drafting Process for the *Earth Constitution*

The *Constitution for the Federation of Earth* was written at a time when scientists initially raised concerns about our collapsing environmental stability and viability. Rachel Carson's book *Silent Spring* appeared in 1962, and it launched a worldwide movement in environmental studies. Four years earlier in 1958, the World Constitution and Parliament Association (WCPA) was founded by Philip and Margaret Isely and other concerned global citizens. Early WCPA literature exhibited a very clear awareness of the environmental crisis.

Philip Isely was a visionary philanthropist, and he set up the WCPA world headquarters in a two-story building in Denver, Colorado. He and Margaret hired employees to coordinate a global effort to bring world citizens together in the process of writing the *Earth Constitution.* This process included many international meetings of legal experts, judges, philosophers, and social science scholars from around the world. It was a sustained and well-coordinated effort. The high points of the process included four "Constituent Assemblies," which took place as follows:

At the First Constituent Assembly in Interlaken, Switzerland in 1968, the delegates chose a drafting committee of twenty-five persons, chaired by Dr. Reinhart Ruge from Mexico. Among the drafting committee were those who became the five primary authors: Philip Isely, Secretary-General of WCPA from the U.S.; Dr. Terence Amerasinghe, an international lawyer from Sri Lanka and Co-President of WCPA; S. M. Hussain, who soon became a Supreme Court Justice in Bangladesh; D. M. Spencer, a professor of law from Mumbai, India; and Dr. Max Habicht, a renowned international lawyer from Switzerland. At this meeting, the Assembly started the drafting process by detailing what should be included within the planetary constitution.

The first draft of the *Earth Constitution* emerged in 1972, and it was circulated worldwide several times to the participants. In 1977, at the Second Constituent Assembly in Innsbruck, Austria, the entire draft was examined and debated paragraph by paragraph. In 1979, the Third Constituent Assembly met in Colombo, Sri Lanka, and the assembly made an important declaration stating that the people of Earth have the right and duty to ratify the *Earth Constitution* regardless of whether this process is led by nation-states.

At the Fourth Assembly in Troia, Portugal in 1991, after a number of final changes, the *Earth Constitution* was declared finished and ready for ratification by the peoples and nations of Earth under the democratic procedures set forth in Article 17. In total, the drafting process took twenty-three years, hundreds of mailings, thousands of telephone calls, and many sub-meetings.

The drafters intended that the *Earth Constitution* replace the unworkable Charter of the United Nations. The *Constitution* explicitly states that all relevant agencies of the U.N. will be incorporated into the Earth Federation government. A serious effort was made to launch the *Earth Constitution* at the United Nations, with citations to the pending environmental collapse. For example, a 1994 publication of WCPA written by Philip Isely, called "A Bill of Particulars Why the UN Must Be Replaced," contained a list of "Environmental Destruction" action items:

> Despite much attention given to global environmental deterioration by the U.N. General Assembly, by the United Nations Environmental Program and by U.N. sponsored conferences, the U.N. has been unable to implement actions necessary to reverse major environmental damages and to sustain a good livable environment on Earth.
>
> Although it has been known for many years that the rain forests of Earth are needed to recycle 50% or more of the Earth's oxygen supply, and to store excess carbon dioxide, the U.N. has been unable to stop continued destruction of the rain forests at very rapid rates, and at current rates most of the rain forest will be gone within two generations. ...
>
> Although it has been known for many years that the burning of fossil fuels is raising the carbon dioxide level in the atmosphere so that resultant heat trapping will cause disastrous climatic changes, nothing has been done by the U.N. to stop oil and coal production and burning for fuel.
>
> Although the technical feasibility for safe, sustainable and plentiful energy supplies from solar and hydrogen sources has been known for many years, no intensive global "crash" program has been launched to develop such sources rapidly to replace oil and coal.
>
> Although the reduction of carbon dioxide emissions by 20% has been encouraged at various conventions, this will not stop the other

80% from continuing to cause a rise in CO_2 levels in the atmosphere, and the U.N. has no way to achieve even the 20% reduction. ....

To reverse the catastrophic climatic changes ... requires a very massive and globally coordinated program of many interrelated parts, which will cost many hundreds of billions of dollars per year for many years if human civilization on Earth is to be saved, but the U.N. is totally unprepared and unable to launch or administer such a program.

The most drastic result of climatic changes, following upon imbalances of carbon dioxide in the atmosphere and heat trapping, will be agricultural failures worldwide and consequent global starvation of a magnitude reaching into billions of people, but this problem is not even mentioned seriously at U.N. conferences or in the U.N. General Assembly or Security Council. ...

When taking up the issue of the ownership and development of the oceans and seabeds as the common heritage of humanity, the decision made at the U.N. sponsored "Law of the Seas" conferences was to give 200 miles offshore to each nation with a seacoast, which is the 200 miles containing the most accessible resources of the common heritage of humanity, and is also the areas needing the most protection by global intervention from pollution.

Although radioactive wastes and residues from the production of nuclear power ... are a deadly threat to human life for thousands and tens of thousands of years, the U.N. has done nothing to stop the production of nuclear power with the resulting accumulations of radioactive poisons, despite the additional fact that there are no safe disposal procedures known for the accumulating life-threatening nuclear wastes. ...

Since 1945, enormous quantities of other toxic wastes have been accumulating from a great many industrial processes, which are dumped in the oceans or shipped from the "advanced" industrial nations to "less advanced" countries, and the U.N. has no program for safe disposal or control over this global problem.

Dozens of other urgent and extreme environmental problems continue to proliferate and become worse ... and the U.N. is unable to do anything except make studies, collect documentation, establish commissions to study the problems, and hold conferences which cannot make any binding decisions to solve the problems. ... The foregoing is only a partial listing of the global environmental problems with which the U.N. has been unable to cope.

Clearly, this 1994 WCPA document was far ahead of its time. It represents the comprehensive and expert thinking behind the *Earth Constitution*, and it includes all the ideas raised by current climate scientists and economists. Another WCPA document from that period entitled "Who Speaks for Humanity?" also provided a list of interrelated world problems, including: the spread of nuclear and high-tech weapons; threat of war in many places and the breeding of dictatorships and revolts; massive climate changes that threaten global starvation, more hurricanes, and earthquakes; use of fossil fuels and the need for global transition to safe and sustainable energy supplies; agricultural soil erosion, pest control, and desertification; inadequate water supplies; ozone holes resulting from CFC gasses and oxygen depletion; paying for the maintenance of the global commons for healthy living; population growth exceeding the Earth's carrying capacity and the need for family planning; capital intensive high-tech production vs. labor-intensive production resulting in wealth inequality; control of multi-national corporations; protection of human rights and safeguarding democracy, cultural diversity and civil liberties; and providing the education necessary to live in an interdependent world and acknowledge social responsibility.

From this list it should be clear that the framers of the *Earth Constitution* understood we live in a single, inter-dependent world system that requires planetary holism and coordination on all these issues. The U.N. is simply not designed to address any of the environmental or other global problems listed above. It is a merely a "trade association" under the control of the powerful Security Council nations. It has no binding authority and it is therefore impotent when it comes to planetary issues.

> *Today, a new generation of activists manage the world-wide programs of WCPA and ECI, and they are getting the word out that the Earth Constitution is a simple and effective way of transcending the current worldwide chaos.*

Today, the World Constitution and Parliament Association (WCPA), in coordination with the Earth Constitution Institute (ECI), continues to promote the *Constitution for the Federation of Earth.* Together, WCPA and ECI organize Provisional World Parliaments that ratify and pass legislation under the auspices of the *Earth Constitution.* A new generation of activists now manage the worldwide programs of WCPA and ECI, getting the word out to the people of Earth that a simple and effective way of transcending the current world chaos exists in the form of the *Earth Constitution.*

The holistic paradigm that has emerged in the 20th and 21st centuries is embodied in the *Earth Constitution.* It is a federation model *designed* to address the entire nexus of global problems efficiently and effectively. The hundreds of world citizen thinkers, international lawyers, futurists, and cultural creatives who drafted and today promote the *Earth Constitution* understand that it is organized to holistically and synergistically address these problems. Now, let's delve into the document in more detail.

7.4 Design Features of the *Earth Constitution* for Establishing a Sustainable Future

The *Earth Constitution* is a coherent and comprehensive document organized to catalyze civilization into becoming a democratic community living in peace, justice, and sustainability. As such, its design features bearing on sustainability, ending war, creating justice, and promoting general human well-being are all intertwined. The following discussion of these design features also reveals the framers' intent to actualize civilization as a universal human family.

7.4.1 A Founded World Society

This "founded world system" aspect is an extremely important feature because the *Earth Constitution* recognizes that a sustainable world system requires a biospheric or planetary consciousness. The race to sustainability is pushing humanity to a new level of self-awareness. Calls to cognitive, spiritual, and moral growth are not new, but the difference now is that the very survival of humanity is at stake. The New Age movement calls this "conscious evolution."

The economic-political model of the Second Industrial Revolution era involved the development of giant hierarchical corporate systems of extraction, production, transportation, consumption, and waste disposal – linear structures created by people utterly self-confident in their success and practices (Rifkin 2013). Elites often saw this system as the apex of human civilization. Second rate thinkers, such as Francis Fukuyama (1992) saw this system as a triumph of civilization and "the end of history." It was similarly extolled as the epitome of freedom by economic ideologists such as Friedrich von Hayek, Joseph Schumpeter, Milton Friedman, and Ayn Rand. This entire system was powered by fossil fuels (primarily coal, oil, and gas), and it was facilitated by new technologies, such as the telephone, railroad, internal combustion engines, and later airline transport.

As we have seen, the corporate system interfaced with the system of nation-states and the United States' dream of empire that emerged on a global scale after World War II (Grandin 2007; Ferguson 2004). This world system presupposed by ideologists included the rivalry of militarized nations, fragmenting the world into a multiplicity of competing units. The resulting disorganization and hegemony of the allied and so-called "free world" was thought to be supreme. The public face of nationalism bespoke "democracy, freedom, and prosperity" for the world, while its top-secret agents spoke of controlling "the world's wealth" and "vital resources" through the use of "straight power concepts" (see Chomsky 1996).

These assumptions denied the reality of a world system broken down by two World Wars using industrial scale technology to wipe out some 100 million members of our human family. Thereafter, the war system continued via interminable smaller internecine conflicts around the planet. Thoughtful world citizens – often called "world federalists" – began raising their voices over the irrationality and destructive nature of international politics since the close of World War I, yet the dominant elites and the thoughtless masses refused to listen.

The world heard, for example, from Emery Reves (1945), Albert Camus (1946), Errol E. Harris (1966), and Albert Einstein (1968). Even before the climate crisis, these thinkers understood the utter madness of the fragmented world system and its suicidal trajectory. Reves observed, "A picture of the world pieced together like a mosaic from its various national components is a picture that never and under no circumstances can have any relation to reality, unless we deny that such a thing as reality exists" (1945, 22). Camus declared, "The most striking feature of the world we live in is that most of its inhabitants ... are cut off from the future" (1946, 27). Camus observed that a world predicated on murder, nation-state violence, and the background threat of nuclear holocaust obliterates and denies a credible, coherent future for humanity. In reality, he affirmed humanity is one and Earth is one, which means sovereign nation-states bear no relation to this truth.

Similarly, Harris stated, "The restriction of sovereignty to national limits undermines the conditions of human welfare and frustrates the ends of civilized living" (1966, 188). Civilized living means the flourishing of human beings in peace, justice, and sustainability, that is, a world governed by principles focused on the common good of all, not on nation-state power, international rivalry, or military dominance. And Einstein affirmed, "The human race will cease to exist unless a world government capable of enforcing world law is established by peaceful means. Only in this way can war be averted, and the peace and plenty we all desire for humanity can be made possible. The choice is indeed between one world or none" (1968, 421).

These four astute thinkers and many others understood the immense significance of civilization breaking down in two World Wars and the threat of planetary omnicide posed by nuclear weapons in the hands of militarized sovereign nation-states (cf. Glover 1999). More recently, since approximately the 1960s, millions more voices have been raised, including scientists who have been urgently warning that this Second Industrial Revolution model destructively conflicts with the ecosystem of our planet and must be rapidly changed through "third industrial revolution" technologies in conjunction with a new biospheric consciousness (Rifkin 2013).

Thus, critics of this self-destructive world political system have long been calling for a new "planetary consciousness" and world citizenship. In addition, critics of the self-destructive Second Industrial Revolution economic system have been calling for a new ecological economy "beyond growth" and predicated on planetary sustainability. All these voices have been urging humankind to move into a new and higher level of self-awareness.

Even by the 19th century, advanced thinkers such as Karl Marx had begun calling for a transformation of human consciousness away from ego and self-interest to a cooperative, integral awareness of our common "species-being." Concerning Marx's understanding of what was needed in human history, Roslyn Wallach Bologh affirms:

> Marx formulates history from within a form of life characterized by the possibility of self-conscious community. ... He reads history in terms of repressed community (capitalism) versus natural community (pre-capitalism) and self-conscious community (post-capitalism). ... This is how I interpret Marx's concept of socialism – a self-consciously social mode of (re)production, (comm)unity as a historical accomplishment not conceived as external to the members and their activity. (1979, pp. 237 & 239)

The self-centered model of being human *represses* our more fundamental commonality and interdependence, and it must be replaced by a conscious community model in which human beings begin living from their common species-being. This growth in self-awareness, Marx believed, would result in a cooperative economics characterized in his famous slogan "from each according to his abilities, to each according to his needs." Since that time, 20th and 21st century scholars have produced the stunning congruence of research regarding the stages of human growth in maturity that buttresses Marxist philosophy.

Earlier we saw that the stages of growth toward human maturity reflect a broad consensus from such diverse specialists as Erich Fromm, Lawrence Kohlberg, Carol Gilligan, Jürgen Habermas, Clare Graves, and Ken Wilber. The basic pattern of growth moves from a childlike egocentric orientation to a conventionally based ethnocentric orientation, to an autonomous and universalized worldcentric orientation, to a spiritually awakened and deeply aware cosmocentric orientation. One key to this progressive development is the increase in self-other awareness.

A worldcentric person has a planetary consciousness. A person entering the cosmocentric level is developing a biospheric consciousness. He or she is becoming aware of the vast intelligence manifested in the ecological networks of the biosphere and of the need to live in harmony with nature's holarchy. People at these highest levels are capable of more rapid growth because they now comprehend the nature of the growth process itself. They have entered what Beck and Cowan label as "Second Tier" consciousness (2006, Chaps. 15 & 16). Prior to this level of self-awareness, ideas of planetary or biospheric consciousness appear as mere empty phrases without meaningful existential content.

Yet human survival now depends on widespread rapid growth toward cosmocentric consciousness. The noosphere must become a holosphere, patterning itself on the cosmic holism manifest everywhere in the universe and in the evolutionary process. Ratifying the *Earth Constitution* becomes an absolutely necessary step in conscious human evolution. In short, humanity must unite within a democratic holarchy if we are to survive and flourish much longer on Earth.

The *intentional founding* of a unified global society via ratification of the *Earth Constitution* would represent the decision on the part of humanity to adopt a world system *designed* for sustainability. To try to *evolve* such a system by encouraging green capitalism or working to implement arms reduction agreements for weapons of mass destruction would take too long and ensure the suicide of the human project. The Covid-19 pandemic highlights our planetary vulnerability, lack of cohesion in the face of global crisis, and lack of holarchical unity. Our window of opportunity to secure a future for humankind diminishes daily.

If parties were to come to the table to discuss writing a new world constitution, it would be because the *Earth Constitution* is too honest, too well-designed, and too transformative for the power elite. Just imagine the slow and ponderous set of qualifications and compromises the imperial nations, powerful corporations, military-industrial complex, and super-wealthy elitists would attempt to insert (and probably successfully inject) into that document. The result would be a dark version of world federation, the globalization plan commonly referred to as the "New World Order."

The Provisional World Parliament has facilitated the process of ratifying the *Earth Constitution* by formulating rules for a Founding Ratification Convention. These rules guard against subversion of the *Earth Constitution* by any of the elitist forces now dominating our world system. We need nations to ratify the *Earth Constitution* as it is now – without factionalism, corrupt compromise, or watering-down its transformative and liberating features. The time for editing the *Earth Constitution* (if needed) is *after* the initiation of the Earth Federation government. There is a powerful amendment provision in the *Earth Constitution*, but now is not the time to tinker; now is the time for organized global action.

Of course, there have been other founded societies in the past. For example, the Constitution for the United States was drafted by committee and led to its founding in 1787. In 1947, India similarly was founded upon a new constitution, and a number of African countries during the 1960s were founded under like means. A "founded society" is to be distinguished from an evolved society by its decision to base the new political entity on an intentionally written and ratified constitution. Such a constitution often includes important principles of justice, freedom, etc.

> *The process of ratifying the Earth Constitution is organized and monitored by the Provisional World Parliament (PWP) in cooperation with nations that have approved preliminary ratification. The PWP already has passed procedural legislation for this process. No further constituent assemblies are necessary because PWP already has adopted rules for a Founding Ratification Convention.*

The attempt to found such an egalitarian society within today's world system is a doomed exercise, since the current global elite system dominates the form and practice of every national entity. Indeed, no nation can escape the grip of the current imperial and corporate powers. Nations that attempt to escape – like Cuba in 1959, Chile in 1970, Nicaragua in the 1980s, Yugoslavia in the 1990s, or Venezuela today – are severely punished, overthrown, bombed into oblivion, or economically crippled.

Consequently, the only way to found a new system based on principles of peace, justice, and sustainability is to do so at the world level. This act would transform the entire world system from fragmentation to intentional holism, from a war-system of nations recognizing no enforceable laws above themselves to a holarchical peace system predicated on the principle of unity in diversity. Passing the *Earth Constitution* would actualize the world-system design necessary for our living planet.

If we want planetary sustainability in time for the survival of humanity, then we must organize a world system for that objective. The best way to achieve this would be to ratify the already well-designed *Earth Constitution.* The people of Earth are today – for the first time in history – in a position to *found* a truly intentional world system based on peace, justice, and sustainability.

Moreover, if enough mature people ratify the *Earth Constitution* – perhaps only 10% of humanity may be required – the result of such unification would be the moral stimulation and rapid ethical and spiritual growth toward worldcentric maturity among the remaining 90% of humanity. Changing the structural design for the governance of our planet would achieve the transformative economic and political goals that follow from the organs of the *Earth Constitution*. And the feature of the *Earth Constitution* that supports sustainability is its intentional design to transform not only the consequences of human actions for our planetary biosystem but human consciousness as well.

7.4.2 A Holistic World System

The *Earth Constitution* is founded on the unity in diversity of the Earth's peoples and the principles that "war shall be outlawed and peace prevail" and "Earth's total resources shall be equitably used for human welfare." It therefore conceptualizes and embraces the holistic vision of a sustainable noosphere and links this with the highest planetary principles. In fact, the entire Preamble reads as a contrast between holism and fragmentation (Martin 2010a, Chap. 4).

Following World War II, major thinkers recognized that the design of the U.N. was fundamentally flawed and incapable of dealing with lethal world problems. Just prior to the war, a number of prominent intellectuals published a document called "The City of Man," which recognized the fundamental role of the principle of unity in diversity:

> Diversity in unity and unity in diversity will be the symbols of federal peace in universal democracy. Universal and total democracy is the principle of liberty and life which the dignity of man opposes to the principle of slavery and spiritual death represented by totalitarian autocracy. No other system can be proposed to the dignity of man, since democracy alone combines the fundamental characteristics of law, equality, and justice. (Agar, *et.al.* 1941, 27-28)

This holistic principle is clearly articulated in the Preamble to the *Earth Constitution*. Global democracy is not one political option among others; it is the principle of law, equality, and justice projected to its proper domain – "federal peace in universal democracy." Articles 21 and 28 of the U.N. Universal Declaration of Human Rights agree with these principles ... in part. Yet it is this relationship among all human beings that grounds the ecological consciousness necessary for sustainability. It also makes possible the synergistic coordination among humans necessary for real worldwide resilience. As R. Buckminster Fuller affirms:

> Today, all human existence depends on the swift, world-around intercommunication system operating at 186,000 miles per second. We have transformed reality from Newton's "at rest" norm to an Einstein's

186,000 miles-per-second norm. ... The once noble and essential but now obsolete nations belonged to the rooted socioeconomic land-capitalism era of humanity. In reality, humanity is now uprooted kinetically and occupying the whole planet. ... The world's economic accounting system, if properly entered into the world's computers, will quickly indicate that comprehensive economic success for all humanity is now realizable within a Design Science Decade. All it takes is shifting from weaponry to livingry production. (1983, 78 & 80)

Beginning with the Preamble and the "broad functions" for our collective human project articulated in Article 1 and empowered today in Article 19, the *Earth Constitution* actually mandates a new "Design Science Decade." Upon ratification – when this holistic design will give rise to transformed economic, political, communications, production, and transport systems – the *Earth Constitution* is designed to work on behalf of prosperity, justice, and sustainability for all humanity (see Article 17). The Preamble articulates the needed transformation from "weaponry to livingry" and the steps necessary to move into an era of "comprehensive economic success for all humanity."

7.4.3 Focus on Global Problems and Crises Beyond the Scope of Nation-States

Article 1 outlines six functions to end war, disarm the nations, protect universal human rights, diminish social differences and poverty, regulate global systems of trade and communications, protect the global environment for healthy living for both present and future generations, and address all the other problems beyond the scope of nation-states. The U.N. Charter, by contrast, does not mention any of these global problems except war. Regarding war, the U.N. reaffirms the 350-year-old system of militarized sovereign nations by declaring it will "maintain the sovereign integrity of its member nations." The U.N. Security Council (the victors in World War II), ostensibly keeps the peace through its power to make war against any nation deemed threatening. The militaristic Security Council also may veto any and all decisions and resolutions that come from the General Assembly. This bizarre and undemocratic design therefore cripples the U.N. from effectively dealing with global problems.

Article 4 of the *Earth Constitution* gives the Earth Federation government binding authority over all global problems that are beyond the capacity of nation-states to address, while leaving the nations authority over their internal affairs (Articles 14). Consequently, the *Earth Constitution* does not keep the peace through the threat of war, as does the U.N. system. Instead, it keeps the peace through an integrated set of non-military institutions *designed* to do precisely that.

Thus, the *Earth Constitution* keeps the peace by design, with agencies making judicial decisions, promoting civil resolution of disputes, and ensuring global justice, fairness, and equity. There have been some 150 wars around the world since the U.N. was founded, resulting in tens of millions of deaths, most of them civilian. The U.N. system maintains the global war-system of "might makes right." The *Earth Constitution* replaces this deadly and immoral system with the rule of effective democratic law. The same is true for all other global problems including climate change: The *Earth Constitution* regards our planet as an interconnected system.

7.4.4   Explicit Focus on Sustainability

By now, it is obvious that the climate crisis requires holistic transformation of our world system. The *Earth Constitution* specifically focuses on sustainability and ecological resilience in more than forty of its articles, including:

Articles Related to Ecology			
1.5	Primary Goal: Protect ecological fabric of life	8.6.1.4	Monitor the global environment
4.10	Address all ecological disruptions	8.6.1.5	Make ecology recommendations
4.12	Define planetary environmental standards	8.6.1.7	Enlist universities in ecology protection
4.18	Recycling of natural resources	13.9	Natural environment as human right
4.20	Soil conservation & pest control	13.10	Natural resources as human right
4.21	Limit population growth by non-coercive means	13.11	Pure air and water as human rights
4.22	Protect planetary water supplies	13.14	Eliminate environmental dangers
4.23	Conservation of our planet's oceans	13.15	Ecology protection as human right
4.24	Protect the atmosphere of Earth	17.3.10.8	Emergency Earth Rescue Admin.
4.27	Ecologically sound energy supplies	17.3.10.9	Create safe world energy system
4.28	Prevent damage from mining extraction	17.3.10.13	Oceans and seabed protection
4.29	Control over nuclear energy	17.3.12.1	Earth Rescue Administration
4.30	Protect natural resources of Earth	17.3.12.3	Global clean energy system
4.31	Assess technology for environ. impact	17.3.12.4	Global sustainable agriculture
4.32	Develop alternative technologies	17.3.12.6	Deal with problem of nuclear waste
4.33	Technology to safeguard environment	17.3.12.8	Clean air & water programs
4.36	World Service Corps for restoring Earth	17.3.12.9	Conservation & recycling
4.38	Develop world parks & wilderness areas	17.3.12.10	Non-coercive population control
7.3	World Executive environmental departs.	17.4.11.1	Oceans/seas as common heritage
8.4	Climate crisis is one world problem	17.4.11.2	Polar caps/arctic as human right
8.5	Climate crisis is one world problem	17.4.11.4	Food supply system for Earth
8.6	Agency for Environment & Technology Assessment	19.5.1.2	Provisional World Government agenda, including climate crisis

If we put together all these environmental features of the *Earth Constitution,* we see a federated government designed to establish a sustainable world system and restore our badly damaged environment. We find a comprehensive vision, addressed from multiple angles through various agencies, with rights, functions, and initiatives all working synergistically for planetary success.

7.4.5   Focus on Integrating the Diversity of Humanity

All governing bodies and departments of the Earth Federation system – including the House of the Peoples, House of Nations, World Executive, World Judiciary, and World Ombudsmus – all mandated to represent the diversity of humanity. Why? Because all these agencies and organs of the federated system will elect leaders from all continents on Earth, and the system is designed so that no single nation, group, or person can predominate or grab control. Think of it: *For the first time in human history, we will be one united people, one united planet.*

More specifically, each *Earth Constitution* agency is headed by a group of between five and ten members from diverse planetary regions. No agency or institution within the Earth Federation will be able to fragment along ethnocentric or religious lines. Up until now, our "legal communities" have been illusory, never representing the whole. But when the *Earth Constitution* is ratified, a new day will dawn.

The quality of our communities is very much related to our freedom and ability to flourish. But the one community that makes possible the true flourishing and fulfillment of our human project – our planetary community – has yet to be actualized.

History tells us that "positive freedom" comes when people flourish within their community. "Negative freedom," on the other hand, focuses on the struggle for individual autonomy, usually against the backdrop of an interfering and oppressive society. (See Martin, *Global Democracy and Human Self-Transcendence*). Social scientists have progressively understood the intrinsically social nature of human beings and that our freedom and flourishing as persons is not separable from society. Indeed, the quality of the community within which we are embedded is directly related to the quality of our life experience.

The one community capable of making possible the true flourishing and fulfillment of our human project – the planetary community – has yet to be actualized. Instead, the current world system fractures our common community on the basis of race, religion, culture, nationality, wealth, power, and many other factors. Many people still fail to recognize that our common humanity and our bond to the natural world supersede all such superficial divisions.

Fortunately, most people have awakened to the climate crisis and understand that unless we meet the challenge of global warming, we almost certainly will face our own extinction. As Jürgen Habermas concludes, "The welfare-state mass democracies in the Western world now face the end of a 200-year development process that began with the revolutionary birth of the modern nation-state. ... Today developments summarized under the term 'globalization' have put this entire constellation into question" (2001, 60).

Because the *Earth Constitution* is predicated on the unity in diversity of the human community, it can actualize, for the first time in history, the synergistic potential of true community, the kind of mutual freedom and empowerment most of us seek. Contemporary philosopher of law John Finnis writes:

> We must not take the pretensions of the modern state at face value. Its legal claims are founded, as I remarked, on its self-interpretation as a complete and self-sufficient community. But there are relationships between men which transcend the boundaries of all *poleis*, realms, or states. ... If it now appears that the good of individuals can only be fully secured and realized in the context of international community, we must conclude that the claim of the national state to be a complete community is unwarranted and the postulate of the national legal order, that it is supreme and comprehensive and an exclusive source of legal obligation, is increasingly what lawyers would call a 'legal fiction.' (1980, 149-150)

The sovereign nation-state is becoming a "legal fiction" because it is unable to maintain a successful community in which equitable law empowers the freedom and common good of all its citizens. Truly, "community" is becoming planetary, as philosopher Harris concludes:

> At the present time, the two fundamental principles justifying national sovereignty ... the rule of law and the pursuit of the common good of its subjects, have been undermined, so that national sovereignty, as such, has become obsolete. ... In the world today, the only form of democracy that could aspire to the ideals of the traditional philosophical conception would have to be global, one that could legislate to implement global measures to deal with global problems (as sovereign states cannot) and could maintain the Rule of Law world-wide (which the exercise of sovereign rights by independent nations prevents). (2008, 134-35)

By institutionalizing the global community in the design of all its institutions and agencies, the *Earth Constitution* actualizes universal human rights, true freedom, and worldwide peace. It also provides the legitimate authority necessary to address climate crisis, an authority no assembly of sovereign nations can possibility achieve. Why? Because sovereign states are now "put into question" (Habermas), seen as "legal fictions" (Finnis), and viewed as "obsolete" (Harris). The global community of unity in diversity actualizes both the legal form and the synergy needed to effectively address our climate crisis.

Furthermore, by establishing Earth as a unified democratic legal community, the *Earth Constitution* binds us together into a moral community as well. Philosopher of law Ronald Dworkin (1986) writes that sound constitutional law creates a framework of "respect and concern," while another philosopher of law, Lon Fuller (1969), adds that democratic constitutional law establishes a bilateral "moral relationship" between the governed and those in representative power. In sum, the world will blossom as a community when the well-being of each person comingles with the well-being of all.

7.4.6   Recognizes the People of Earth as Sovereign

Article 4 of the *Earth Constitution* places the atmosphere of Earth, the oceans and seabeds, and the other "essential resources" of our planet under the protection of the people of Earth. Hence, it is the first document that, by design, recognizes the global commons belong to all the people and must be preserved, restored, and protected in their name. In other words, it institutionalizes humanity as a holarchy under a federated Earth.

Recall that the U.N. treaty called the "Law of the Seas" grants seacoast nations two hundred miles offshore for the exploitation of natural resources for their own national benefit. As WCPA founder Philip Isely points out, these resources are the "common heritage of humanity" and are "areas needing the most protection." Yet the U.N. chose to protect national sovereignty and give away our global commons for plunder by countries and their private profit-making corporations.

The U.N. Sustainable Development Goals – which are supposed to guide us toward a sustainable future by 2030 – follow this same archaic design. Recall that the U.N.'s "2030 Agenda for Sustainable Development" states: "We reaffirm that every State has, and shall freely exercise, full permanent sovereignty over all its wealth, natural resources and economic activity" (Introduction item 18). I previously noted the disasters that follow from this bizarre assumption. The very design of the U.N. Charter threatens the future of our planet.

Conversely, the *Earth Constitution* is structured to provide democratic and cooperative governance and elicit the conscious participation from people everywhere in taking responsibility for the common good of our planet and its future. It can do this because it recognizes that our global commons belong to the people of Earth, not to sovereign nation-states. It therefore institutionalizes a custodial function in Articles 2, 4, and elsewhere. Take, for example, these four items from Article 4, which grant to the people of Earth the right to:

Article 4: Rights of the People	
4.23	Own, administer, and supervise the development and conservation of the oceans and sea-beds of Earth and all resources thereof, and protect them from damage.
4.24	Protect, control, and supervise the uses of the atmosphere of Earth.
4.27	Develop, operate, and coordinate transnational power systems, or networks of small units, integrating into these systems the power derived from the sun, wind, water, tides, heat differentials, magnetic forces, and any other source of safe, ecologically sound, and continuing energy supply.
4.30	Control essential natural resources which may be limited or unevenly distributed about Earth, and implement ways to reduce wastes and minimize disparities when development or production is insufficient to supply everybody with all that may be needed.

These and other provisions of the *Earth Constitution* place the global commons under the sovereign authority of the people of Earth, thereby making it possible to establish an environmentally sustainable planetary civilization. It is time we collectively assume responsibility for our planet by creating a global public authority representing the people of Earth. To fail in this responsibility would be madness when the *Earth Constitution* possesses a simple and effective solution for the climate crisis and acknowledges the sovereignty of all people to govern our planet.

7.4.7 Democratic, Cooperative, and Collaborative Structure

The *Earth Constitution* establishes one thousand World Electoral Districts energized by People's Assemblies to elicit global grassroots participation. If we want the people of Earth to assume a transformed planetary consciousness, then what better way could there possibly be than to empower them to participate in governing our planet?

In his 2019 book *The Green New Deal,* Jeremy Rifkin recognizes that government at nation-state levels has created "figurative families" and "larger collectivities" that generate common human empathy. Nation-states, he says, have been critical in helping us move toward planetary biospheric consciousness. Yet, when it comes to "thinking like a species" and establishing a planetary empathic civilization with a biosphere consciousness, he implies that the technical infrastructure of the Third Industrial Revolution alone (along with our present system of sovereign nation-states) is sufficient to do this.

He does not suggest the obvious: If national governments created figurative, empathic national families, then democratic world government would clearly expand this transformation of consciousness to the species level he deems necessary for a global green new deal. If we design a truly holistic world system, then "thinking like a species" will automatically follow. On the other hand, if we fail to unite the world, then it will be nearly impossible to overcome nationalistic identities within the shrinking timeframe we are facing.

Again, the *Earth Constitution* is constructed to elicit the participation of the people of Earth in global democracy. In doing so, it fosters the conversion of the fragmented planetary noosphere to an organized and eco-centric holosphere. In addition, the Provisional World Parliament has passed World Legislative Act number 26, called "The Education Act." Under this act, all schools connected with the Earth Federation will need to adapt their curricula to include the study of global issues, such as planetary quality of life, requirements for world peace, the principle of unity in diversity, and what constitutes good government. All these areas of study foster planetary consciousness and therefore promote a world in which human activities will sustainably conform to the resources, ecosystems and limitations of our planetary ecology.

7.4.8   Establishes Planetary Public Banking

Banking should empower local communities with low-cost or free loans and grants to develop sustainable infrastructure and cooperative institutions. In the *Earth Constitution*, public banks will establish loans and grants requiring no collateral or previous assets, so access by poor countries and persons is open and encouraged. The Earth Federation government will assume the international debts of developing countries and dispose of them in ways to prevent hardship for all parties. The significance of global public banking cannot be over-estimated. In *Web of Debt,* attorney Ellen H. Brown writes:

It is here that we find the real cause of global scarcity: somebody is paying interest on most of the money in the world all of the time. A dollar accruing interest at 5 percent, compounded annually, becomes two dollars in about 14 years. At that rate, banks siphon off as much in interest every 14 years as there was money in the entire world 14 years earlier. That explains why M3 has increased by 100 percent or more every 14 years since the Federal Reserve first started tracking it in 1959. According to the Fed chart titled "M3 Money Stock" [M3 is the total money supply as calculated by the US Federal Reserve], M3 was about $300 billion in 1959. In 1973, 14 years later, it had grown to $900 billion. In 1987, 14 years after that, it was $3500 billion; and in 2001, 14 years after that, it was $7200 billion. To meet the huge interest burden required to service all this money-built-on-debt, the money supply must continually expand; and for that to happen, borrowers must continually go deeper into debt, merchants must continually raise their prices, and the odd men out in the banker's game of musical chairs must continue to lose their property to the banks. Wars, competition and strife are the inevitable results of this scarcity-driven system. (2007, 342)

Private banking cartels, including the U.S. Federal Reserve, have been fleecing the people of the world, since most of the global monetary supply is created as debt everywhere on Earth (Brown 2007; Zarlinga 2002). Brown writes: "The obvious solution is to eliminate the parasitic banking scheme that is feeding on the world's prosperity" (ibid.). That is the function of global public banking. It takes control of the Earth's debt-based monetary system on behalf of the common good. Article 8 of the *Earth Constitution* establishes the Planetary Banking System, which is empowered to rectify the many evils of the current debt-based system.

7.4.9   Establishes Debt-Free Money Creation

Article 8 of the *Earth Constitution* also creates the World Financial Administration, which is designed to fundamentally transform the planet's monetary system. The current system grants huge private banks immense legal rights that result in economic inequality and undemocratic power everywhere on Earth. Debt-free money creation will replace the old "scarcity system" of hierarchical power and inequality of wealth. The new economic design features are founded on sustainable prosperity and reasonable equity for all peoples. The immense productive power of the people of Earth will no longer be siphoned off for the enrichment of private banking cartels.

Government has always had the authority to create money, even while the propaganda machine of the bankers has convinced them to cede this right to the private banking system (Brown, 343). There may be graduated income taxes to help restore social equity and move world citizens out of poverty. The money saved by ending wars will be used for rescuing the climate of Earth from disaster. For virtually the first time in history, the democratic representatives of the people of Earth will have control over money creation on behalf of truly sustainable development and the common good. Moreover, the Earth Federation government will spend money on a global civil service administration employing millions of people to replant forests, restore the soils, clean up the waters, and rejuvenate the biosphere.

Mainstream environmental thinkers such as Sally Groener, *et. al.* (2008) and Jeremy Rifkin (2019) appear to think that our only credible hope is the "greening" of capitalism. They describe green investment initiatives and speculate that these efforts could catch-on in ways that transform Earth to a sustainable planet. While they each keep government in the background to regulate and encourage this process, they assume "government" means nation-states. They also assume the primary role of government is mere regulation of the "free market." Therefore, they don't appear to understand the actual design and purposes of capitalism. Unlike Ellen Brown above, who totally grasps the situation, most mainstream economists still subscribe to the current corrupt banking system. (Private banking is still legal under the *Earth Constitution,* but will likely wither away.)

The global economic system must include market socialism by adopting "fair trade" in place of so-called "free trade," and by transforming private profit loans into non-exploitative development loans or grants for the sustainable common good. Harris explains, "The conception of profit must be transformed: It must be socialized rather than individualized" (2000, 107). When banking and money creation belong to the people of Earth and is used for the common good, then the power of planetary economics will solve the environmental crisis, as well. Proper system design makes this happen. No longer will the future of civilization depend on the private whims of investment capitalists or banking cartels.

7.4.10   A Networked System to Monitor Planetary Health

Within the Integrative Complex of the *Earth Constitution,* several agencies monitor the health of the planet and coordinate local communities with the knowledge, technology, and support necessary for democratically distributed prosperity and resilience. The local and the global thus become networked in a cooperative and collaborative union that promises to develop immense synergy for transformation and sustainability. Here are some examples of the design features which network the planet:

Design Feature: Networking People and Planet
Sessions of the World Parliament rotate between 5 world capitals on 5 different continents.
House of Peoples contains 1000 elected members from 1000 electoral districts around the world determined by population.
House of Counselors is comprised of 200 members with 10 members from each of the 20 different world regions.
House of Nations includes 1, 2, or 3 representatives from each country depending on its population.
The World Judiciary contains 8 different benches for addressing different categories of world law.
World Executive is run by a Presidium of 5 persons, one from each continent.
The enforcement system similarly is directed by a team of 5 World Attorneys General and 20 regional World Attorneys from each of the 20 world regions.
There will be 20 World Police Captains serving civil harmony within these 20 regions.
World Ombudsmus will be dedicated to protecting global human rights and serving as a watchdog on the world government, and the office will be directed by a team of 5 World Ombudsmen, one from each continent, and shall be assisted by 20 regional World Advocates in 20 districts around the world.
The 7 agencies of the Integrative Complex shall each be directed by a commission of 10 to 12 members who will be selected globally to reflect unity in diversity.

The Earth Federation government will not be headquartered in a distant continent or permanent host nation. The government will be dispersed and truly worldwide, with regional offices supporting the people of Earth in various capacities. As additional safety measures:

Design Feature:
Assessment and Security Systems

World Civil Service Authority will see that highly qualified people staff all government positions.

World Boundaries and Elections Administration to ensure fair and free elections take place worldwide.

Institute on Governmental Procedures and World Problems will educate bureaucratic officials and give courses in how to address world problems with the latest information, technology, organization, and cooperation.

Agency for Research and Planning will serve the World Parliament with the latest information and keep parliamentarians informed with data on all global and environmental issues. Its research and reports will make the Earth Federation government directly responsive to the health of the biosphere and what is happening on our planet.

Agency for Technological and Environmental Assessment will assess technology for environmental impact and maintain a global network to monitor the health of the planet. It also will collaborate with public and private universities in this effort, keeping the government informed and responsive to variations in the planet's health.

World Financial Administration will protect and monitor the economic health of the planet using global public banking and debt-free money creation. It also will operate a planetary insurance system to protect people everywhere from calamities and disasters.

Commission for Legislative Review will assess the effects of all legislation enacted by the World Parliament and will integrate and harmonize the existing body of international laws with the new legislation to make sure all legislation is beneficial to the health of the biosphere.

Together, all these features of the *Earth Constitution* provide a networked, holarchical world system responsive to the needs of human beings and the environment that supports us. This ingenious design for our living planet is summarized in the chart below:

Figure 7.2
EARTH CONSTITUTION OVERVIEW

World Parliament
Three Chambers with approximately 1500 Representatives

House of Peoples	House of Counselors	House of Nations
1000 reps.	200 reps.	About 300 reps.

World Integrative Complex

World Executive	World Judiciary	World Police and Attorneys	World Ombudsmus
no veto power	8 benches	no military	watchdog on govt.

World Financial Administration
Global Public Banking for Interest-free Development Grants or Loans
Debt-free money creation for worldwide sustainable development

Worldwide Offices of People's Representatives in the World Parliament
Global People's Assemblies generate world citizen input to
their representatives in the World Parliament (WLA 29)

Worldwide Grassroots Development Programs
Funded by Debt-Free Earth Federation Currency

1. Employment of millions in worldwide grassroots ecological restoration and protection, such as planting trees and restoring clean water. Conversion of all military personnel to the global civilian service corps.
2. Empowerment of women and families in rural villages and city slums (including guaranteed monthly income with agreement to limit number of children to two).
3. Housing, income, education, and health care provided for all world citizens, including those now homeless and destitute.
4. Support for local initiatives, coops, and other democratically run projects that promote the common good and restore unity, synergy, and vision.
5. Education of local communities to become self-sustaining with clean energy and technology, organic agriculture, and free web-based information.

7.4.11 Bills of Rights with Environmental Sustainability Rights

Article 12 of the *Earth Constitution* provides an impressive list of freedoms and rights that come into effect with the First Operative Stage in the ratification process. These rights are extended impartially to all world citizens, with no discrimination on the basis of race, color, caste, nationality, sex, religion, political affiliation, or social status (Article 12.1). Freedoms are likewise extended to all and include: freedom of thought, speech, press, communication, publication, and broadcasting (Article 12.3). The only exception is speech or conduct that physically harms others or incites violence.

Article 13 directs the Earth Federation government to "secure certain other rights for all inhabitants of the Federation of Earth." These rights include: free and adequate public education (Article 13.4), free public health care (Article 13.5), protection of the natural environment which is the common heritage of humanity (Article 13.9), conservation of Earth's natural resources (Article 13.10), adequate housing, nutritious food supplies, safe and adequate water supplies (Article 13.12), and social security for everyone (Article 13.13).

The World Ombudsmus is the agency charged with protecting and promoting these rights and freedoms. The mandate of the World Ombudsmus includes: protecting the people of Earth against violations or neglect of universal human rights (Article 11.1.1), and reviewing the functioning of the agencies of the World Government to ascertain whether they are adequately fulfilling their purposes and serving the welfare of humanity (Article 11.1.10).

One can see that the *Earth Constitution* was elegantly drafted and contains a deep recognition of what constitutes "healthy living for all." It also directs the World Ombudsmus to protect and promote ecology-related rights, a unique design feature with great implications for climate protection and restoration. Several coordinated central agencies in the Integrative Complex of the Earth Federation government, in addition to the World Ombudsmus, would therefore be working to protect the "natural environment which is the common heritage of humanity." (See the action agenda of the initial federation, Art. 17.3.12, p. 288.)

7.4.12   Design to Progressively Minimize Levels of Conflict

The *Earth Constitution* repudiates the use of violence everywhere on Earth, including from the World Police. Not only does it disarm the nations (at its second operative stage), but the World Police and World Ombudsmus are explicitly tasked to reduce violence in all law enforcement compliance practices, and to implement non-coercive methods of conflict resolution.

For example: enforcement actions may not violate the civil and human rights guaranteed under the constitution (Article 10.1.3); enforcement of world law applies directly to individuals (Article 10.1.1); agents of the enforcement system will be equipped only with appropriate weapons to aid in the apprehension of dangerous individuals (Articles 10.1.5 and 10.3.5); and enforcement will only serve peaceful human needs and no weapons of mass destruction may be used (Article 10.1.6). A basic condition for preventing outbreaks of violence is the

Earth Constitution's guarantee of a fair hearing for any person or group having a grievance, with due regard for the rights and welfare of all concerned (Article 10.4.4).

Recall that Article 1 set forth the first broad function of the Earth Federation as ending war and disarming nations. The *Earth Constitution* goes on to create a non-military, civilian-based legal framework designed to reduce violence, perceived injustices, group prejudices, and ignorance. Hence, the Earth Federation government is premised on genuine equality, dignity, and freedom of all persons. This elegantly founded framework will produce the results for which it was designed. Humanity, by reducing violence to an absolute minimum, will be moved to a higher level of civilization.

7.4.13 A Living Document Mandating Periodic Reexamination and Modification

In order to achieve its grand objectives, the *Earth Constitution* required a constitutional framework that allows for learning, growth, and conscious evolution. In addition, world citizens must be able to modify it to ensure adaptation to a changing global environment, updated science and technology, and improved processes and best practices. John Finnis and others have pointed out that the purpose of legitimate government is the common good and "flourishing" of the citizens (1980, 221). The purpose of all legitimate government is the empowering of citizens to live meaningful and fulfilling lives, lives they deem to be good and well-lived.

To maximize this function of legitimate government, any constitutional authority must ensure a stable, legally consistent framework, permeated by the "respect and concern" for the well-being of all citizens equally (Dworkin 1986, Chap. 6). To do this in the best way possible, a constitution must be open to further development by adapting to changing environmental conditions, social forces, technological developments, and economic realities. This capacity for change within a stable framework was built into the *Earth Constitution* by careful design. It is a living document dedicated to the common good of the whole of humanity, whose fate is linked to the health of the planetary biosphere and the survival of the other forms of life on this planet.

Article 18 places such learning processes into the open structure of the Earth Federation government. It is relatively easy to make amendments to the *Earth Constitution*, and a complete constitutional review must take place within ten years after the first operative stage has been implemented and every twenty years thereafter. Thus, the document can be altered whenever necessary to achieve its mission of a peaceful, prosperous, just, free, and sustainable human civilization. It was designed to be a living, evolving document for a living, evolving planet.

7.4.14 The "Council of Elders"

The World Parliament includes among its three houses the House of Counselors (an affirmative nod to the indigenous concept of a "Council of Elders"). Many indigenous peoples still live under a system of elder governance, acknowledging that wisdom comes with age and experience. For this reason, the parliamentary system under the *Earth Constitution* blends the broad diversity of Earth's population in the House of Peoples with the wise guidance of experts and scholars concerned with the collective common good in the House of Counselors. Pursuant to its holistic vision, the *Earth Constitution* also includes the existing nation-states within its House of Nations. The three Houses – Peoples, Nations, and Counselors – offer an elegant prescription for planetary success.

Together, these Houses reflect a comprehensive governance model and a 21st century democratic vision. We have seen numerous philosophers point out that the *raison d'etre* of legitimate government is the common good and the flourishing of all citizens equally before the law. The composition of the World Parliament is intended to unite the diverse components of Earth in the best way possible to ensure that common good. Thus, it includes the direct, grassroots participation of the people of Earth, representatives from all the nations, with the extra element of wise and highly educated people in the Parliament. With any luck, His Holiness the Dalai Lama will be in the House of Counselors!

Figure 7.3
Presentation of the *Earth Constitution* to His Holiness the Dalai Lama by Swami Agnivesh, WCPA Distinguished Advisor and Vice-President

7.4.15 Simple Ratification Process

The *Earth Constitution* can be easily launched from its current "provisional world government stage" to its "first operative stage," and then be on its way to the second and final operative stages. Articles 17 and 19 lay out concrete actions and operating procedures to make ratification and implementation both practical and possible. The first operative stage of the Earth Federation government requires ratification by a mere twenty-five nations or the equivalent from within World Electoral Districts.

What will happen after the first operative stage is reached? If twenty-five nations affirm the *Earth Constitution*, how will the other nations of the world react? One or two of the big imperialist nations may strenuously object, assuming as they do that their quest for empire should supersede the common good of humanity and perhaps even the need for humanity to survive climate collapse.

However, the majority of nations will likely see that the founding nations have discovered a way out of the "New World Order" nightmare created by the more powerful and militarized sovereign nation-states. They also will discern that nations in the Earth Federation have more prosperity, security, and sustainability through public banking and cooperative economics. They will witness greater freedom and flourishing among the federated populations. They will admire the more sustainable, green societies being built. And rapidly, they will want to be part of this planetary success story.

The fragmentation of today's anarchic economic system and militarized nation-states continues to block human flourishing. But once the initial phase of Earth Federation is up and running, the rest of humanity will witness what can happen through constitutionally mandated political and economic cooperation premised on peace, equity, and sustainability. After a tipping point, nations will flock to become part of the emerging Earth Federation. Soon, everyone will understand that ratification of the *Earth Constitution* is the best path for addressing the climate crisis and the many other global challenges we face.

7.4.16 Earth Federation Already Has Begun!

In 1991, the 4th Constituent Assembly determined that the *Earth Constitution* was a completed document. Since then, the World Constitution and Parliament Association (WCPA) has been encouraging humanity to assume the right and duty to begin world government. Article 17 outlines the criteria for ratification, but there is no need to wait for full ratification because the *Earth Constitution* provides another path for implementation. Therefore, our vision of unity in diversity and a viable planetary future can manifest itself *here and now*.

If just a few thousand people with vision, integrity, and enthusiasm join the grassroots movement to activate the Earth Constitution, humanity will reach a higher level of consciousness and begin to actually build the world we seek – one based on global peace, justice and sustainability.

Article 19 provides for a "Provisional World Parliament" while ratification is pending. Thus, the *Earth Constitution* lays out concrete steps to actually begin world government. Herein lies the key to *immediate transformative praxis*, and the process already has begun! For the past five decades, the Provisional World Parliament has been organizing and meeting. Though still a small movement, it could grow quickly if enough people awaken and seize the immense opportunity to activate the *Earth Constitution*. In short, Article 19 provides humanity with *a tool to change the world*. If just a few thousand people with vision and integrity seize hold of this tool, humanity will soon reach a higher level of existence in which global peace, justice, and sustainability are normalized and protected mainstream values.

As the Preamble to the *Earth Constitution* declares, the justification for granting this extraordinary opportunity to initiate the Provisional World Parliament is that the world faces multiple existential threats, including nuclear holocaust and climate collapse. What are we waiting for? Since the *Earth Constitution* is a rational plan for the next phase of our common human journey, and because it embodies the moral authority of our collective wisdom, it is both more legitimate and more authoritative than the antiquated model perpetuated by militarized sovereign nations and corrupt conglomerates.

At the Second Constituent Assembly in 1979, the Parliament declared: *The people of Earth have the right and responsibility to create world government since every one of the world's nation-states, without exception, had abdicated their duty in this regard.* Since then, WCPA has sent the *Earth Constitution* to the heads of all nations and their ambassadors at the U.N. Over the years, we have gained the support of some so-called third-world nations, but to date no nation has seized the opportunity to ratify the *Earth Constitution* for fear of retaliation by the imperial powers, including the fear of invasion or overthrow. However, NGOs and individuals can quickly empower global transformative change by joining our movement and continuing to activate the Provisional World Parliament under Article 19. We the people can and must lead the way.

As we have seen, the only credible solution for our multiple crises is democratic world law uniting humanity. Our choice is one world or no world at all, a phrase that Albert Einstein used when he called for democratic world government. Harris also used this phrase as the title for his 1993 book about why we need to ratify the *Earth Constitution*. Article 19 gives us the right and duty to start the Earth Federation government now, and we are calling on all activists and people of good conscience to assist in this sacred mission.

Article 19 also outlines the emergence of provisional world government through the formation of "preparatory commissions" that address the multiple initiatives necessary, including a Ratification Commission, World Elections Commission, World Problems Commission, Finance Commission, etc. These commissions are organized and promoted through the Provisional World Parliament (PWP), the primary organ within the emerging Earth Federation government.

The PWP will elect its own Presidium and be responsible for the formation of the Provisional World Executive. It then is authorized to begin programs dealing with the most urgent world problems (Article 19.5.1). Article 19 goes on to state, "Insofar as considered appropriate and feasible, the Provisional World Parliament and Provisional World Executive may undertake some of the actions specified under Article 17.3.12, for the first operative stage of World Government" (Article 19.5.4). With regard to climate change, Article 17.3.12.1 provides for the organization of an Emergency Earth Rescue Administration, which is tasked with all aspects of climate change and the climate crisis.

Figure 7.4
1996 Meeting of WCPA Executive Council members in Denver, Colorado
(2nd from left) **Dr. Terence Amerasinghe, Co-President WCPA;**
(5th from left) **Prof. Philip Isely, Secretary General;** (7th from left) **Dr. Reinhart Ruge, then Co-President;** (8th from left) **Yogi Shanti Swaroop, Spiritual Advisor and Executive Cabinet member;** (far right) **Sarwar Alam, Youth Coordinator;** (2nd from right) **author, Dr. Glen Martin**

To date, WCPA and the Earth Constitution Institute (ECI) have achieved only limited success due to lack of funding and the people power needed to propel this movement into the mainstream. Despite these challenges, we already have initiated several features of the Provisional World Parliament, and all prior parliaments have been conducted to the highest professionalized standards of legislative expertise. This has largely been due to the hard work of Dr. Eugenia Almand, who serves as Secretary of the Parliament and Chair of the Legislative Review Commission. In total, we have held fourteen sessions of the Provisional World Parliament in cities and countries around the world, and we have passed over sixty World Legislative Acts that have been implemented to the extent possible given our organizational and financial resources. The 15th Session of the Parliament is scheduled to convene within the coming year.

A brief history of the Provisional World Parliament is in order. The 1st Session of the PWP was held in 1982 at the famous Royal Pavilion in Brighton, England. It included delegates from twenty-five nations and six continents. It was presided over by Sir Chaudry Mohammed Zafrullah Kahn of Pakistan, former President of the U.N. General Assembly and former Foreign Minister of Pakistan. The 2nd PWP opened at the renowned Constitution Club in New Delhi, India, by then President of India, Zail Singh. The 3rd PWP met for ten days at the Fontainebleau Hotel in Miami Beach, Florida; the 4th PWP was held in Barcelona, Spain; the 5th PWP took place on the island of Malta; and the 6th PWP was in Bangkok, Thailand. The most recent 14th Session of the PWP was held in 2014 at the Ramakrishna Mission Institute of Culture in Kolkata, India. Altogether, these fourteen sessions have passed a substantial body of provisional world law that illuminates the mature progress we can make under the authority of the *Earth Constitution* (www.EarthConstitution.world/world-legislative-acts/).

Figure 7.5
Second Session of the Provisional World Parliament opened at the
Constitution Club in New Delhi (1985), chaired by then President of India,
His Excellency Giani Zail Singh (center with turban)

These sessions constitute the World Parliament in action, in rudimentary form and without the benefit of significant resources, but provisional world government nevertheless. The next World Parliament is being planned for December 2021 or 2022 in India, and the implications couldn't be greater. World citizenship on planet Earth has started, but time is of the essence. We need your help to scale this movement for the betterment of humanity.

By becoming a signatory to the *Earth Constitution* and attending a session of the World Parliament as a delegate (observers are welcome too), you become a provisional government official. We invite you to do so and please consider volunteering for one of the PWP ministries or standing commissions (i.e., World Judiciary, World Attorney's General, World Police, World Ombudsmus, etc.). We need people of integrity and stature to head these burgeoning agencies and make it clear that world law under the *Earth Constitution* is actually emerging (cf., Almand and Martin, eds., 2009).

For example, the provisional world judiciary was started at the 12th Session of the PWP in 2010 with the passage of the Collegium of World Judges Act (WLA 48). Activating the provisional Collegium of World Judges was a great accomplishment, since this body is working to develop the World Court system under the authority of the *Earth Constitution.* Respected high court justices from various countries, both sitting and retired, have signed up as members of the Collegium. This initiative is headed by Peruvian High Court Justice David Quispe Salsavilca. Next, the first Bench of the World Supreme Court can then be created. Please remember, at this moment the world has no binding court system due to the refusal of sovereign nation-states to submit to such jurisdiction. This is true of both the International Court of Justice (ICJ) and International Criminal Court (ICC) in The Hague.

The text of the *Earth Constitution* does not mention its sponsors, though WCPA and ECI are the two Earth Federation NGOs spearheading the movement. In addition, Article 19 states that all past PWP delegates remain official delegates for future sessions. For example, from the 14th Session of the PWP, Gandhian activist E.P. Menon and leading Malaysian advocate Puan Saraswathy Devi remain recognized as governmental law makers. Also, original signatories to the *Earth Constitution*, such as Dr. Roger Kotila and Dr. Dauji Gupta (both Vice-Presidents of WCPA) remain recognized lawmakers. The elected President of the last PWP session is myself, and the Executive Secretary is Dr. Eugenia Almand.

To summarize this section, the key principle is that the provisional world government already is operating, and Article 19 of the *Earth Constitution* authorizes world citizens to actualize the global institutions and actions needed to address global problems. In addition, we wish to expand local WCPA chapters around the world. The first step in the process of becoming a representative of the PWP is to sign the Pledge of Allegiance to the *Earth Constitution* contained in the Appendix. For more information on how to get involved, visit the ECI website: www.EarthConstitution.world.

We are not merely visionaries but also sacred activists, helping people around the world awaken to the significance of our movement and the collective power provided by the *Earth Constitution*. Together, with both pride and courage, we must rapidly move forward to evolve and scale the Provisional World Parliament and ratify the *Earth Constitution*. The future of our planet hangs in the balance; it depends on what we do today.

7.4.17   Current Projects under Authority of Article 19

The emerging Earth Federation movement is wide open, and we invite all readers to join us. Currently, WCPA and ECI are engaged in the following projects, and we hope you will contribute time and/or treasure to one or more of them:

1.   Provisional World Parliament (PWP): World citizens who are signatories of the *Earth Constitution* have come together fourteen times and are currently developing a 15th Session of the PWP as we have seen. During PWP sessions, world citizens come together to jump-start the process of self-governance for the people of Earth. Delegates typically represent groups that support the *Earth Constitution* or nations concerned with the future of our planet. The announcement for the next session of the Provisional World Parliament can be found at the ECI website (www.EarthConstitution.world).

2.   Collegium of World Judges: As mentioned above, the PWP already has activated a Collegium of highly qualified judges who are prepared to staff the benches of the World Court System. The Collegium is open to qualified judges everywhere, especially retired judges who no longer have responsibilities directly related to their home country. This project was activated in 2010 at the 12th Session of the PWP in Kolkata, India (WLA 48).

3.   Global Cooperative Communities Empowerment. The PWP has activated the Ministry of Habitats and Settlements, and it has created the Agency for Cooperative Communities to foster truly sustainable local empowerment and self-sufficiency (WLA 63). We are in the process of defining transferrable development protocols for the all the peoples and nations of Earth.

4.   Ministry of Health and Nutrition: This agency has been activated to protect and equitably provide food and nutrition resources for the people of Earth. The objective is to ensure sustainable healthy living for all persons. It is headed by Guruji Arun Kumar and currently operating in India.

5. Ministry of the Environment: This ministry also has been activated with the goal of replanting forests, restoring organic soil fertility, and addressing climate change in other respects (WLA 9). For instance, we appointed a leading environmentalist as Director for Green Economic Development. We also have a leading advocate for cooperative rural economic empowerment serving as Earth Federation Minister for Habitat and Settlements.

What is needed, of course, is a comprehensive global program unifying all nations and peoples to combat climate collapse. In line with this, the 2nd PWP assembly created the Emergency Earth Rescue Administration (EERA; WLA 6). EERA's mission is to spearhead the gigantic task of restoring Earth's environment once the first operative stage of world government has been activated. The EERA needs funding and resources to expand its environmental measures worldwide. Other legislative acts protecting the environment include: the World Hydrogen Energy Authority for renewable clean energy (WLA 10); the Hydrocarbon Resource Act to conserve, regulate, and utilize the world's remaining hydrocarbon resources (WLA 16); and the Water Act, which recognizes clean water as a human right and takes steps to protect our diminishing water resources, restore sources of fresh water, and democratically apportion these resources to all world citizens (WLA 30). Major initiatives will be needed to implement all these acts, and millions of currently unemployed people can be hired by the Earth Federation for these purposes, thereby diminishing global poverty and activating the global economy for the benefit of everyone.

6. WCPA Global Schools Project: This project was initiated in Assam, India, with the vision of public schools developing worldwide based on the *Earth Constitution* and the principle of unity in diversity of humanity.

Figure 7.6
Ninth Session of the Provisional World Parliament in Tripoli, Libya (2006).
The group in the foreground is a children's delegation petitioning the parliament
for mosquito netting to combat the spread of malaria in Africa and Asia.

7. Model World Parliament Project: This project encourages educational institutions everywhere to abandon the useless "Model U.N. pageants" and initiate truly progressive and educational model world parliaments based on the *Earth Constitution*.

8. Global Headquarters and Communications Centers: ECI and WCPA have developed internet platforms and social media venues for the promotion study, translation, and ratification of the *Earth Constitution*. Our U.S. headquarters is the "Peace Pentagon" – a conference and training center located at the Oracle Institute campus in Independence, VA (www.PeacePentagon.net). Resting along the New River in southern Virginia, the Peace Pentagon serves as a countermeasure to the War Pentagon in northern Virginia. ECI Board Member Rev. Laura George, J.D. manages Oracle Campus, which is open to the public.

9. Development of the World Electoral Districts: A project of paramount importance is the creation of one thousand World Electoral Districts – the voting districts for the people of Earth to elect representatives in the House of Peoples. Currently, we are seeking demographers and world geographers to finalize these electoral and administrative districts. Dr. Eugenia Almand, Secretary-General of WCPA, and Mr. Kalani James Evans have developed a "World District Template" that describes specific procedures and requirements for the creation of the electoral districts. Mr. Lucío Martins Rodrigues has joined our team to head up this project. Once established, the World Electoral Districts will be utilized for voting, administrative, and educational purposes.

10. World Parliament University: The *Earth Constitution* calls for a World University System, and Article 19 empowers us to begin that system. The Education Act passed in 2004 describes the broad requirements for educational institutions receiving funding or support from the Earth Federation government (WLA 26). Its legal precursor was the Graduate School of World Problems established in 1982 at the very 1st Session of the PWP in Brighton, England (WLA 4). This was followed by the Institute on World Problems (2003 to 2019), and then the Earth Constitution Institute (ECI). The university is a project of ECI, our educational charity, and it will offer online and in-person courses on issues related to the *Earth Constitution*. Both WCPA and the WPU are funded and supported by ECI. Preparing world citizens to work in the provisional and final agencies of the Earth Federation government will be the university's top priority.

11. WCPA Chapters: The World Constitution and Parliament Association was created to advance the *Earth Constitution*, organize the Provisional World Parliaments, and plan other relevant international meetings. Currently, the largest WCPA chapters are in the U.S. and India; the latter has its own website maintained by Prof. Narasimha Murthy, WCPA Communications Coordinator for India and Southeast Asia (see: www.WCPAindia.org). Our coordinator for Latin American is Leopoldo Cook Antonorsi in Venezuela. WCPA Chapters are non-profit associations and NGOs that support the study, promotion, and ratification of the *Earth Constitution* (while World District organizing committees are citizen-staffed governmental bodies established for electoral and administrative purposes; see *Figure 7.2* above). We wish to develop additional WCPA Chapters and invite you to start a chapter in your region of the world.

In addition, WCPA taskforce groups have emerged to tackle world problems and craft *Earth Constitution* solutions. While some of these projects are newly begun, their progress is impressive. The list of projects below shows that the *Earth Constitution*, by design, is a living document for a living planet. Virtually anything constructive, cooperative, positive, and beneficial to humanity can be undertaken under any one of its many Articles – all authorized by Article 19, which conveys our authority to immediately start worldwide government.

Current WCPA Projects (by Article)	
7.3.1 Disarmament and War Prevention	7.3.14 Science and Technology
7.3.2 Population and Youth Welfare	7.3.17 Social Welfare
7.3.3 Agriculture and Food Processing	7.3.16 Employment and Labor
7.3.4 Water Resources	7.3.18 Commerce and Industry
7.3.5 Healthcare	7.3.21 Media and Communication
7.3.6 Global Education	7.3.22 Human and Animal Rights
7.3.7 Language, Art History, and Culture	7.3.26 Exterior Relations & Foreign Affairs
7.3.9 Environment and Ecology	7.3.27 Democratic Procedures

The *Earth Constitution* has the power to transform the world system. If you are doing good work – whether social work, health care, peacebuilding, or environmental protection – your personal mission is embodied in the *Earth Constitution*, a visionary and holistic plan for democratic world government. Please join us and help promote this vision so more people will become conscious of the goal implicit in our human ontological project from the very beginning. Indeed, Article 19 makes this our right and our duty.

In this chapter, we have seen that our "design for a living planet" is the *Earth Constitution* itself. There can be no partial solution that targets environmental issues alone. The world system as a whole requires a comprehensive design for success. We will resolve all the crises of our human condition concurrently by founding a decent and inclusive world system, or we will solve none of them. The *Earth Constitution* addresses all these issues – sustainability, peace, justice, freedom – and brings civilization into harmony with our planetary biosphere.

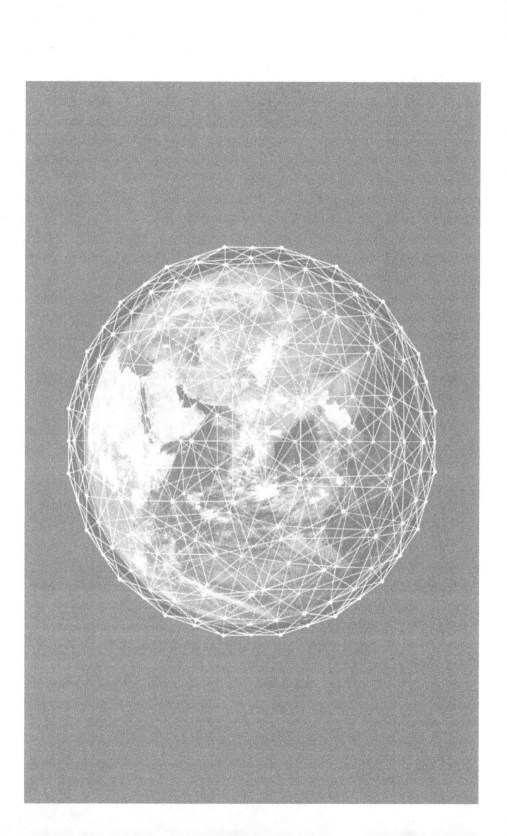

CHAPTER EIGHT
Conclusions

*Every act, seen from the perspective not of the agent
but of the process in whose framework it occurs and
whose automatism it interrupts, is a "miracle" – that is,
something which could not be expected. If it is true that
action and beginning are essentially the same, it follows
that a capacity for performing miracles must likewise be
within the range of human faculties. ...*

*Hence, it is not in the least superstitious, it is even
a counsel of realism, to look for the unforeseeable and
unpredictable, to be prepared for and to expect "miracles"
in the political realm. And the more heavily the scales are
weighted in favor of disaster, the more miraculous will the
deed done in freedom appear ....*

*The decisive difference between the "infinite
improbabilities" on which the reality of our earthly life
rests and the miraculous character in those events which
establish historical reality is that, in the realm of human
affairs, we know the author of the "miracles." It is men
who perform them – men who because they have received
the twofold gift of freedom and action can establish a
reality of their own.*

Hannah Arendt

8.1   Five Dimensions of Sustainability

Systems have consequences and designs have predictable results. This much
we have seen throughout this book. The *Earth Constitution* is an integrated,
comprehensive document *designed* to transform our biosphere to a sustainable
world system and our noosphere to planetary consciousness – all within the very
short time yet available to us. These elegant features of the *Earth Constitution*
work together synergistically to create a democratic and resilient world system.

Its design also leads to other positive consequences for Earth and for humanity. It changes us; it elevates us. Its holistic design can help make us whole and comprehend the power of unity in diversity.

We will attain holistic results if we begin to think holistically. The *Earth Constitution* was designed by holistic thinkers to help humankind think holistically. Here is physicist David Bohm explaining how systems thinking works:

> If he thinks of the totality as constituted of independent fragments, then that is how his mind will tend to operate, but if he can include everything coherently and harmoniously in an overall whole that is undivided, unbroken, and without a border (for every border is a division or break) then his mind will tend to move in a similar way, from this will flow an orderly action within the whole." (1980, xiii)

Barbara Marx Hubbard writes, "Through conscious evolution we realize, for the first time as a species, that it is our responsibility to proactively design social systems that are in alignment with the tendency in nature toward higher consciousness, greater freedom, and more synergistic order" (1998, 71). Robert Ornstein and Paul Ehrlich declare, "The time has come to take our own evolution into our hands and create a *new* evolutionary process, a process of conscious evolution" (2018, 12). The *Earth Constitution* provides humanity with a concrete tool that supports conscious evolution and synthesizes what climate scientists already understand in relation to sustainability.

Article 18 of the *Earth Constitution* requires reexamination of the entire document within the first ten years of its ratification and at least every twenty years thereafter, making it a living document that can be updated when needed. It incorporates and makes possible the five fundamental features of sustainable design discussed in Chapter One:

1. **Internalize Holism.** The Preamble to the *Earth Constitution* underscores, through its emphasis on holism, the interdependence and interrelation-ship of the geosphere, biosphere, and noosphere of Earth. We explored this holistic awareness in relation to contemporary science in Chapters One, Two, and Three. The *Earth Constitution* simultaneously points to and promotes this awareness. It also empowers local communities to restore and sustain all the dimensions of our richly diverse and nurturing ecosystem.

2. **Establish Ecological Economics.** The *Earth Constitution* is predicated on debt-free money and global public banking, and is designed for the planetary common good. We also have seen that it makes possible

financial programs for restoring Earth and employing millions of people in a coordinated effort to revitalize our living planet. As Buckminster Fuller affirms: "Our wealth is inherently common wealth and our common wealth can only increase, and it is increasing at a constantly self-accelerating synergetic rate" (1972, 89).

3. **Promote Spiritual Growth.** Revolutions require spiritual and moral authority to succeed, and the *Earth Constitution* clearly promotes growth toward worldcentric and cosmocentric modes of awareness. For example, the Global Education Act of the Provisional World Parliament (WLA 26) will help people around the world understand how and why Earth Federation will ensure humanity's continued evolutionary journey.

4. **Institutionalize Planetary Applied Science.** We need a design for studying and monitoring Earth as a whole and for responding to and maintaining ecosystem stability. We have seen how the Integrative Complex created under the *Earth Constitution* is designed to do exactly this. Under Article 13.9, the people of Earth enjoy "Protection of the natural environment which is the common heritage of humanity against pollution, ecological destruction or damage which could imperil life or lower the quality of life." This goal can only be attained by empowering communities around the globe to restore the health of their local ecosystems by using clean energy.

5. **Give Earth a Brain.** The noosphere is another term for the world's mind-sphere – the coordinating center or brain of Earth. We must acquire the global intelligence needed to strategize, plan, and implement programs that create a stable, prosperous, and regenerative future for all. The World Parliament and corresponding agencies comprise a neural-like network framed through democratic world law. The *Earth Constitution* is nothing short of an ingenious hub for the wheel of life.

These five design features are built into numerous passages in the *Earth Constitution*. For instance, Article 8 on the Integrative Complex describes the scope of the Agency for Technological and Environmental Assessment. This following passage represents holism in action, integrating human beings with their planetary environment. It also mandates ecological economics, uses planetary applied science, and clearly facilitates the flow of information, acting as a brain for Earth. Indeed, this passage also promotes spiritual growth since the nexus of qualities compels us to reflect and improve:

> To examine, analyze and assess environmental problems, in particular the environmental and ecological problems which may result from any intrusions or changes of the environmental or ecological relationships which may be caused by technological innovations, processes of resource development, patterns of human settlements, the production of energy, patterns of economic and industrial development, or other man-made intrusions and changes of the environment, or which may result from natural causes. (§8.6.1.3)

Human reason, love, and intuition function together in the Earth Federation movement. Now is the time for us to actualize our place in the cosmos as parts of the whole, microcosms within a macrocosm. Our immense human dignity needs to become manifest in the way we design our common life on Earth. Uniting under the *Constitution for the Federation of Earth* will allow us to achieve these five essential design features of a livable planet. Moreover, ratifying the *Earth Constitution* will animate our human dignity. We will become who we are meant to be.

8.2   The United Nations AND the *Earth Constitution*?

Transformation to a sustainable world system cannot take place if we insist on retaining the unworkable United Nations system of militarized sovereign nation-states. Viable U.N. agencies will be transferred to their counterparts under the *Earth Constitution*, which must replace the U.N. Charter – a charter based on centuries-old, outdated economic and political assumptions. Under the *Earth Constitution*, nation-states agree to stop all wars and forego their ability to use their nuclear, biological, and other weapons of mass destruction. Instead, those military storehouses would be equitably and transparently dismantled by the world government and the nations become governmental units within the Earth Federation.

Indeed, the U.N. could easily provide the infrastructure necessary to begin the Earth Federation process and ensure its smooth implementation, as provided for in Article 17.4.8. The U.N. General Assembly would become the House of Nations in the World Parliament. The International Court of Justice and International Criminal Court would become two benches of the World Supreme Court system. Similarly, the U.N. High Commission for Human Rights would become the nucleus for the World Ombudsmus. In addition, U.N. Peacekeeping forces would be retrained as civilian World Police, and the U.N. Secretariat would form the initial World Executive Branch. Something similar could be done for all the agencies of the U.N. That way, the many highly competent and well-meaning people now staffing the U.N. would no longer feel helpless. Rather, they finally would be empowered as legal representatives of the people of Earth to actualize global peace, justice, and sustainability.

At the 14th Session of the Provisional World Parliament, two World Legislative Acts were passed (WLA 64 and WLA 65). Those acts refine the process by which the U.N. can be integrated within the Earth Federation, and they have been bound into a book entitled *Our Common Human Future: The U.N. as an Effective Peace and Sustainability System* (Martin, ed. 2016). WCPA has sent this book to the heads of all U.N. Agencies in Vienna, Geneva, and New York. To date, there has been no response.

The rules governing a Founding Ratification Convention already have been passed by the Provisional World Parliament (PWP). These rules are inclusive, democratic, and designed to make founding possible without fear of the *Earth Constitution* being compromised by the many forces of corruption, delay, and chaos that would inevitably seek to destroy the integrity of the document. The Founding Convention would make decisions regarding how to include the U.N. in the newly formed world government. Just to be clear: The U.N. is not an alternative government because it is not a government at all, only a treaty between nations. However, its infrastructure would be invaluable for launching real, legitimate government for the people of Earth.

The international law regime within which the U.N. is immersed and entangled would be integrated into actual world law (Articles 4.8, 5.1.3, and 8.8.1). International law would not be abandoned but strengthened by measured stages into binding statutes over all persons and nations. The powerful nations that today ignore international law with impunity will lose their totalitarian privilege. Under the spurious notion of absolute sovereignty, nations are equal in theory but not in practice, with a handful exercising imperial economic, political, and military powers that harm humanity and our environment. Under the *Earth Constitution*, all the nations of the world would become truly equal. Errol E. Harris reminds us:

> The fact is that health, welfare, and the very survival of all peoples the world over now depend on what happens and what is done in any and every country, for that has effects on soil, water, and air, the food supply and the climate over the entire globe. *Unless and until this common destiny of all nations is widely recognized and is reflected in political institutions*, the hope of resolving the difficulties and removing the environmental perils now threatening mankind will continue to be very dim. (2014, 114)

8.3 The Gaia Principle and the Golden Rule

We saw in Chapter One how the Gaia principle – the idea that Earth as it evolved over 4.6 billion years has formed one all-encompassing, living ecosystem – is rapidly becoming understood by large numbers of people. Professor of Environmental Systems Erle C. Ellis writes, "With Gaia, climate stability in the face of a warming sun and other self-regulating behaviors came to be understood as complex, system-level processes emerging from interactions of the Earth's component systems" (2018, 20).

Earth system science has engendered a growing awareness of the need for protection and restoration of our planet's ecosystem. Phenomena such as global warming, melting of the polar ice caps, depletion of the ozone layer, collapsing of entire ocean fisheries, acidification of the oceans and rising water levels, rapid extinction of species on a daily basis, global diseases and pandemics, increased planetary disasters and super-storms, and possible inversions of major ocean currents and weather patterns are becoming well understood (Lovelock 1991).

Today, thoughtful human beings understand that human life is inseparable from the web of life on Earth. They understand that we must rapidly alter our economic, social, and political practices to bring human civilization into harmony with the planetary web of life that sustains us. They understand all development must be sustainable and support human life in ways that guarantee the thrivability of future generations. In short, life must be seen as primarily qualitative rather than quantitative.

Geologist Martin Redfern writes that "we are no longer the victims of our planet, we are the custodians of it" (2003, 132). Yet today, virtually all societies are living at the expense of future generations, both of humans and other species (Caldicott 1992; Daly 1996; Speth 2004). Meaningless maximization of our life-prospects diminishes the life-prospects of all other species. At the current rate of destruction, it is possible that we will reduce planetary life-prospects to zero. We become our own victims if we refuse to recognize the common destiny of all species, all people, and all nations within our shared biosphere.

The vast majority of people on Earth want peace, equality, and a healthy environment. They want lives replete with quality and meaning. They also have an ethical-intuitive longing for dignity, mutual recognition, and community. They therefore do not comprehend why we have chaos instead: war, violence, brutal competition, fear, suspicion, and injustice.

Many people lack a systems theory view of our global disorder, the "big picture" of why the present world system *will never* bring peace, justice, or sustainability. They do not discern that it is our anachronistic world system that keeps in place powerful and elite financial, political, corporate, and military forces that benefit from the present system and *prevent* civilization from taking steps leading to coherence, order, community, and the democratic rule of law. Currently, the people of Earth are still dominated and derailed by regressive economic and political institutions. Until the masses see this big picture, our prospects for transformative world government look dim.

Nevertheless, a new noosphere of mental constructs, archetypes, and images is coalescing around Earth. We are starting to build a "brain" for Earth, a collective selfhood that represents the aspirations of burgeoning world citizens. This global neural network must integrate a planetary consciousness, a holistic pattern fractally reflected in its constituent parts around the world. We need to adopt planetary laws and customs that reflect our most cherished principle, universally expressed by every culture on Earth. Commonly called the Golden Rule, this standard calls for human reciprocity informed and enhanced by kindness and empathy. The Golden Rule is the matrix, the starting point for yet higher forms of compassion and love.

> *We need to adopt planetary laws and customs that reflect our most cherished and universal principle, commonly called the Golden Rule.*

How do we unite people democratically so as to give expression to the Golden Rule and to our yearning for peace, justice, and equality for all before the law? There is only one practical way to unite humanity into a global community that expresses the common good of all, and that way is the *Earth Constitution*. It is an effective instrument for actualizing the transformational world system we seek. It contains a framework for expressing the will of our planetary majority, giving the people of Earth a brain, and voicing our desire for solidarity and sustainability. In the language of Ken Wilber, we need "integral ecology," a master plan for sustainability, which will materialize once civilization as a whole integrates values, economics, and politics within a higher, more mature level of consciousness:

> The basic idea is simple: anything less than an integral or comprehensive approach to environmental issues is doomed to failure. ... *Exterior* environmental sustainability is clearly needed; but without growth and development in the *interior* domains to world-centric levels of values and consciousness, the environment remains gravely at risk. Those focusing only on exterior solutions are contributing to the problem. Self, culture, and nature must be liberated together or not at all. How to do so is the focus of integral ecology. (2007, 100)

The *Earth Constitution* is the master plan, and the Provisional World Parliament is dedicated to rectifying the horrific consequences of the disordered current world system. WCPA and the Earth Constitution Institute (ECI) are committed to actualizing the decent world system dreamt of by the majority of human beings. This exemplary constitution integrates economics, resource conservation, regeneration, technological monitoring and development, education, moral growth, and beneficial legislation. All these aspects of planetary life are synthesized and harmonized in the *Earth Constitution* so humanity may achieve sustainability and embark on the next phase of our collective journey.

This is "integral ecology" in action. For the first time in history, the true will of the majority can prevail. No more class distinctions: masters over slaves, lords over serfs, men over women, bourgeoisie over proletariat, corporations over working people, rich over poor, one caste over another caste, one race over another race, imperial nations over weaker nations ... the list goes on and on. Hannah Arendt says that democratic mutual recognition of people with one another is constitutive of society itself (1977, 236-38). Jürgen Habermas says we "could not be scandalized" by the horror of the present world disorder "if we did not know that these shameful conditions might *also* be *different*" (2003, 63).

Yet it can only be different and our mutual recognition of one another can only be enabled if we finally take the design step necessary to address the root causes of the chaos. Law Professor Peter Gabel declares, "The world as it really is is suffused with moral longing that pulls upon the conscience of humanity to elevate ourselves from the limitations of what is to what ought to be" (2013, 48). He argues that law should play a key role in bringing us toward what ought to be. The *Earth Constitution* provides the legal framework, economic mechanisms, and people-power to make this happen with reasonable speed.

Additionally, the *Earth Constitution* contains the assumptions necessary to make the economy a subset of the planetary ecosystem and not a forever expanding engine of destruction based on purely quantitative premises. Ultimately, our future on Earth depends not only on the practical steps provided by the *Earth Constitution*, but also on our immense human potentialities for self-actualization and self-transcendence. We all know that these shameful conditions could be different. For instance, in a civilized world, poverty would not prevail, women would not be subjugated, and we would not struggle with global pandemics. In the case of Covid-19, the Earth Federation Ministry of Health could have brought the outbreak quickly under control – a perfect example of the efficiencies of world government.

As we have seen, the *Earth Constitution* functions both as a means and an end. It is an end because it embodies the very real dignity, equality, and community of civilization for the first time in human history. In this sense it represents a significant end or goal of history within the human quest for fulfillment. It would prove an immense success within our cosmic adventure. In the words of Panikkar, we need "to awaken our sense of belonging to the adventure of reality as a whole" (2013, 343). In the words of Nolan Pliny Jacobson, "Man is the juncture where the effects of past causes lose the fateful cast of necessity and provide opportunity for man to free himself to move onto an entirely new level of being" (1974, 121). For the first time, the people of Earth would democratically govern themselves.

This cosmic adventure takes its next step through the emergent unity of humankind via ratification of the *Earth Constitution*. True democracy, true self-government for humanity, can only happen at the global level, when all are embraced and all invited to participate. The *Earth Constitution* allows us to realize our "ontological vocation" and become ever more fully human and ever more fully expressive of the ground of Being. It makes possible our ascent to an entirely new level of being. Thus, it also is the means, a path leading to the further growth and actualization of the human project and our collective destiny. As both means and end, ratification of the *Earth Constitution* is absolutely necessary.

Can this be done in time? The task is huge but it is not "infinitely improbable," as Hannah Arendt reminds us in the quote opening this chapter. She challenges us to recognize that human beings can "establish a reality of their own" and perform "miracles" that may appear infinitely improbable but are in fact only the actualization of a reality already imminent. If our conscious awareness discerns our Oneness with one another and the cosmos, then the renaissance of the world is preparing to emerge. The people of Earth will just say, "Enough!"

Enough of absurd wealth inequality. Enough of endless wars and weapons of mass destruction. Enough of corporate greed and rivalry of militarized nations. Enough domination by undemocratic concentrations of wealth and power. Enough of polluted air in our cities, global pandemics, homeless people in the streets, and other solvable social problems. And most definitely, we have had enough of corrupt politicians! The people of Earth have the power to stop all these existential threats rapidly through ratification of the *Earth Constitution*.

The reader may rightly ask: *Why would things be so different under the Earth Constitution?* Answer: Because human beings would be moving to a higher level of existence, a higher level of civilizational maturity, within a framework designed to help them coalesce around and codify the Golden Rule, and beyond this to integral forms of love, thereby building a new world based on compassion, justice, and democracy. As Einstein and many others have said, we cannot solve a fundamental problem on the same level that originated the problem. We must move to a higher level from which the problem disappears of its own accord. Embracing holism is not merely an intellectual act. It must be a genuinely transformative existential act.

Father Pierre Teilhard de Chardin was the great thinker who initiated humanity to the language of geosphere, biosphere, and noosphere. He envisioned the growing unity of humankind and the integration of the noosphere in a process of ever-greater coherence and love. He declared:

> Only love can bring individual beings to their perfect completion, as individuals, by uniting them one with another, because only love takes possession of them and unites them by what lies deepest within them. ... Humanity, the spirit of the earth, the synthesis of individuals and peoples, the paradoxical conciliation of the element with the whole, of the one with the many: all these are regarded as utopian fantasies, yet they are biologically necessary; and if we would see them made flesh in the world what more need we do than imagine our power to love growing and broadening till it can embrace the totality of men and the earth (1969, 145)

De Chardin posited a biological imperative integrated with the love at the heart of holism. He envisioned diverse peoples becoming one within the womb of Earth and the wholeness of the noosphere. Indeed, holistic thought bridges the gap between body, mind, and spirit, and between individuality and community, compromise and freedom. Also implicit in the noosphere is the holism of the biosphere and the holism of the geosphere. Human beings are expressions of this nested holarchy and through their love will acquire true unity within diversity.

Humans are drawn forth and guided by reason and love. We are capable of moving to a higher level of consciousness that will largely dissolve our most fundamental problems. This interdependency of all things on Earth – including life and the unifying love emphasized by de Chardin – cry out for a corresponding unity in responding to the global climate crisis. "Without life, there would not be the feedback mechanisms on atmospheric composition that have, so far at least, kept the climate bearable" states climate scientist Martin Redfern (2003, 131-32). Begun by natural life processes, we need to continue the beneficent functions and expand the evolutionary impulse though a loving transformative praxis that actively monitors and modulates life conditions on Earth.

Life itself is most eminently made possible by the *Constitution for the Federation of Earth*. The trajectory of evolution is greater consciousness and the self-realization of the noosphere, therefore a key step in this process is made possible by the *Earth Constitution*. Ervin Laszlo observes:

> Teilhard de Chardin agreed ... that consciousness strives toward ever greater freedom of expression, and also agreed that a divine spirit drives its evolution. Consciousness is pulled from the future rather than pushed from the past. Through the action of love, an evolved consciousness fuses the elements of the mind into a higher unity. Attaining this "spiritual evolution" is the meaning of existence. (2017, 41)

To be "pulled from the future" means we must begin thinking in terms of practical utopian visions that live in our awareness but we have been trained to ignore or repress. As I elaborated in *Global Democracy and Human Self-Transcendence*, this practical utopian dimension functions as a fundamental component of our human reality. Dreams are made manifest every day by courageous, dedicated people. Our collective dream is for a united humanity, a resilient planet, and global peace. Conscious evolution and sustainability necessarily belong together.

We have seen throughout this book that "love" is a key word for this uniting process, and love united with reason gives us democratic world law. Democratic world law is the 21st century form of love. The *Earth Constitution* explicitly protects the ecological fabric of life on Earth at all levels, respecting and contextualizing the Gaia principle and the Golden Rule. It therefore holds the key to our collective future. Our best and most practical option for the continued evolution of our ourselves and our planet is the *Earth Constitution*.

8.4 Practical Utopia as Eminently Practical

Our future depends on the entire human community uniting around shared principles of integral social science and climate science. To unite all people under non-military democratic world government, we must accept the Gaia principle and the Golden rule as guiding principles for all human political, economic, and social advancement. Human democracy is thus inseparable from natural ecology. What democratic world government will do for humanity, the Gaia principle does for nature.

We have the very real capacity to actualize what I have called "practical utopia" and what Erich Fromm calls "awake utopia." If we adopt such a vision, then the transformation of our world system will happen, we will become more fully human, and we will actualize our common human dignity in ways that fulfill our human potential. Philosopher Frederick L. Polak explains:

Man cannot become fully man, and attain the summit of human dignity – known in platonic Antiquity and rediscovered in the Renaissance – cannot evolve toward his final maturity in the Kantian sense, if he cannot simultaneously elaborate and refine his mental image of a different and future world. Eschatological or utopian, this image of the future, infusing man with the foreknowledge of happiness and harmony to come, haunts him and challenges him to work for its realization. (1967, 287)

The *Earth Constitution* not only addresses how to overcome climate crisis, it represents a true opportunity for us to become more fully human, to become what we are meant to be. Our human task, Buckminster Fuller declares, is to realize that "the most ideal is the most realistically practical" (1972, 79). The design of a world system that simultaneously solves the climate crisis and human self-realization is literally a dream come true.

Recall that human self-realization is not limited to a process of "internal" awakening. Like the cosmos, humans are a nested holarchy comprised of our bodies, our senses, our minds, our spirit, and the human community with which we are all intrinsically involved. Panikkar declares "that the awareness of our complex being discloses to us the harmony between all 'components' of our beings, not only among themselves, but also with the entire universe" (2013, 327). In short, we need one another. Our job is to fully awaken to this truth, to holism, and to the depth of our collective journey.

As we have seen, planetary sustainability can only happen if it is integrated into economic, political, cultural, and educational institutions designed for a living planet. The *Earth Constitution* joins these structures together to create a truly ecological and sustainable world order. Our choice is between climate change and system change. We also must choose between remaining selfish, egoistic, and immature – all of which will lead to our probable extinction – or becoming mature human beings who routinely express love, compassion, and deep intelligence. The *Earth Constitution* invites our affirmation on many levels.

Resistance and criticism of the present system, including "extinction rebellion," while necessary and appropriate, are not powerful enough to unite humanity. Personal conservation and recycling are important practices, but they will not solve the climate crisis. Withdrawal into small, self-sustaining communities won't get the job done either. We need a concrete and positive planetary blueprint that we can support and actualize, collectively. Since the design of any system produces the consequences inherently designed into it, the *Earth Constitution* – which was carefully designed to produce a sustainable global civilization – provides the best possible blueprint for system-wide change and for founding a truly sustainable and democratic world civilization.

Buckminster Fuller famously stated, "You never change things by fighting the existing reality. To change something, build a new model that makes the existing model obsolete." That is exactly what the framers of the *Earth Constitution* did. They built a new world model that makes the U.N. Charter obsolete. In fact, Dr. Robert Muller, who was the Assistant Secretary General of the United Nations for eighteen years and whom I knew personally, declared after reading the *Earth Constitution*, "God bless this Constitution!" The document is simple to read (see Appendix III), and it has been translated into some twenty-three languages.

Today, we have an ingenious model designed to address our many planetary crises. Adoption of the *Earth Constitution* may take a miracle, but miracles happen every day to many people. If we act swiftly with focus and commitment, this miracle can happen. Indeed, the more we humans expand our consciousness, the more everything in existence appears miraculous, and the more we realize our unique position and purpose within the holistic structure of the cosmos.

All of us were born on this living planet and each of us has a stake in its integral well-being. We therefore need a properly designed world government to mirror holism on all levels: the matrix of our geosphere, the web of life in our biosphere, and the wisdom contained in our emerging noosphere. The *Constitution for the Federation of Earth* provides this level of integration, and it represents the most viable and realistic opportunity for humanity to become one self-aware, loving and law-biding community. It is entirely up to us. Let us choose to activate both our reason and our love, embrace this sacred global vision, and make it happen for ourselves and our precious planet Earth.

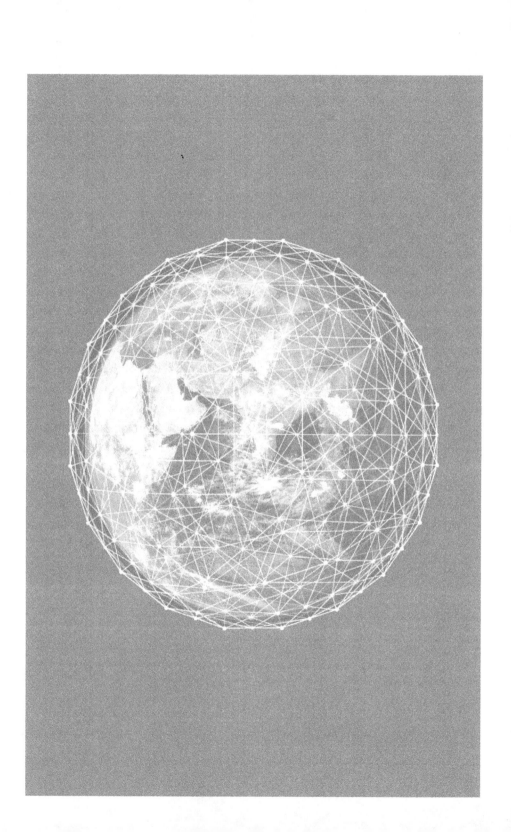

APPENDIX I

Earth Constitution Mandala

WORLD PARLIAMENT = ALL THREE

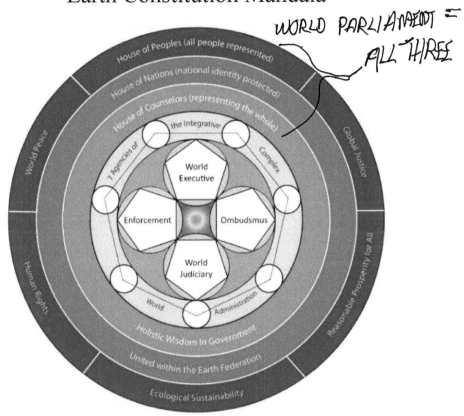

Figure A-1

The Earth Constitution Mandala symbolizes the holistic nature of a federated world system based on the *Earth Constitution*.

➤ **House of Peoples:** draws from 1000 World Electoral Districts.
➤ **House of Nations:** 1, 2, or 3 representatives from each nation.
➤ **House of Counselors:** includes 200 scholars from around the world.
➤ **Integrated Complex:** 7 agencies serve with expertise and the latest information.

The inner quadrant of the Mandala depicts the four main agencies responsible to the World Parliament and the people of Earth.

➤ **World Executive:** 5 Presidents with no military powers.
➤ **World Police:** only weapons needed to apprehend individual persons.
➤ **Ombudsmus:** watchdog on government and protector of universal human rights.
➤ **World Judiciary:** upholds world laws and settles international disputes.

APPENDIX II

A Pledge of Allegiance to the
Constitution for the Federation of Earth

*I pledge allegiance to the
Constitution for the Federation of Earth,
And to the Republic of free world citizens for which it stands,*

*One Earth Federation, protecting by law
the rich diversity of the Earth's citizens,*

*One Earth Federation, protecting
the precious ecology of our planet.*

*I pledge allegiance to the World Parliament
representing all nations and peoples,
And to the democratic processes by which it proceeds,*

*One law for the Earth, with freedom and equality for all,
One standard of justice, with a bill of rights protecting each.*

*I pledge allegiance to the future generations
protected by the Earth Constitution,
And to the unity, integrity, and beauty of humankind,
living in harmony on the Earth,*

*One Earth Federation, conceived in love, truth, and hope,
with peace and prosperity for all.*

This Pledge of Allegiance was ratified at the
Seventh Session of the Provisional World Parliament,
which met in Chennai, India from December 26-30, 2003.

APPENDIX III
Constitution for the Federation of Earth

PREAMBLE

Realizing that Humanity today has come to a turning point in history and that we are on the threshold of a new world order which promises to usher in an era of peace, prosperity, justice and harmony;

Aware of the interdependence of people, nations and all life;

Aware that man's abuse of science and technology has brought Humanity to the brink of disaster through the production of horrendous weaponry of mass destruction and to the brink of ecological and social catastrophe;

Aware that the traditional concept of security through military defense is a total illusion both for the present and for the future;

Aware of the misery and conflicts caused by ever increasing disparity between rich and poor;

Conscious of our obligation to posterity to save Humanity from imminent and total annihilation;

Conscious that Humanity is One despite the existence of diverse nations, races, creeds, ideologies and cultures and that the principle of unity in diversity is the basis for a new age when war shall be outlawed and peace prevail; when the earth's total resources shall be equitably used for human welfare; and when basic human rights and responsibilities shall be shared by all without discrimination;

Conscious of the inescapable reality that the greatest hope for the survival of life on earth is the establishment of a democratic world government;

We, citizens of the world, hereby resolve to establish a world federation to be governed in accordance with this constitution for the Federation of Earth.

ARTICLE 1: Broad Functions of the Earth Federation

The broad functions of the Federation of Earth shall be:

1.1 To prevent war, secure disarmament, and resolve territorial and other disputes which endanger peace and human rights.

1.2 To protect universal human rights, including life, liberty, security, democracy, and equal opportunities in life.

1.3 To obtain for all people on earth the conditions required for equitable economic and social development and for diminishing social differences.

1.4 To regulate world trade, communications, transportation, currency, standards, use of world resources, and other global and international processes.

1.5 To protect the environment and the ecological fabric of life from all sources of damage, and to control technological innovations whose effects transcend national boundaries, for the purpose of keeping Earth a safe, healthy and happy home for humanity.

1.6 To devise and implement solutions to all problems which are beyond the capacity of national governments, or which are now or may become of global or international concern or consequence.

ARTICLE 2: Basic Structure of Earth Federation

2.1 The Federation of Earth shall be organized as a universal federation, to include all nations and all people, and to encompass all oceans, seas and lands of Earth, inclusive of non-self governing territories, together with the surrounding atmosphere.

2.2 The World Government for the Federation of Earth shall be non-military and shall be democratic in its own structure, with ultimate sovereignty residing in all the people who live on Earth.

2.3 The authority and powers granted to the World Government shall be limited to those defined in this Constitution for the Federation of Earth, applicable to problems and affairs which transcend national boundaries, leaving to national governments jurisdiction over the internal affairs of the respective nations but consistent with the authority of the World Government to protect universal human rights as defined in this World Constitution.

2.4 The basic direct electoral and administrative units of the World Government shall be World Electoral and Administrative Districts. A total of not more than 1000 World Electoral and Administrative Districts shall be defined, and shall be nearly equal in population, within the limits of plus or minus ten percent.

2.5 Contiguous World Electoral and Administrative Districts shall be combined as may be appropriate to compose a total of twenty World Electoral and Administrative Regions for the following purposes, but not limited thereto: for the election or appointment of certain world government officials; for administrative purposes; for composing various organs of the world government as enumerated in Article 4; for the functioning of the Judiciary, the Enforcement System, and the Ombudsmus, as well as for the functioning of any other organ or agency of the World Government.

2.6 The World Electoral and Administrative Regions may be composed of a variable number of World Electoral and Administrative Districts, taking into consideration geographic, cultural, ecological and other factors as well as population.

2.7 Contiguous World Electoral and Administrative Regions shall be grouped together in pairs to compose Magna-Regions.

2.8 The boundaries for World Electoral and Administrative Regions shall not cross the boundaries of the World Electoral and Administrative Districts, and shall be common insofar as feasible for the various administrative departments and for the several organs and agencies of the World Government. Boundaries for the World Electoral and Administrative Districts as well as for the Regions need not conform to existing national boundaries, but shall conform as far as practicable.

2.9 The World Electoral and Administrative Regions shall be grouped to compose at least five Continental Divisions of the Earth, for the election or appointment of certain world government officials, and for certain aspects of the composition and functioning of the several organs and agencies of the World Government as specified hereinafter. The boundaries of Continental Divisions shall not cross existing national boundaries as far as practicable. Continental Divisions may be composed of a variable number of World Electoral and Administrative Regions.

ARTICLE 3: Organs of the Earth Federation

The organs of the World Government shall be:

3.1   The World Parliament
3.2   The World Executive
3.3   The World Administration
3.4   The Integrative Complex
3.5   The World Judiciary
3.6   The Enforcement System
3.7   The World Ombudsmus

SEE PAGE 253

ARTICLE 4: Grant of Specific Powers to the Earth Federation

The powers of the World government to be exercised through its several organs and agencies shall comprise the following:

4.1   Prevent wars and armed conflicts among the nations, regions, districts, parts and peoples of Earth.

4.2   Supervise disarmament and prevent re-armament; prohibit and eliminate the design, testing, manufacture, sale, purchase, use and possession of weapons of mass destruction, and prohibit or regulate all lethal weapons which the World Parliament may decide.

4.3   Prohibit incitement to war, and discrimination against or defamation of conscientious objectors.

4.4   Provide the means for peaceful and just solutions of disputes and conflicts among or between nations, peoples, and/or other components within the Federation of Earth.

4.5   Supervise boundary settlements and conduct plebiscites as needed.

4.6   Define the boundaries for the districts, regions and divisions which are established for electoral, administrative, judicial and other purposes of the World Government.

4.7   Define and regulate procedures for the nomination and election of the members of each House of the World Parliament, and for the nomination, election, appointment and employment of all World Government officials and personnel.

4.8   Codify world laws, including the body of international law developed prior to adoption of the world constitution, but not inconsistent therewith, and which is approved by the World Parliament.

4.9   Establish universal standards for weights, measurements, accounting and records.

4.10   Provide assistance in the event of large scale calamities, including drought, famine, pestilence, flood, earthquake, hurricane, ecological disruptions and other disasters.

4.11   Guarantee and enforce the civil liberties and the basic human rights which are defined in the Bill of Rights for the Citizens of Earth which is made a part of this World Constitution under Article 12.

4.12   Define standards and promote the worldwide improvement in working conditions, nutrition, health, housing, human settlements, environmental conditions, education, economic security, and other conditions defined under Article 13 of this World Constitution.

4.13   Regulate and supervise international transportation, communications, postal services, and migrations of people.

4.14   Regulate and supervise supra-national trade, industry corporations, businesses, cartels, professional services, labor supply, finances, investments and insurance.

4.15   Secure and supervise the elimination of tariffs and other trade barriers among nations, but with provisions to prevent or minimize hardship for those previously protected by tariffs.

4.16   Raise the revenues and funds, by direct and/or indirect means, which are necessary for the purposes and activities of the World Government.

4.17   Establish and operate world financial, banking, credit and insurance institutions designed to serve human needs; establish, issue and regulate world currency, credit and exchange.

4.18   Plan for and regulate the development, use, conservation and re-cycling of the natural resources of Earth as the common heritage of Humanity; protect the environment in every way for the benefit of both present and future generations.

4.19   Create and operate a World Economic Development Organization to serve equitably the needs of all nations and people included within the World Federation.

4.20   Develop and implement solutions to transnational problems of food supply, agricultural production, soil fertility, soil conservation, pest control, diet, nutrition, drugs and poisons, and the disposal of toxic wastes.

4.21   Develop and implement means to control population growth in relation to the life-support capacities of Earth, and solve problems of population distribution.

4.22   Develop, protect, regulate and conserve the water supplies of Earth; develop, operate and/or coordinate transnational irrigation and other water supply and control projects; assure equitable allocation of trans-national water supplies, and protect against adverse trans-national effects of water or moisture diversion or weather control projects within national boundaries.

4.23   Own, administer and supervise the development and conservation of the oceans and sea-beds of Earth and all resources thereof, and protect from damage.

4.24   Protect from damage, and control and supervise the uses of the atmosphere of Earth.

4.25   Conduct inter-planetary and cosmic explorations and research; have exclusive jurisdiction over the Moon and over all satellites launched from Earth.

4.26   Establish, operate and/or coordinate global air lines, ocean transport systems, international railways and highways, global communication systems, and means for interplanetary travel and communications; control and administer vital waterways.

4.27   Develop, operate and/or coordinate transnational power systems, or networks of small units, integrating into the systems or networks power derived from the sun, wind, water, tides, heat differentials, magnetic forces, and any other source of safe, ecologically sound and continuing energy supply.

4.28   Control the mining, production, transportation and use of fossil sources of energy to the extent necessary to reduce and prevent damages to the environment and the ecology, as well as to prevent conflicts and conserve supplies for sustained use by succeeding generations.

4.29   Exercise exclusive jurisdiction and control over nuclear energy research and testing and nuclear power production, including the right to prohibit any form of testing or production considered hazardous.

4.30   Place under world controls essential natural resources which may be limited or unevenly distributed about the Earth. Find and implement ways to reduce wastes and find ways to minimize disparities when development or production is insufficient to supply everybody with all that may be needed.

4.31   Provide for the examination and assessment of technological innovations which are or may be of supranational consequence, to determine possible hazards or perils to humanity or the environment; institute such controls and regulations of technology as may be found necessary to prevent or correct widespread hazards or perils to human health and welfare.

4.32   Carry out intensive programs to develop safe alternatives to any technology or technological processes which may be hazardous to the environment, the ecological system, or human health and welfare.

4.33   Resolve supra-national problems caused by gross disparities in technological development or capability, capital formation, availability of natural resources, educational opportunity, economic opportunity, and wage and price differentials. Assist the processes of technology transfer under conditions which safeguard human welfare and the environment and contribute to minimizing disparities.

4.34   Intervene under procedures to be defined by the World Parliament in cases of either intra-state violence and intra-state problems which seriously affect world peace or universal human rights.

4.35   Develop a world university system. Obtain the correction of prejudicial communicative materials which cause misunderstandings or conflicts due to differences of race, religion, sex, national origin or affiliation.

4.36   Organize, coordinate and/or administer a voluntary, non-military World Service Corps, to carry out a wide variety of projects designed to serve human welfare.

4.37   Designate as may be found desirable an official world language or official world languages.

4.38   Establish and operate a system of world parks, wild life preserves, natural places, and wilderness areas.

4.39   Define and establish procedures for initiative and referendum by the Citizens of Earth on matters of supra-national legislation not prohibited by this World Constitution.

4.40   Establish such departments, bureaus, commissions, institutes, corporations, administrations, or agencies as may be needed to carry out any and all of the functions and powers of the World Government.

4.41   Serve the needs of humanity in any and all ways which are now, or may prove in the future to be, beyond the capacity of national and local governments.

ARTICLE 5: The World Parliament

5.1   Functions and Powers of the World Parliament

The functions and powers of the World Parliament shall comprise the following:

5.1.1   To prepare and enact detailed legislation in all areas of authority and jurisdiction granted to the World Government under Article 4 of this World Constitution.

5.1.2   To amend or repeal world laws as may be found necessary or desirable.

5.1.3   To approve, amend or reject the international laws developed prior to the advent of World Government, and to codify and integrate the system of world law and world legislation under the World Government.

5.1.4   To establish such regulations and directions as may be needed, consistent with this world constitution, for the proper functioning of all organs, branches, departments, bureaus, commissions, institutes, agencies or parts of the World Government.

5.1.5   To review, amend and give final approval to each budget for the World Government, as submitted by the World Executive; to devise the specific means for directly raising funds needed to fulfill the budget, including taxes, licenses, fees, globally accounted social and public costs which must be added into the prices for goods and services, loans and credit advances, and any other appropriate means; and to appropriate and allocate funds for all operations and functions of the World Government in accordance with approved budgets, but subject to the right of the Parliament to revise any appropriation not yet spent or contractually committed.

5.1.6 To create, alter, abolish or consolidate the departments, bureaus, commissions, institutes, agencies or other parts of the World Government as may be needed for the best functioning of the several organs of the World Government, subject to the specific provisions of this World Constitution.

5.1.7 To approve the appointments of the heads of all major departments, commissions, offices, agencies and other parts of the several organs of the World Government, except those chosen by electoral or civil service procedures.

5.1.8 To remove from office for cause any member of the World Executive, and any elective or appointive head of any organ, department, office, agency or other part of the World Government, subject to the specific provisions in this World Constitution concerning specific offices.

5.1.9 To define and revise the boundaries of the World Electoral and Administrative Districts, the World Electoral and Administrative Regions and Magna Regions, and the Continental Divisions.

5.1.10 To schedule the implementation of those provisions of the World Constitution which require implementation by stages during the several stages of Provisional World Government, First Operative Stage of World Government, Second Operative Stage of World Government, and Full Operative Stage of World Government, as defined in Articles 17 and 19 of this World Constitution.

5.1.11 To plan and schedule the implementation of those provisions of the World Constitution which may require a period of years to be accomplished.

5.2 Composition of the World Parliament

5.2.1 The World Parliament shall be composed of three houses, designated as follows: The House of Peoples, to represent the people of Earth directly and equally; The House of Nations, to represent the nations which are joined together in the Federation of Earth; and a House of Counselors with particular functions to represent the highest good and best interests of humanity as a whole.

P. 253

5.2.2 All members of the World Parliament, regardless of House, shall be designated as Members of the World Parliament.

5.3 The House of Peoples

5.3.1 The House of Peoples shall be composed of the peoples delegates directly elected in proportion to population from the World Electoral and Administrative Districts, as defined in Article 2.4.

5.3.2 Peoples delegates shall be elected by universal adult suffrage, open to all persons of age 18 and above.

5.3.3 One peoples delegate shall be elected from each World Electoral and Administrative District to serve a five-year term in the House of Peoples. Peoples delegates may be elected to serve successive terms without limit. Each peoples delegate shall have one vote.

5.3.4 A candidate for election to serve as a peoples delegate must be at least 21 years of age, a resident for at least one year of the electoral district from which the candidate is seeking election, and shall take a pledge of service to humanity.

5.4   The House of Nations
5.4.1   The House of Nations shall be composed of national delegates elected or appointed by procedures to be determined by each national government on the following basis:
      5.4.1.1   One national delegate from each nation of at least 100,000 population, but less than 10,000,000 population.
      5.4.1.2   Two national delegates from each nation of at least 10,000,000 population, but less than 100,000,000 population.
      5.4.1.3   Three national delegates from each nation of 100,000,000 population or more.
5.4.2   Nations of less than 100,000 population may join in groups with other nations for purposes of representation in the House of Nations.
5.4.3   National delegates shall be elected or appointed to serve for terms of five years, and may be elected or appointed to serve successive terms without limit. Each national delegate shall have one vote.
5.4.4   Any person to serve as a national delegate shall be a citizen for at least two years of the nation to be represented, must be at least 21 years of age, and shall take a pledge of service to humanity.

5.5   The House of Counselors
5.5.1   The House of Counselors shall be composed of 200 counselors chosen in equal numbers from nominations submitted from the twenty World Electoral and Administrative Regions, as defined in Article 2.5 and 2.6, ten from each Region.
5.5.2   Nominations for members of the House of Counselors shall be made by the teachers and students of universities and colleges and of scientific academies and institutes within each world electoral and administrative region. Nominees may be persons who are off campus in any walk of life as well as on campus.
5.5.3   Nominees to the House of Counselors from each World Electoral and Administrative Region shall, by vote taken among themselves, reduce the number of nominees to no less than two times and no more than three times the number to be elected.
5.5.4   Nominees to serve as members of the House of Counselors must be at least 25 years of age, and shall take a pledge of service to humanity. There shall be no residence requirement, and a nominee need not be a resident of the region from which nominated or elected.
5.5.5   The members of the House of Counselors from each region shall be elected by the members of the other two houses of the World Parliament from the particular region.
5.5.6   Counselors shall be elected to serve terms of ten years. One-half of the members of the House of Counselors shall be elected every five years. Counselors may serve successive terms without limit. Each Counselor shall have one vote.

5.6   Procedures of the World Parliament
5.6.1   Each house of the World Parliament during its first session after general elections shall elect a panel of five chairpersons from among its own members, one from each of five Continental Divisions. The chairpersons shall rotate annually so that each will serve for one year as chief presiding officer, while the other four serve as vice-chairpersons.
5.6.2   The panels of Chairpersons from each House shall meet together, as needed, for the purpose of coordinating the work of the Houses of the World Parliament, both severally and jointly.
5.6.3   Any legislative measure or action may be initiated in either House of Peoples or House of Nations or both concurrently, and shall become effective when passed by a simple majority vote of both the House of Peoples and of the House of Nations, except in those

cases where an absolute majority vote or other voting majority is specified in this World Constitution.

5.6.4   In case of deadlock on a measure initiated in either the House of Peoples or House of Nations, the measure shall then automatically go to the House of Counselors for decision by simple majority vote of the House of Counselors, except in the cases where other majority vote is required in this World Constitution. Any measure may be referred for decision to the House of Counselors by a concurrent vote of the other two houses.

5.6.5   The House of Counselors may initiate any legislative measure, which shall then be submitted to the other two houses and must be passed by simple majority vote of both the House of Peoples and House of Nations to become effective, unless other voting majority is required by some provision of this World Constitution.

5.6.6   The House of Counselors may introduce an opinion or resolution on any measure pending before either of the other two houses; either of the other houses may request the opinion of the House of Counselors before acting upon a measure.

5.6.7   Each house of the World Parliament shall adopt its own detailed rules of procedure, which shall be consistent with the procedures set forth in this World Constitution, and which shall be designed to facilitate coordinated functioning of the three houses.

5.6.8   Approval of appointments by the World Parliament or any house thereof shall require simple majority votes, while removals for cause shall require absolute majority votes.

5.6.9   After the full operative stage of World Government is declared, general elections for members of the World Parliament to the House of Peoples shall be held every five years. The first general elections shall be held within the first two years following the declaration of the full operative stage of World Government.

5.6.10   Until the full operative stage of World Government is declared, elections for members of the World Parliament to the House of Peoples may be conducted whenever feasible in relation to the campaign for ratification of this World Constitution.

5.6.11   Regular sessions of the House of Peoples and House of Nations of the World Parliament shall convene on the second Monday of January of each and every year.

5.6.12   Each nation, according to its own procedures, shall appoint or elect members of the World Parliament to the House of Nations at least thirty days prior to the date for convening the World Parliament in January.

5.6.13   The House of Peoples together with the House of Nations shall elect the members of the World Parliament to the House of Counselors during the month of January after the general elections. For its first session after general elections, the House of Counselors shall convene on the second Monday of March, and thereafter concurrently with the other two houses.

5.6.14   Bi-elections to fill vacancies shall be held within three months from occurrence of the vacancy or vacancies.

5.6.15   The World Parliament shall remain in session for a minimum of nine months of each year. One or two breaks may be taken during each year, at times and for durations to be decided by simple majority vote of the House of Peoples and House of Nations sitting jointly.

5.6.16   Annual salaries for members of the World Parliament of all three houses shall be the same, except for those who serve also as members of the Presidium and of the Executive Cabinet.

5.6.17   Salary schedules for members of the World Parliament and for members of the Presidium and of the Executive Cabinet shall be determined by the World Parliament.

ARTICLE 6: The World Executive

6.1   Functions and Powers of the World Executive
6.1.1   To implement the basic system of world law as defined in the World Constitution and in the codified system of world law after approval by the World Parliament.
6.1.2   To implement legislation enacted by the World Parliament.
6.1.3   To propose and recommend legislation for enactment by the World Parliament.
6.1.4   To convene the World Parliament in special sessions when necessary.
6.1.5   To supervise the World Administration and the Integrative Complex and all of the departments, bureaus, offices, institutes and agencies thereof.
6.1.6   To nominate, select and remove the heads of various organs, branches, departments, bureaus, offices, commissions, institutes, agencies and other parts of the World Government, in accordance with the provisions of this World Constitution and as specified in measures enacted by the World Parliament.
6.1.7   To prepare and submit annually to the World Parliament a comprehensive budget for the operations of the World Government, and to prepare and submit periodically budget projections over periods of several years.
6.1.8   To define and propose priorities for world legislation and budgetary allocations.
6.1.9   To be held accountable to the World Parliament for the expenditures of appropriations made by the World Parliament in accordance with approved and longer term budgets, subject to revisions approved by the World Parliament.

6.2   Composition of the World Executive
The World Executive shall consist of a Presidium of five members, and of an Executive Cabinet of from twenty to thirty members, all of whom shall be members of the World Parliament.

6.3   The Presidium
6.3.1   The Presidium shall be composed of five members, one to be designated as President and the other four to be designated as Vice Presidents. Each member of the Presidium shall be from a different Continental Division.
6.3.2   The Presidency of the Presidium shall rotate each year, with each member in turn to serve as President, while the other four serve as Vice Presidents. The order of rotation shall be decided by the Presidium.
6.3.3   The decisions of the Presidium shall be taken collectively, on the basis of majority decisions.
6.3.4   Each member of the Presidium shall be a member of the World Parliament, either elected to the House of Peoples or to the House of Counselors, or appointed or elected to the House of Nations.
6.3.5   Nominations for the Presidium shall be made by the House of Counselors. The number of nominees shall be from two to three times the number to be elected. No more than one-third of the nominees shall be from the House of Counselors or from the House of Nations, and nominees must be included from all Continental Divisions.
6.3.6   From among the nominees submitted by the House of Counselors, the Presidium shall be elected by vote of the combined membership of all three houses of the World Parliament in joint session. A plurality vote equal to at least 40 percent of the total membership of the World Parliament shall be required for the election of each member to the Presidium, with successive elimination votes taken as necessary until the required plurality is achieved.

6.3.7   Members of the Presidium may be removed for cause, either individually or collectively, by an absolute majority vote of the combined membership of the three houses of the World Parliament in joint session.

6.3.8   The term of office for the Presidium shall be five years and shall run concurrently with the terms of office for the members as Members of the World Parliament, except that at the end of each five year period, the Presidium members in office shall continue to serve until the new Presidium for the succeeding term is elected. Membership in the Presidium shall be limited to two consecutive terms.

6.4   The Executive Cabinet

6.4.1   The Executive Cabinet shall be composed of from twenty to thirty members, with at least one member from each of the ten World Electoral and Administrative Magna Regions of the world.

6.4.2   All members of the Executive Cabinet shall be Members of the World Parliament.

6.4.3   There shall be no more than two members of the Executive Cabinet from any single nation of the World Federation. There may be only one member of the Executive Cabinet from a nation from which a Member of the World Parliament is serving as a member of the Presidium.

6.4.4   Each member of the Executive Cabinet shall serve as the head of a department or agency of the World Administration or Integrative Complex, and in this capacity shall be designated as Minister of the particular department or agency.

6.4.5   Nominations for members of the Executive Cabinet shall be made by the Presidium, taking into consideration the various functions which Executive Cabinets members are to perform. The Presidium shall nominate no more than two times the number to be elected.

6.4.6   The Executive Cabinet shall be elected by simple majority vote of the combined membership of all three houses of the World Parliament in joint session.

6.4.7   Members of the Executive Cabinet either individually or collectively may be removed for cause by an absolute majority vote of the combined membership of all three houses of the World Parliament sitting in joint session.

6.4.8   The term of office in the Executive Cabinet shall be five years, and shall run concurrently with the terms of office for the members as Members of the World Parliament, except that at the end of each five year period, the Cabinet members in office shall continue to serve until the new Executive Cabinet for the succeeding term is elected. Membership in the Executive Cabinet shall be limited to three consecutive terms, regardless of change in ministerial position.

6.5   Procedures of the World Executive

6.5.1   The Presidium shall assign the ministerial positions among the Cabinet members to head the several administrative departments and major agencies of the Administration and of the Integrative Complex. Each Vice President may also serve as a Minister to head an administrative department, but not the President. Ministerial positions may be changed at the discretion of the Presidium. A Cabinet member or Vice President may hold more than one ministerial post, but no more than three, providing that no Cabinet member is without a Ministerial post.

6.5.2   The Presidium, in consultation with the Executive Cabinet, shall prepare and present to the World Parliament near the beginning of each year a proposed program of world legislation. The Presidium may propose other legislation during the year.

6.5.3　The Presidium, in consultation with the Executive Cabinet, and in consultation with the World Financial Administration, (see Article 8, Sec. 7.1.9.) shall be responsible for preparing and submitting to the World Parliament the proposed annual budget, and budgetary projections over periods of years.

6.5.4　Each Cabinet Member and Vice President as Minister of a particular department or agency shall prepare an annual report for the particular department or agency, to be submitted both to the Presidium and to the World Parliament.

6.5.5　The members of the Presidium and of the Executive Cabinet at all times shall be responsible both individually and collectively to the World Parliament.

6.5.6　Vacancies occurring at any time in the World Executive shall be filled within sixty days by nomination and election in the same manner as specified for filling the offices originally.

6.6　Limitations on the World Executive

6.6.1　The World Executive shall not at any time alter, suspend, abridge, infringe or otherwise violate any provision of this World Constitution or any legislation or world law enacted or approved by the World Parliament in accordance with the provisions of this World Constitution.

6.6.2　The World Executive shall not have veto power over any legislation passed by the World Parliament.

6.6.3　The World Executive may not dissolve the World Parliament or any House of the World Parliament.

6.6.4　The World Executive may not act contrary to decisions of the World Courts.

6.6.5　The World Executive shall be bound to faithfully execute all legislation passed by the World Parliament in accordance with the provisions of this World Constitution, and may not impound or refuse to spend funds appropriated by the World Parliament, nor spend more funds than are appropriated by the World Parliament.

6.6.6　The World Executive may not transcend or contradict the decisions or controls of the World Parliament, the World Judiciary or the Provisions of this World Constitution by any device of executive order or executive privilege or emergency declaration or decree.

ARTICLE 7: The World Administration

7.1　Functions of the World Administration

7.1.1　The World Administration shall be organized to carry out the detailed and continuous administration and implementation of world legislation and world law.

7.1.2　The World Administration shall be under the direction of the World Executive, and shall at all times be responsible to the World Executive.

7.1.3　The World Administration shall be organized so as to give professional continuity to the work of administration and implementation.

7.2　Structure and Procedures of the World Administration

7.2.1　The World Administration shall be composed of professionally organized departments and other agencies in all areas of activity requiring continuity of administration and implementation by the World Government.

7.2.2　Each Department or major agency of the World Administration shall be headed by a Minister who shall be either a member of the Executive Cabinet or a Vice President of the Presidium.

7.2.3   Each Department or major agency of the World Administration shall have as chief of staff a Senior Administrator, who shall assist the Minister and supervise the detailed work of the Department or agency.

7.2.4   Each Senior Administrator shall be nominated by the Minister of the particular Department or agency from among persons in the senior lists of the World Civil Service Administration, as soon as senior lists have been established by the World Civil Service Administration, and shall be confirmed by the Presidium. Temporary qualified appointments shall be made by the Ministers, with confirmation by the Presidium, pending establishment of the senior lists.

7.2.5   There shall be a Secretary General of the World Administration, who shall be nominated by the Presidium and confirmed by absolute majority vote of the entire Executive Cabinet.

7.2.6   The functions and responsibilities of the Secretary General of the World Administration shall be to assist in coordinating the work of the Senior Administrators of the several Departments and agencies of the World Administration. The Secretary General shall at all times be subject to the direction of the Presidium, and shall be directly responsible to the Presidium.

7.2.7   The employment of any Senior Administrator and of the Secretary General may be terminated for cause by absolute majority vote of both the Executive Cabinet and Presidium combined, but not contrary to civil service rules which protect tenure on grounds of competence.

7.2.8   Each Minister of a Department or agency of the World Administration, being also a Member of the World Parliament, shall provide continuous liaison between the particular Department or agency and the World Parliament, shall respond at any time to any questions or requests for information from the Parliament, including committees of any House of the World Parliament.

7.2.9   The Presidium, in cooperation with the particular Ministers in each case, shall be responsible for the original organization of each of the Departments and major agencies of the World Administration.

7.2.10   The assignment of legislative measures, constitutional provisions and areas of world law to particular Departments and agencies for administration and implementation shall be done by the Presidium in consultation with the Executive Cabinet and Secretary General, unless specifically provided in legislation passed by the World Parliament.

7.2.11   The Presidium, in consultation with the Executive Cabinet, may propose the creation of other departments and agencies to have ministerial status; and may propose the alteration, combination or termination of existing Departments and agencies of ministerial status as may seem necessary or desirable. Any such creation, alteration, combination or termination shall require a simple majority vote of approval of the three houses of the World Parliament in joint session.

7.2.12   The World Parliament by absolute majority vote of the three houses in joint session may specify the creation of new departments or agencies of ministerial status in the World Administration, or may direct the World Executive to alter, combine, or terminate existing departments or agencies of ministerial status.

7.2.13   The Presidium and the World Executive may not create, establish or maintain any administrative or executive department or agency for the purpose of circumventing control by the World Parliament.

7.3   Departments of the World Administration

Among the Departments and agencies of the World Administration of ministerial status, but not limited thereto and subject to combinations and to changes in descriptive terminology, shall be those listed under this Section. Each major area of administration shall be headed by a Cabinet Minister and a Senior Administrator, or by a Vice President and a Senior Administrator.

7.3.1   Disarmament & War Prevention.
7.3.2   Population.
7.3.3   Food and Agriculture.
7.3.4   Water Supplies and Waterways.
7.3.5   Health and Nutrition.
7.3.6   Education.
7.3.7   Cultural Diversity and the Arts.
7.3.8   Habitat and Settlements.
7.3.9   Environment and Ecology.
7.3.10   World Resources.
7.3.11   Oceans and Seabeds.
7.3.12   Atmosphere and Space.
7.3.13   Energy.
7.3.14   Science and Technology.
7.3.15   Genetic Research & Engineering.
7.3.16   Labor and Income.
7.3.17   Economic & Social Development.
7.3.18   Commerce & Industry
7.3.19   Transportation and Travel.
7.3.20   Multi-National Corporations.
7.3.21   Communications & Information.
7.3.22   Human Rights.
7.3.23   Distributive Justice.
7.3.24   World Service Corps.
7.3.25   World Territories, Capitals & Parks.
7.3.26   Exterior Relations.
7.3.27   Democratic Procedures.
7.3.28   Revenue.

ARTICLE 8: The Integrative Complex

8.1   Definition

8.1.1   Certain administrative, research, planning and facilitative agencies of the World Government which are particularly essential for the satisfactory functioning of all or most aspects of the World Government, shall be designated as the Integrative Complex. The Integrative Complex shall include the agencies listed under this Section, with the proviso that other such agencies may be added upon recommendation of the Presidium followed by decision of the World Parliament.

8.1.1.1   The World Civil Service Administration.
8.1.1.2   The World Boundaries and Elections Administration.
8.1.1.3   The Institute on Governmental Procedures and World Problems.
8.1.1.4   The Agency for Research and Planning.

8.1.1.5   The Agency for Technological and Environmental Assessment.

8.1.1.6   The World Financial Administration.

8.1.1.7   Commission for Legislative Review.

8.1.2   Each agency of the Integrative Complex shall be headed by a Cabinet Minister and a Senior Administrator, or by a Vice President and a Senior Administrator, together with a Commission as provided hereunder. The rules of procedure for each agency shall be decided by majority decision of the Commission members together with the Administrator and the Minister or Vice President.

8.1.3   The World Parliament may at any time define further the responsibilities, functioning and organization of the several agencies of the Integrative Complex, consistent with the provisions of Article 8 and other provisions of the World Constitution.

8.1.4   Each agency of the Integrative Complex shall make an annual report to the World Parliament and to the Presidium.

8.2   The World Civil Service Administration

8.2.1   The functions of the World Civil Service Administration shall be the following, but not limited thereto:

8.2.1.1   To formulate and define standards, qualifications, tests, examinations and salary scales for the personnel of all organs, departments, bureaus, offices, commissions and agencies of the World Government, in conformity with the provisions of this World Constitution and requiring approval by the Presidium and Executive Cabinet, subject to review and approval by the World Parliament.

8.2.1.2   To establish rosters or lists of competent personnel for all categories of personnel to be appointed or employed in the service of the World Government.

8.2.1.3   To select and employ upon request by any government organ, department, bureau, office, institute, commission, agency or authorized official, such competent personnel as may be needed and authorized, except for those positions which are made elective or appointive under provisions of the World Constitution or by specific legislation of the World Parliament.

8.2.2   The World Civil Service Administration shall be headed by a ten member commission in addition to the Cabinet Minister or Vice President and Senior Administrator. The Commission shall be composed of one commissioner from each of ten World Electoral and Administrative Magna-Regions. The persons to serve as Commissioners shall be nominated by the House of Counselors and then appointed by the Presidium for five year terms. Commissioners may serve consecutive terms.

8.3   The World Boundaries and Elections Administration

8.3.1   The functions of the World Boundaries and Elections Administration shall include the following, but not limited thereto:

8.3.1.1   To define the boundaries for the basic World Electoral and Administrative Districts, the World Electoral and Administrative Regions and Magna-Regions, and the Continental Divisions, for submission to the World Parliament for approval by legislative action.

8.3.1.2   To make periodic adjustments every ten or five years, as needed, of the boundaries for the World Electoral and Administrative Districts, the World Electoral and Administrative Regions and Magna-Regions, and of the Continental Divisions, subject to approval by the World Parliament.

8.3.1.3   To define the detailed procedures for the nomination and election of Members of the World Parliament to the House of Peoples and to the House of Counselors, subject to approval by the World Parliament.

8.3.1.4   To conduct the elections for Members of the World Parliament to the House of Peoples and to the House of Counselors.

8.3.1.5   Before each World Parliamentary Election, to prepare Voters' Information Booklets which shall summarize major current public issues, and shall list each candidate for elective office together with standard information about each candidate, and give space for each candidate to state his or her views on the defined major issues as well as on any other major issue of choice; to include information on any initiatives or referendums which are to be voted upon; to distribute the Voter's Information Booklets for each World Electoral District, or suitable group of Districts; and to obtain the advice of the Institute on Governmental Procedures and World Problems, the Agency for Research and Planning, and the Agency for Technological and Environmental Assessment in preparing the booklets.

8.3.1.6   To define the rules for world political parties, subject to approval by the World Parliament, and subject to review and recommendations of the World Ombudsmus.

8.3.1.7   To define the detailed procedures for legislative initiative and referendum by the Citizens of Earth, and to conduct voting on supra- national or global initiatives and referendums in conjunction with world parliamentary elections.

8.3.1.8   To conduct plebiscites when requested by other Organs of the World Government, and to make recommendations for the settlement of boundary disputes.

8.3.1.9   To conduct a global census every five years, and to prepare and maintain complete demographic analyses for Earth.

8.3.2   The World Boundaries and Elections Administration shall be headed by a ten member commission in addition to the Senior Administrator and the Cabinet Minister or Vice President. The commission shall be composed of one commissioner each from ten World Electoral and Administrative Magna-Regions. The persons to serve as commissioners shall be nominated by the House of Counselors and then appointed by the World Presidium for five year terms. Commissioners may serve consecutive terms.

8.4   Institute on Governmental Procedures and World Problems

8.4.1   The functions of the Institute on Governmental Procedures and World Problems shall be as follows, but not limited thereto:

8.4.1.1   To prepare and conduct courses of information, education and training for all personnel in the service of the World Government, including Members of the World Parliament and of all other elective, appointive and civil service personnel, so that every person in the service of the World Government may have a better understanding of the functions, structure, procedures and inter-relationships of the various organs, departments, bureaus, offices, institutes, commissions, agencies and other parts of the World Government.

8.4.1.2   To prepare and conduct courses and seminars for information, education, discussion, updating and new ideas in all areas of world problems, particularly for Members of the World Parliament and of the World Executive, and for the chief personnel of all organs, departments and agencies of the World Government, but open to all in the service of the World Government.

8.4.1.3   To bring in qualified persons from private and public universities, colleges and research and action organizations of many countries, as well as other qualified persons, to lecture and to be resource persons for the courses and seminars organized by the Institute on Governmental Procedures and World Problems.

8.4.1.4   To contract with private or public universities and colleges or other agencies to conduct courses and seminars for the Institute.

8.4.2   The Institute on Governmental Procedures and World Problems shall be supervised by a ten member commission in addition to the Senior Administrator and Cabinet Minister or Vice President. The commission shall be composed of one commissioner each to be named by the House of Peoples, the House of Nations, the House of Counselors, the Presidium, the Collegium of World Judges, The World Ombudsmus, The World Attorneys General Office, the Agency for Research and Planning, the Agency for Technological and Environmental Assessment, and the World Financial Administration. Commissioners shall serve five year terms, and may serve consecutive terms.

8.5   The Agency for Research and Planning

8.5.1   The functions of the Agency for Research and Planning shall be as follows, but not limited thereto:

8.5.1.1   To serve the World Parliament, the World Executive, the World Administration, and other organs, departments and agencies of the World Government in any matter requiring research and planning within the competence of the agency.

8.5.1.2   To prepare and maintain a comprehensive inventory of world resources.

8.5.1.3   To prepare comprehensive long-range plans for the development, conservation, re-cycling and equitable sharing of the resources of Earth for the benefit of all people on Earth, subject to legislative action by the World Parliament.

8.5.1.4   To prepare and maintain a comprehensive list and description of all world problems, including their inter-relationships, impact time projections and proposed solutions, together with bibliographies.

8.5.1.5   To do research and help prepare legislative measures at the request of any Member of the World Parliament or of any committee of any House of the World Parliament.

8.5.1.6   To do research and help prepare proposed legislation or proposed legislative programs and schedules at the request of the Presidium or Executive Cabinet or of any Cabinet Minister.

8.5.1.7   To do research and prepare reports at the request of any other organ, department or agency of the World Government.

8.5.1.8   To enlist the help of public and private universities, colleges, research agencies, and other associations and organizations for various research and planning projects.

8.5.1.9   To contract with public and private universities, colleges, research agencies and other organizations for the preparation of specific reports, studies and proposals.

8.5.1.10   To maintain a comprehensive World Library for the use of all Members of the World Parliament, and for the use of all other officials and persons in the service of the World Government, as well as for public information.

8.5.2   The Agency for Research and Planning shall be supervised by a ten member commission in addition to the Senior Administrator and Cabinet Minister or Vice President. The commission shall be composed of one commissioner each to be named by the House of Peoples, the House of Nations, the House of Counselors, the Presidium, the Collegium of World Judges, the Office of World Attorneys General, World Ombudsmus, the Agency for Technological and Environmental Assessment, the Institute on Governmental Procedures and World Problems, and the World Financial Administration. Commissioners shall serve five year terms, and may serve consecutive terms.

8.6   The Agency for Technological and Environmental Assessment
8.6.1   The functions of the agency for Technological and Environmental Assessment shall include the following, but not limited thereto:

8.6.1.1   To establish and maintain a registration and description of all significant technological innovations, together with impact projections.

8.6.1.2   To examine, analyze and assess the impacts and consequences of technological innovations which may have either significant beneficial or significant harmful or dangerous consequences for human life or for the ecology of life on Earth, or which may require particular regulations or prohibitions to prevent or eliminate dangers or to assure benefits.

8.6.1.3   To examine, analyze and assess environmental and ecological problems, in particular the environmental and ecological problems which may result from any intrusions or changes of the environment or ecological relationships which may be caused by technological innovations, processes of resource development, patterns of human settlements, the production of energy, patterns of economic and industrial development, or other man-made intrusions and changes of the environment, or which may result from natural causes.

8.6.1.4   To maintain a global monitoring network to measure possible harmful effects of technological innovations and environmental disturbances so that corrective measures can be designed.

8.6.1.5   To prepare recommendations based on technological and environmental analyses and assessments, which can serve as guides to the World Parliament, the World Executive, the World Administration, the Agency for Research and Planning, and to the other organs, departments and agencies of the World Government, as well as to individuals in the service of the World Government and to national and local governments and legislative bodies.

8.6.1.6   To enlist the voluntary or contractual aid and participation of private and public universities, colleges, research institutions and other associations and organizations in the work of technological and environmental assessment.

8.6.1.7   To enlist the voluntary or contractual aid and participation of private and public universities and colleges, research institutions and other organizations in devising and developing alternatives to harmful or dangerous technologies and environmentally disruptive activities, and in devising controls to assure beneficial results from technological innovations or to prevent harmful results from either technological innovations or environmental changes, all subject to legislation for implementation by the World Parliament.

8.6.2   The Agency for Technological and Environmental Assessment shall be supervised by a ten member commission in addition to the Senior Administrator and Cabinet Minister or Vice President. The commission shall be composed of one commissioner from each of ten World Electoral and Administrative Magna-Regions. The persons to serve as commissioners shall be nominated by the House of Counselors, and then appointed by the World Presidium for five year terms. Commissioners may serve consecutive terms.

8.7   The World Financial Administration

8.7.1   The functions of the World Financial Administration shall include the following, but not limited thereto:

8.7.1.1   To establish and operate the procedures for the collection of revenues for the World Government, pursuant to legislation by the World Parliament, inclusive of taxes, globally accounted social and public costs, licenses, fees, revenue sharing arrangements, income derived from supra-national public enterprises or projects or resource developments, and all other sources.

8.7.1.2   To operate a Planetary Accounting Office, and thereunder to make cost/benefit studies and reports of the functioning and activities of the World Government and of its several organs, departments, branches, bureaus, offices, commissions, institutes, agencies and other parts or projects. In making such studies and reports, account shall be taken not only of direct financial costs and benefits, but also of human, social, environmental, indirect, long-term and other costs and benefits, and of actual or possible hazards and damages. Such studies and reports shall also be designed to uncover any wastes, inefficiencies, misapplications, corruptions, diversions, unnecessary costs, and other possible irregularities.

8.7.1.3   To make cost/benefit studies and reports at the request of any House or committee of the World Parliament, and of the Presidium, the Executive Cabinet, the World Ombudsmus, the Office of World Attorneys General, the World Supreme Court, or of any administrative department or any agency of the Integrative Complex, as well as upon its own initiative.

8.7.1.4   To operate a Planetary Comptroller's Office and thereunder to supervise the disbursement of the funds of the World Government for all purposes, projects and activities duly authorized by this World Constitution, the World Parliament, the World Executive, and other organs, departments and agencies of the World Government.

8.7.1.5   To establish and operate a Planetary Banking System, making the transition to a common global currency, under the terms of specific legislation passed by the World Parliament.

8.7.1.6   Pursuant to specific legislation enacted by the World Parliament, and in conjunction with the Planetary Banking System, to establish and implement the procedures of a Planetary Monetary and Credit System based upon useful productive capacity and performance, both in goods and services. Such a monetary and credit system shall be designed for use within the Planetary Banking System for the financing of the activities and projects of the World Government, and for all other financial purposes approved by the World Parliament, without requiring the payment of interest on bonds, investments or other claims of financial ownership or debt.

8.7.1.7 To establish criteria for the extension of financial credit based upon such considerations as people available to work, usefulness, cost/benefit accounting, human and social values, environmental health and esthetics, minimizing disparities, integrity, competent management, appropriate technology, potential production and performance.

8.7.1.8 To establish and operate a Planetary Insurance System in areas of world need which transcend national boundaries and in accordance with legislation passed by the World Parliament.

8.7.1.9 To assist the Presidium as may be requested in the technical preparation of budgets for the operation of the World Government.

8.7.2 The World Financial Administration shall be supervised by a commission of ten members, together with a Senior Administrator and a Cabinet Minister or Vice President. The commission shall be composed of one commissioner each to be named by the House of Peoples, the House of Nations, the House of Counselors, the Presidium, the Collegium of World Judges, the Office of Attorneys General, the World Ombudsmus, the Agency for Research and Planning, the Agency for Technological and Environmental Assessment, and the Institute on Governmental Procedures and World Problems. Commissioners shall serve terms of five years, and may serve consecutive terms.

8.8 Commission for Legislative Review

8.8.1 The functions of the Commission for Legislative Review shall be to examine World Legislation and World Laws which the World Parliament enacts or adopts from the previous Body of International Law for the purpose of analyzing whether any particular legislation or law has become obsolete or obstructive or defective in serving the purposes intended; and to make recommendations to the World Parliament accordingly for repeal or amendment or replacement.

8.8.2 The Commission for Legislative Review shall be composed of twelve members, including two each to be elected by the House of Peoples, the House of Nations, the House of Counselors, the Collegium of World Judges, the World Ombudsmus and the Presidium. Members of the Commission shall serve terms of ten years, and may be re-elected to serve consecutive terms. One half of the Commission members after the Commission is first formed shall be elected every five years, with the first terms for one half of the members to be only five years.

ARTICLE 9: The World Judiciary

9.1 Jurisdiction of the World Supreme Court

9.1.1 A World Supreme Court shall be established, together with such regional and district World Courts as may subsequently be found necessary. The World Supreme Court shall comprise a number of benches.

9.1.2 The World Supreme Court, together with such regional and district World Courts as may be established, shall have mandatory jurisdiction in all cases, actions, disputes, conflicts, violations of law, civil suits, guarantees of civil and human rights, constitutional interpretations, and other litigations arising under the provisions of this World Constitution, world legislation, and the body of world law approved by the World Parliament.

9.1.3 Decisions of the World Supreme Court shall be binding on all parties involved in all cases, actions and litigations brought before any bench of the World Supreme Court for settlement. Each bench of the World Supreme Court shall constitute a court of highest appeal, except when matters of extra-ordinary public importance are assigned or transferred to the Superior Tribunal of the World Supreme Court, as provided in Section 5 of Article 9.

9.2   Benches of the World Supreme Court

The benches of the World Supreme Court and their respective jurisdictions shall be as follows:

9.2.1   Bench for Human Rights: To deal with issues of human rights arising under the guarantee of civil and human rights provided by Article 12 of this World Constitution, and arising in pursuance of the provisions of Article 13 of this World Constitution, and arising otherwise under world legislation and the body of world law approved by the World Parliament.

9.2.2   Bench for Criminal Cases: To deal with issues arising from the violation of world laws and world legislation by individuals, corporations, groups and associations, but not issues primarily concerned with human rights.

9.2.3   Bench for Civil Cases: To deal with issues involving civil law suits and disputes between individuals, corporations, groups and associations arising under world legislation and world law and the administration thereof.

9.2.4   Bench for Constitutional Cases: To deal with the interpretation of the World Constitution and with issues and actions arising in connection with the interpretation of the World Constitution.

9.2.5   Bench for International Conflicts: To deal with disputes, conflicts and legal contest arising between or among the nations which have joined in the Federation of Earth.

9.2.6   Bench for Public Cases: To deal with issues not under the jurisdiction of another bench arising from conflicts, disputes, civil suits or other legal contests between the World Government and corporations, groups or individuals, or between national governments and corporations, groups or individuals in cases involving world legislation and world law.

9.2.7   Appellate Bench: To deal with issues involving world legislation and world law which may be appealed from national courts; and to decide which bench to assign a case or action or litigation when a question or disagreement arises over the proper jurisdiction.

9.2.8   Advisory Bench: To give opinions upon request on any legal question arising under world law or world legislation, exclusive of contests or actions involving interpretation of the World Constitution. Advisory opinions may be requested by any House or committee of the World Parliament, by the Presidium, any Administrative Department, the Office of World Attorneys General, the World Ombudsmus, or by any agency of the Integrative Complex.

9.2.9   Other benches may be established, combined or terminated upon recommendation of the Collegium of World Judges with approval by the World Parliament; but benches number one through eight may not be combined nor terminated except by amendment of this World Constitution.

9.3   Seats of the World Supreme Court

9.3.1   The primary seat of the World Supreme Court and all benches shall be the same as for the location of the Primary World Capital and for the location of the World Parliament and the World Executive.

9.3.2   Continental seats of the World Supreme Court shall be established in the four secondary capitals of the World Government located in four different Continental Divisions of Earth, as provided in Article 15.

9.3.3   The following permanent benches of the World Supreme Court shall be established both at the primary seat and at each of the continental seats: Human Rights, Criminal Cases, Civil Cases, and Public Cases.

9.3.4   The following permanent benches of the World Supreme Court shall be located only at the primary seat of the World Supreme Court: Constitutional Cases, International Conflicts, Appellate Bench, and Advisory Bench.

9.3.5   Benches which are located permanently only at the primary seat of the World Supreme Court may hold special sessions at the other continental seats of the World Supreme Court when necessary, or may establish continental circuits if needed.

9.3.6   Benches of the World Supreme Court which have permanent continental locations may hold special sessions at other locations when needed, or may establish regional circuits if needed.

9.4   The Collegium of World Judges

9.4.1   A Collegium of World Judges shall be established by the World Parliament. The Collegium shall consist of a minimum of twenty member judges, and may be expanded as needed but not to exceed sixty members.

9.4.2   The World Judges to compose the Collegium of World Judges shall be nominated by the House of Counselors and shall be elected by plurality vote of the three Houses of the World Parliament in joint session. The House of Counselors shall nominate between two and three times the number of world judges to be elected at any one time. An equal number of World Judges shall be elected from each of ten World Electoral and Administrative Magna-Regions, if not immediately then by rotation.

9.4.3   The term of office for a World Judge shall be ten years. Successive terms may be served without limit.

9.4.4   The Collegium of World Judges shall elect a Presiding Council of World Judges, consisting of a Chief Justice and four Associate Chief Justices. One member of the Presiding Council of World Judges shall be elected from each of five Continental Divisions of Earth. Members of the Presiding Council of World Judges shall serve five year terms on the Presiding Council, and may serve two successive terms, but not two successive terms as Chief Justice.

9.4.5   The Presiding Council of World Judges shall assign all World Judges, including themselves, to the several benches of the World Supreme Court. Each bench for a sitting at each location shall have a minimum of three World Judges, except that the number of World Judges for benches on Continental Cases and International Conflicts, and the Appellate Bench, shall be no less than five.

9.4.6   The member judges of each bench at each location shall choose annually a Presiding Judge, who may serve two successive terms.

9.4.7   The members of the several benches may be reconstituted from time to time as may seem desirable or necessary upon the decision of the Presiding Council of World Judges. Any decision to re-constitute a bench shall be referred to a vote of the entire Collegium of World Judges by request of any World Judge.

9.4.8   Any World Judge may be removed from office for cause by an absolute two thirds majority vote of the three Houses of the World Parliament in joint session.

9.4.9   Qualifications for Judges of the World Supreme Court shall be at least ten years of legal or juristic experience, minimum age of thirty years, and evident competence in world law and the humanities.

9.4.10   The salaries, expenses, remunerations and prerogatives of the World Judges shall be determined by the World Parliament, and shall be reviewed every five years, but shall not be changed to the disadvantage of any World Judge during a term of office. All members of the Collegium of World Judges shall receive the same salaries, except that additional compensation may be given to the Presiding Council of World Judges.

9.4.11   Upon recommendation by the Collegium of World Judges, the World Parliament shall have the authority to establish regional and district world courts below the World Supreme Court, and to establish the jurisdictions thereof, and the procedures for appeal to the World Supreme Court or to the several benches thereof.

9.4.12   The detailed rules of procedure for the functioning of the World Supreme Court, the Collegium of World Judges, and for each bench of the World Supreme Court, shall be decided and amended by absolute majority vote of the Collegium of World Judges.

9.5   The Superior Tribunal of the World Supreme Court

9.5.1   A Superior Tribunal of the World Supreme Court shall be established to take cases which are considered to be of extra-ordinary public importance. The Superior Tribunal for any calendar year shall consist of the Presiding Council of World Judges together with one World Judge named by the Presiding Judge of each bench of the World Court sitting at the primary seat of the World Supreme Court. The composition of the Superior Tribunal may be continued unchanged for a second year by decision of the Presiding Council of World Judges.

9.5.2   Any party to any dispute, issue, case or litigation coming under the jurisdiction of the World Supreme Court, may apply to any particular bench of the World Supreme Court or to the Presiding Council of World Judges for the assignment or transfer of the case to the Superior Tribunal on the grounds of extra-ordinary public importance. If the application is granted, the case shall be heard and disposed of by the Superior Tribunal. Also, any bench taking any particular case, if satisfied that the case is of extra-ordinary public importance, may of its own discretion transfer the case to the Superior Tribunal.

ARTICLE 10: The Enforcement System

10.1   Basic Principles

10.1.1   The enforcement of world law and world legislation shall apply directly to individual, and individuals shall be held responsible for compliance with world law and world legislation regardless of whether the individuals are acting in their own capacity or as agents or officials of governments at any level or of the institutions of governments, or as agents or officials of corporations, organizations, associations or groups of any kind.

10.1.2   When world law or world legislation or decisions of the world courts are violated, the Enforcement System shall operate to identify and apprehend the individuals responsible for violations.

10.1.3   Any enforcement action shall not violate the civil and human rights guaranteed under this World Constitution.

10.1.4   The enforcement of world law and world legislation shall be carried out in the context of a non-military world federation wherein all member nations shall disarm as a condition for joining and benefiting from the world federation, subject to Article 17, Sec. 3.8 and 4.6. The Federation of Earth and World Government under this World Constitution shall neither keep nor use weapons of mass destruction.

10.1.5   Those agents of the enforcement system whose function shall be to apprehend and bring to court violators of world law and world legislation shall be equipped only with such weapons as are appropriate for the apprehension of the individuals responsible for violations.

10.1.6   The enforcement of world law and world legislation under this World Constitution shall be conceived and developed primarily as the processes of effective design and administration of world law and world legislation to serve the welfare of all people on Earth, with equity and justice for all, in which the resources of Earth and the funds and the credits of the World Government are used only to serve peaceful human needs, and none used for weapons of mass destruction or for war making capabilities.

10.2   The Structure for Enforcement: World Attorneys General

10.2.1   The Enforcement System shall be headed by an Office of World Attorneys General and a Commission of Regional World Attorneys.

10.2.2   The Office of World Attorneys General shall be comprised of five members, one of whom shall be designated as the World Attorney General and the other four shall each be designated an Associate World Attorney General.

10.2.3   The Commission of Regional World Attorneys shall consist of twenty Regional World Attorneys.

10.2.4   The members to compose the Office of World Attorneys General shall be nominated by the House of Counselors, with three nominees from each Continental Division of Earth. One member of the Office shall be elected from each of five Continental Divisions by plurality vote of the three houses of the World Parliament in joint session.

10.2.5   The term of office for a member of the Office of World Attorneys General shall be ten years. A member may serve two consecutive terms. The position of World Attorney General shall rotate every two years among the five members of the Office. The order of rotation shall be decided among the five members of the Office.

10.2.6   The Office of World Attorneys General shall nominate members for the Commission of twenty Regional World Attorneys from the twenty World Electoral and Administrative Regions, with between two and three nominees submitted for each Region. From these nominations, the three Houses of the World Parliament in joint session shall elect one Regional World Attorney from each of the twenty Regions. Regional World Attorneys shall serve terms of five years, and may serve three consecutive terms.

10.2.7   Each Regional World Attorney shall organize and be in charge of an Office of Regional World Attorney. Each Associate World Attorney General shall supervise five Offices of Regional World Attorneys.

10.2.8   The staff to carry out the work of enforcement, in addition to the five members of the Office of World Attorneys General and the twenty Regional World Attorneys, shall be selected from civil service lists, and shall be organized for the following functions:

10.2.8.1 Investigation.

10.2.8.2 Apprehension and arrest.

10.2.8.3 Prosecution.

10.2.8.4 Remedies and correction.

10.2.8.5 Conflict resolution.

10.2.9   Qualifications for a member of the Office of World Attorneys General and for the Regional World Attorneys shall be at least thirty years of age, at least seven years legal experience, and education in law and the humanities.

10.2.10   The World Attorney General, the Associate World Attorneys General, and the Regional World Attorneys shall at all times be responsible to the World Parliament. Any member of the Office of World Attorneys General and any Regional World Attorney can be removed from office for cause by a simple majority vote of the three Houses of the World Parliament in joint session.

10.3   The World Police

10.3.1   That section of the staff of the Office of World Attorneys General and of the Offices of Regional World Attorneys responsible for the apprehension and arrest of violators of world law and world legislation, shall be designated as World Police.

10.3.2   Each regional staff of the World Police shall be headed by a Regional World Police Captain, who shall be appointed by the Regional World Attorney.

10.3.3   The Office of World Attorneys General shall appoint a World Police Supervisor, to be in charge of those activities which transcend regional boundaries. The World Police Supervisor shall direct the Regional World Police Captains in any actions which require coordinated or joint action transcending regional boundaries, and shall direct any action which requires initiation or direction from the Office of World Attorneys General.

10.3.4   Searches and arrests to be made by World Police shall be made only upon warrants issued by the Office of World Attorneys General or by a Regional World Attorney.

10.3.5   World Police shall be armed only with weapons appropriate for the apprehension of the individuals responsible for violation of world law.

10.3.6   Employment in the capacity of World Police Captain and World Police Supervisor shall be limited to ten years.

10.3.7   The World Police Supervisor and any Regional World Police Captain may be removed from office for cause by decision of the Office of World Attorneys General or by absolute majority vote of the three Houses of the World Parliament in joint session.

10.4   The Means of Enforcement

10.4.1   Non-military means of enforcement of world law and world legislation shall be developed by the World Parliament and by the Office of World Attorneys General in consultation with the Commission of Regional World Attorneys, the Collegium of World Judges, the World Presidium, and the World Ombudsmus. The actual means of enforcement shall require legislation by the World Parliament.

10.4.2   Non-military means of enforcement which can be developed may include: Denial of financial credit; denial of material resources and personnel; revocation of licenses, charters, or corporate rights; impounding of equipment; fines and damage payments; performance of work to rectify damages; imprisonment or isolation; and other means appropriate to the specific situations.

10.4.3   To cope with situations of potential or actual riots, insurrection and resort to armed violence, particular strategies and methods shall be developed by the World Parliament and by the Office of World Attorneys General in consultation with the Commission of Regional World Attorneys, the Collegium of World Judges, the Presidium and the World Ombudsmus. Such strategies and methods shall require enabling legislation by the World Parliament where required in addition to the specific provisions of this World Constitution.

10.4.4   A basic condition for preventing outbreaks of violence which the Enforcement System shall facilitate in every way possible, shall be to assure a fair hearing under non-violent circumstances for any person or group having a grievance, and likewise to assure a fair opportunity for a just settlement of any grievance with due regard for the rights and welfare of all concerned.

ARTICLE 11: The World Ombudsmus

11.1 Functions and Powers of the World Ombudsmus

The functions and powers of the World Ombudsmus, as public defender, shall include the following:

11.1.1 To protect the People of Earth and all individuals against violations or neglect of universal human and civil rights which are stipulated in Article 12 and other sections of this World Constitution.

11.1.2 To protect the People of Earth against violations of this World Constitution by any official or agency of the World Government, including both elected and appointed officials or public employees regardless of organ, department, office, agency or rank.

11.1.3 To press for the implementation of the Directive Principles for the World Government as defined in Article 13 of this World Constitution.

11.1.4 To promote the welfare of the people of Earth by seeking to assure that conditions of social justice and of minimizing disparities are achieved in the implementation and administration of world legislation and world law.

11.1.5 To keep on the alert for perils to humanity arising from technological innovations, environmental disruptions and other diverse sources, and to launch initiatives for correction or prevention of such perils.

11.1.6 To ascertain that the administration of otherwise proper laws, ordinances and procedures of the World Government do not result in unforeseen injustices or inequities, or become stultified in bureaucracy or the details of administration.

11.1.7 To receive and hear complaints, grievances or requests for aid from any person, group, organization, association, body politic or agency concerning any matter which comes within the purview of the World Ombudsmus.

11.1.8 To request the Office of World Attorneys General or any Regional World Attorney to initiate legal actions or court proceedings whenever and wherever considered necessary or desirable in the view of the World Ombudsmus.

11.1.9 To directly initiate legal actions and court proceedings whenever the World Ombudsmus deems necessary.

11.1.10 To review the functioning of the departments, bureaus, offices, commissions, institutes, organs and agencies of the World Government to ascertain whether the procedures of the World government are adequately fulfilling their purposes and serving the welfare of humanity in optimum fashion, and to make recommendations for improvements.

11.1.11 To present an annual report to the World Parliament and to the Presidium on the activities of the World Ombudsmus, together with any recommendations for legislative measures to improve the functioning of the World Government for the purpose of better serving the welfare of the People of Earth.

11.2 Composition of the World Ombudsmus

11.2.1 The World Ombudsmus shall be headed by a Council of World Ombudsen of five members, one of whom shall be designated as Principal World Ombudsan, while the other four shall each be designated as an Associate World Ombudsan.

11.2.2 Members to compose the Council of World Ombudsen shall be nominated by the House of Counselors, with three nominees from each Continental Division of Earth. One member of the Council shall be elected from each of five Continental Divisions by plurality vote of the three Houses of the World Parliament in joint session.

11.2.3 The term of office for a World Ombudsan shall be ten years. A World Ombudsan may serve two successive terms. The position of Principal World Ombudsan shall be rotated every two years. The order of rotation shall be determined by the Council of World Ombudsen.

11.2.4 The Council of World Ombudsen shall be assisted by a Commission of World Advocates of twenty members. Members for the Commission of World Advocates shall be nominated by the Council of World Ombudsen from twenty World Electoral and Administrative Regions, with between two and three nominees submitted for each Region. One World Advocate shall be elected from each of the twenty World Electoral and Administrative Regions by the three Houses of the World Parliament in joint session. World Advocates shall serve terms of five years, and may serve a maximum of four successive terms.

11.2.5 The Council of World Ombudsen shall establish twenty regional offices, in addition to the principal world office at the primary seat of the World Government. The twenty regional offices of the World Ombudsmus shall parallel the organization of the twenty Offices of Regional World Attorney.

11.2.6 Each regional office of the World Ombudsmus shall be headed by a World Advocate. Each five regional offices of the World Ombudmus shall be supervised by an Associate World Ombudsan.

11.2.7 Any World Ombudsan and any World Advocate may be removed from office for cause by an absolute majority vote of the three Houses of the World Parliament in joint session.

11.2.8 Staff members for the World Ombudsmus and for each regional office of the World Ombudsmus shall be selected and employed from civil service lists.

11.2.9 Qualifications for World Ombudsan and for World Advocate shall be at least thirty years of age, at least five years legal experience, and education in law and other relevant education.

ARTICLE 12: Bill of Rights for the Citizens of Earth

The inhabitants and citizens of Earth who are within the Federation of Earth shall have certain inalienable rights defined hereunder. It shall be mandatory for the World Parliament, the World Executive, and all organs and agencies of the World Government to honor, implement and enforce these rights, as well as for the national governments of all member nations in the Federation of Earth to do likewise. Individuals or groups suffering violation or neglect of such rights shall have full recourse through the World Ombudsmus, the Enforcement System and the World Courts for redress of grievances. The inalienable rights shall include the following:

12.1 Equal rights for all citizens of the Federation of Earth, with no discrimination on grounds of race, color, caste, nationality, sex, religion, political affiliation, property, or social status.

12.2 Equal protection and application of world legislation and world laws for all citizens of the Federation of Earth.

12.3 Freedom of thought and conscience, speech, press, writing, communication, expression, publication, broadcasting, telecasting, and cinema, except as an overt part of or incitement to violence, armed riot or insurrection.

12.4 Freedom of assembly, association, organization, petition and peaceful demonstration.

12.5 Freedom to vote without duress, and freedom for political organization and campaigning without censorship or recrimination.

12.6 Freedom to profess, practice and promote religion or religious beliefs or no religion or religious belief.

12.7   Freedom to profess and promote political beliefs or no political beliefs.
12.8   Freedom for investigation, research and reporting.
12.9   Freedom to travel without passport or visas or other forms of registration used to limit travel between, among or within nations.
12.10   Prohibition against slavery, peonage, involuntary servitude, and conscription of labor.
12.11   Prohibition against military conscription.
12.12   Safety of person from arbitrary or unreasonable arrest, detention, exile, search or seizure; requirement of warrants for searches and arrests.
12.13   Prohibition against physical or psychological duress or torture during any period of investigation, arrest, detention or imprisonment, and against cruel or unusual punishment.
12.14   Right of habeas corpus; no ex-post-facto laws; no double jeopardy; right to refuse self-incrimination or the incrimination of another.
12.15   Prohibition against private armies and paramilitary organizations as being threats to the common peace and safety.
12.16   Safety of property from arbitrary seizure; protection against exercise of the power of eminent domain without reasonable compensation.
12.17   Right to family planning and free public assistance to achieve family planning objectives.
12.18   Right of privacy of person, family and association; prohibition against surveillance as a means of political control.

ARTICLE 13: Directive Principles for the Earth Federation

It shall be the aim of the World Government to secure certain other rights for all inhabitants within the Federation of Earth, but without immediate guarantee of universal achievement and enforcement. These rights are defined as Directive Principles, obligating the World Government to pursue every reasonable means for universal realization and implementation, and shall include the following:

13.1   Equal opportunity for useful employment for everyone, with wages or remuneration sufficient to assure human dignity.
13.2   Freedom of choice in work, occupation, employment or profession.
13.3   Full access to information and to the accumulated knowledge of the human race.
13.4   Free and adequate public education available to everyone, extending to the pre-university level; Equal opportunities for elementary and higher education for all persons; equal opportunity for continued education for all persons throughout life; the right of any person or parent to choose a private educational institution at any time.
13.5   Free and adequate public health services and medical care available to everyone throughout life under conditions of free choice.
13.6   Equal opportunity for leisure time for everyone; better distribution of the work load of society so that every person may have equitable leisure time opportunities.
13.7   Equal opportunity for everyone to enjoy the benefits of scientific and technological discoveries and developments.
13.8   Protection for everyone against the hazards and perils of technological innovations and developments.
13.9   Protection of the natural environment which is the common heritage of humanity against pollution, ecological disruption or damage which could imperil life or lower the quality of life.
13.10   Conservation of those natural resources of Earth which are limited so that present and future generations may continue to enjoy life on the planet Earth.

13.11   Assurance for everyone of adequate housing, of adequate and nutritious food supplies, of safe and adequate water supplies, of pure air with protection of oxygen supplies and the ozone layer, and in general for the continuance of an environment which can sustain healthy living for all.

13.12   Assure to each child the right to the full realization of his or her potential.

13.13   Social Security for everyone to relieve the hazards of unemployment, sickness, old age, family circumstances, disability, catastrophes of nature, and technological change, and to allow retirement with sufficient lifetime income for living under conditions of human dignity during older age.

13.14   Rapid elimination of and prohibitions against technological hazards and manmade environmental disturbances which are found to create dangers to life on Earth.

13.15   Implementation of intensive programs to discover, develop and institute safe alternatives and practical substitutions for technologies which must be eliminated and prohibited because of hazards and dangers to life.

13.16   Encouragement for cultural diversity; encouragement for decentralized administration.

13.17   Freedom for peaceful self-determination for minorities, refugees and dissenters.

13.18   Freedom for change of residence to anywhere on Earth conditioned by provisions for temporary sanctuaries in events of large numbers of refugees, stateless persons, or mass migrations.

13.19   Prohibition against the death penalty.

ARTICLE 14: Safeguards and Reservations

14.1   **Certain Safeguards**
The World Government shall operate to secure for all nations and peoples within the Federation of Earth the safeguards which are defined hereunder:

14.1.1   Guarantee that full faith and credit shall be given to the public acts, records, legislation and judicial proceedings of the member nations within the Federation of Earth, consistent with the several provisions of this World Constitution.

14.1.2   Assure freedom of choice within the member nations and countries of the Federation of Earth to determine their internal political, economic and social systems consistent with the guarantees and protections given under this World Constitution to assure civil liberties and human rights and a safe environment for life, and otherwise consistent with the several provisions of this World Constitution.

14.1.3   Grant the right of asylum within the Federation of Earth for persons who may seek refuge from countries or nations which are not yet included within the Federation of Earth.

14.1.4   Grant the right of individuals and groups, after the Federation of Earth includes 90 percent of the territory of Earth, to peacefully leave the hegemony of the Federation of Earth and to live in suitable territory set aside by the Federation neither restricted nor protected by the World Government, provided that such territory does not extend beyond five percent of Earth's habitable territory, is kept completely disarmed and not used as a base for inciting violence or insurrection within or against the Federation of Earth or any member nation, and is kept free of acts of environmental or technological damage which seriously affect Earth outside such territory.

14.2   **Reservation of Powers**
The powers not delegated to the World Government by this World Constitution shall be reserved to the nations of the Federation of Earth and to the people of Earth.

ARTICLE 15: World Federal Zones and the World Capitals

15.1   World Federal Zones
15.1.1   Twenty World Federal Zones shall be established within the twenty World Electoral and Administrative Regions, for the purposes of the location of the several organs of the World Government and of the administrative departments, the world courts, the offices of the Regional World Attorneys, the offices of the World Advocates, and for the location of other branches, departments, institutes, offices, bureaus, commissions, agencies and parts of the World Government.
15.1.2   The World Federal Zones shall be established as the needs and resources of the World Government develop and expand. World Federal Zones shall be established first within each of five Continental Divisions.
15.1.3   The location and administration of the World Federal Zones, including the first five, shall be determined by the World Parliament.

15.2   The World Capitals
15.2.1   Five World Capitals shall be established in each of five Continental Divisions of Earth, to be located in each of the five World Federal Zones which are established first as provided in Article 15 of this World Constitution.
15.2.2   One of the World Capitals shall be designated by the World Parliament as the Primary World Capital, and the other four shall be designated as Secondary World Capitals.
15.2.3   The primary seats of all organs of the World Government shall be located in the Primary World Capital, and other major seats of the several organs of the World Government shall be located in the Secondary World Capitals.

15.3   Locational Procedures
15.3.1   Choices for location of the twenty World Federal Zones and for the five World Capitals shall be proposed by the Presidium, and then shall be decided by a simple majority vote of the three Houses of the World Parliament in joint session. The Presidium shall offer choices of two or three locations in each of the twenty World Electoral and Administrative Regions to be World Federal Zones, and shall offer two alternative choices for each of the five World Capitals.
15.3.2   The Presidium in consultation with the Executive Cabinet shall then propose which of the five World Capitals shall be the Primary World Capital, to be decided by a simple majority vote of the three Houses of the World Parliament in joint session.
15.3.3   Each organ of the World Government shall decide how best to apportion and organize its functions and activities among the five World Capitals, and among the twenty World Federal Zones, subject to specific directions from the World Parliament.
15.3.4   The World Parliament may decide to rotate its sessions among the five World Capitals, and if so, to decide the procedure for rotation.
15.3.5   For the first two operative stages of World Government as defined in Article 17, and for the Provisional World Government as defined in Article 19, a provisional location may be selected for the Primary World Capital. The provisional location need not be continued as a permanent location.
15.3.6   Any World Capital or World Federal Zone may be relocated by an absolute two-thirds majority vote of the three Houses of the World Parliament in joint session.
15.3.7   Additional World Federal Zones may be designated if found necessary by proposal of the Presidium and approval by an absolute majority vote of the three Houses of the World Parliament in joint session.

ARTICLE 16: World Territories and Exterior Relations

16.1 World Territory

16.1.1 Those areas of the Earth and Earth's moon which are not under the jurisdiction of existing nations at the time of forming the Federation of Earth, or which are not reasonably within the province of national ownership and administration, or which are declared to be World Territory subsequent to establishment of the Federation of Earth, shall be designated as World Territory and shall belong to all of the people of Earth.

16.1.2 The administration of World Territory shall be determined by the World Parliament and implemented by the World Executive, and shall apply to the following areas:

16.1.2.1 All oceans and seas having an international or supra-national character, together with the seabeds and resources thereof, beginning at a distance of twenty kilometers offshore, excluding inland seas of traditional national ownership.

16.1.2.2 Vital straits, channels, and canals.

16.1.2.3 The atmosphere enveloping Earth, beginning at an elevation of one kilometer above the general surface of the land, excluding the depressions in areas of much variation in elevation.

16.1.2.4 Man-made satellites and Earth's moon.

16.1.2.5 Colonies which may choose the status of World Territory; non-independent territories under the trust administration of nations or of the United Nations; any islands or atolls which are unclaimed by any nation; independent lands or countries which choose the status of World Territory; and disputed lands which choose the status of World Territory.

16.1.3 The residents of any World Territory, except designated World Federal Zones, shall have the right within reason to decide by plebiscite to become a self-governing nation within the Federation of Earth, either singly or in combination with other World Territories, or to unite with an existing nation with the Federation of Earth.

16.2 Exterior Relations

16.2.1 The World Government shall maintain exterior relations with those nations of Earth which have not joined the Federation of Earth. Exterior relations shall be under the administration of the Presidium, subject at all times to specific instructions and approval by the World Parliament.

16.2.2 All treaties and agreements with nations remaining outside the Federation of Earth shall be negotiated by the Presidium and must be ratified by a simple majority vote of the three Houses of the World Parliament.

16.2.3 The World Government for the Federation of Earth shall establish and maintain peaceful relations with other planets and celestial bodies where and when it may become possible to establish communications with the possible inhabitants thereof.

16.2.4 All explorations into outer space, both within and beyond the solar system in which Planet Earth is located, shall be under the exclusive direction and control of the World Government, and shall be conducted in such manner as shall be determined by the World Parliament.

ARTICLE 17: Ratification and Implementation

17.1   Ratification of the World Constitution
This World Constitution shall be submitted to the nations and people of Earth for ratification by the following procedures:

17.1.1   The World Constitution shall be transmitted to the General Assembly of the United Nations Organization and to each national government on Earth, with the request that the World Constitution be submitted to the national legislature of each nation for preliminary ratification and to the people of each nation for final ratification by popular referendum.

17.1.2   Preliminary ratification by a national legislature shall be accomplished by simple majority vote of the national legislature.

17.1.3   Final ratification by the people shall be accomplished by a simple majority of votes cast in a popular referendum, provided that a minimum of twenty-five percent of eligible voters of age eighteen years and over have cast ballots within the nation or country or within World Electoral and Administrative Districts.

17.1.4   In the case of a nation without a national legislature, the head of the national government shall be requested to give preliminary ratification and to submit the World Constitution for final ratification by popular referendum.

17.1.5   In the event that a national government, after six months, fails to submit the World Constitution for ratification as requested, then the global agency assuming responsibility for the worldwide ratification campaign may proceed to conduct a direct referendum for ratification of the World Constitution by the people. Direct referendums may be organized on the basis of entire nations or countries, or on the basis of existing defined communities within nations.

17.1.6   In the event of a direct ratification referendum, final ratification shall be accomplished by a majority of the votes cast whether for an entire nation or for a World Electoral and Administrative District, provided that ballots are cast by a minimum of twenty-five percent of eligible voters of the area who are over eighteen years of age.

17.1.7   For ratification by existing communities within a nation, the procedure shall be to request local communities, cities, counties, states, provinces, cantons, prefectures, tribal jurisdictions, or other defined political units within a nation to ratify the World Constitution, and to submit the World Constitution for a referendum vote by the citizens of the community or political unit. Ratification may be accomplished by proceeding in this way until all eligible voters of age eighteen and above within the nation or World Electoral and Administrative District have had the opportunity to vote, provided that ballots are cast by a minimum of twenty-five percent of those eligible to vote.

17.1.8   Prior to the Full Operative Stage of World Government, as defined under Section 5 of Article 17, the universities, colleges and scientific academies and institutes in any country may ratify the World Constitution, thus qualifying them for participation in the nomination of Members of the World Parliament to the House of Counselors.

17.1.9   In the case of those nations currently involved in serious international disputes or where traditional enmities and chronic disputes may exist among two or more nations, a procedure for concurrent paired ratification shall be instituted whereby the nations which are parties to a current or chronic international dispute or conflict may simultaneously ratify the World Constitution. In such cases, the paired nations shall be admitted into the Federation of Earth simultaneously, with the obligation for each such nation to immediately turn over all weapons of mass destruction to the World Government, and to turn over the conflict or dispute for mandatory peaceful settlement by the World Government.

17.1.10 Each nation or political unit which ratifies this World Constitution, either by preliminary ratification or final ratification, shall be bound never to use any armed forces or weapons of mass destruction against another member or unit of the Federation of Earth, regardless of how long it may take to achieve full disarmament of all the nations and political units which ratify this World Constitution.

17.1.11 When ratified, the Constitution for the Federation of Earth becomes the supreme law of Earth. By the act of ratifying this Earth Constitution, any provision in the Constitution or Legislation of any country so ratifying, which is contrary to this Earth Constitution, is either repealed or amended to conform with the Constitution for the Federation of Earth, effective as soon as 25 countries have so ratified. The amendment of National or State Constitutions to allow entry into World Federation is not necessary prior to ratification of the Constitution for the Federation of Earth.

17.2 Stages of Implementation

17.2.1 Implementation of this World Constitution and the establishment of World Government pursuant to the terms of this World Constitution, may be accomplished in three stages, as follows, in addition to the stage of a Provisional World Government as provided under Article 19:

17.2.1.1 First Operative Stage of World Government.

17.2.1.2 Second Operative Stage of World Government.

17.2.1.3 Full Operative Stage of World Government.

17.2.2 At the beginning and during each stage, the World Parliament and the World Executive together shall establish goals and develop means for the progressive implementation of the World Constitution, and for the implementation of legislation enacted by the World Parliament.

17.3 First Operative Stage of World Government

17.3.1 The first operative stage of World Government under this World Constitution shall be implemented when the World Constitution is ratified by a sufficient number of nations and/or people to meet one or the other of the following conditions or equivalent:

17.3.1.1 Preliminary or final ratification by a minimum of twenty-five nations, each having a population of more than 100,000.

17.3.1.2 Preliminary or final ratification by a minimum of ten nations above 100,000 population, together with ratification by direct referendum within a minimum of fifty additional World Electoral and Administrative Districts.

17.3.1.3 Ratification by direct referendum within a minimum of 100 World Electoral and Administrative Districts, even though no nation as such has ratified.

17.3.2 The election of Members of the World Parliament to the House of Peoples shall be conducted in all World Electoral and Administrative Districts where ratification has been accomplished by popular referendum.

17.3.3 The Election of Members of the World Parliament to the House of Peoples may proceed concurrently with direct popular referendums both prior to and after the First Operative Stage of World Government is reached.

17.3.4 The appointment or election of Members of the World Parliament to the House of Nations shall proceed in all nations where preliminary ratification has been accomplished.

17.3.5 One-fourth of the Members of the World Parliament to the House of Counselors may be elected from nominees submitted by universities and colleges which have ratified the World Constitution.

17.3.6 The World Presidium and the Executive Cabinet shall be elected according to the provisions in Article 6, except that in the absence of a House of Counselors, the nominations shall be made by the members of the House of Peoples and of the House of Nations in joint session. Until this is accomplished, the Presidium and Executive Cabinet of the Provisional World Government as provided in Article 19 shall continue to serve.

17.3.7 When composed, the Presidium for the first operative stage of World Government shall assign or re-assign Ministerial posts among Cabinet and Presidium members, and shall immediately establish or confirm a World Disarmament Agency and a World Economic and Development Organization.

17.3.8 Those nations which ratify this World Constitution and thereby join the Federation of Earth, shall immediately transfer all weapons of mass destruction as defined and designated by the World Disarmament Agency to that Agency. (See Article 19, Sections 19.1.4 and 19.5.5).The World Disarmament Agency shall immediately immobilize all such weapons and shall proceed with dispatch to dismantle, convert to peacetime use, re-cycle the materials thereof or otherwise destroy all such weapons. During the first operative stage of World Government, the ratifying nations may retain armed forces equipped with weapons other than weapons of mass destruction as defined and designated by the World Disarmament Agency.

17.3.9 Concurrently with the reduction or elimination of such weapons of mass destruction and other military expenditures as can be accomplished during the first operative stage of World Government, the member nations of the Federation of Earth shall pay annually to the Treasury of the World Government amounts equal to one-half the amounts saved from their respective national military budgets during the last year before joining the Federation, and shall continue such payments until the full operative stage of World Government is reached. The World Government shall use fifty percent of the funds thus received to finance the work and projects of the World Economic Development Organization.

17.3.10 The World Parliament and the World Executive shall continue to develop the organs, departments, agencies and activities originated under the Provisional World Government, with such amendments as deemed necessary; and shall proceed to establish and begin the following organs, departments and agencies of the World Government, if not already underway, together with such other departments, and agencies as are considered desirable and feasible during the first operative stage of World Government:

17.3.10.1 The World Supreme Court;

17.3.10.2 The Enforcement System;

17.3.10.3 The World Ombudsmus;

17.3.10.4 The World Civil Service Administration;

17.3.10.5 The World Financial Administration;

17.3.10.6 The Agency for Research and Planning;

17.3.10.7 The Agency for Technological and Environmental Assessment;

17.3.10.8 An Emergency Earth Rescue Administration, concerned with all aspects of climate change and related factors;

17.3.10.9 An Integrated Global Energy System, based on environmentally safe sources;

17.3.10.10 A World University System, under the Department of Education;

17.3.10.11 A World Corporations Office, under the Department of Commerce and Industry;

17.3.10.12 The World Service Corps;

17.3.10.13 A World Oceans and Seabeds Administration.

17.3.11 At the beginning of the first operative stage, the Presidium in consultation with the Executive Cabinet shall formulate and put forward a proposed program for solving the most urgent world problems currently confronting humanity.

17.3.12 The World Parliament shall proceed to work upon solutions to world problems. The World Parliament and the World Executive working together shall institute through the several organs, departments and agencies of the World Government whatever means shall seem appropriate and feasible to accomplish the implementation and enforcement of world legislation, world law and the World Constitution; and in particular shall take certain decisive actions for the welfare of all people on Earth, applicable throughout the world, including but not limited to the following:

17.3.12.1 Expedite the organization and work of an Emergency Earth Rescue Administration, concerned with all aspects of climate change and climate crises;

17.3.12.2 Expedite the new finance, credit and monetary system, to serve human needs;

17.3.12.3 Expedite an integrated global energy system, utilizing solar energy, hydrogen energy, and other safe and sustainable sources of energy;

17.3.12.4 Push forward a global program for agricultural production to achieve maximum sustained yield under conditions which are ecologically sound;

17.3.12.5 Establish conditions for free trade within the Federation of Earth;

17.3.12.6 Call for and find ways to implement a moratorium on nuclear energy projects until all problems are solved concerning safety, disposal of toxic wastes and the dangers of use or diversion of materials for the production of nuclear weapons;

17.3.12.7 Outlaw and find ways to completely terminate the production of nuclear weapons and all weapons of mass destruction;

17.3.12.8 Push forward programs to assure adequate and non-polluted water supplies and clean air supplies for everybody on Earth;

17.3.12.9 Push forward a global program to conserve and re-cycle the resources of Earth.

17.3.12.10 Develop an acceptable program to bring population growth under control, especially by raising standards of living.

17.4 Second Operative Stage of World Government

17.4.1 The second operative stage of World Government shall be implemented when fifty percent or more of the nations of Earth have given either preliminary or final ratification to this World Constitution, provided that fifty percent of the total population of Earth is included either within the ratifying nations or within the ratifying nations together with additional World Electoral and Administrative Districts where people have ratified the World Constitution by direct referendum.

17.4.2 The election and appointment of Members of the World Parliament to the several Houses of the World Parliament shall proceed in the same manner as specified for the first operative stage in Section 3.2, 3.3, 3.4, and 3.5 of Article 17.

17.4.3 The terms of office of the Members of the World Parliament elected or appointed for the first operative stage of World Government, shall be extended into the second operative stage unless they have already served five year terms, in which case new elections or appointments shall be arranged. The terms of holdover Members of the World Parliament into the second operative stage shall be adjusted to run concurrently with the terms of those who are newly elected at the beginning of the second operative stage.

17.4.4   The World Presidium and the Executive Cabinet shall be re-constituted or reconfirmed, as needed, at the beginning of the second operative stage of World Government.

17.4.5   The World Parliament and the World Executive shall continue to develop the organs, departments, agencies and activities which are already underway from the first operative stage of World Government, with such amendments as deemed necessary; and shall proceed to establish and develop all other organs and major departments and agencies of the World Government to the extent deemed feasible during the second operative stage.

17.4.6   All nations joining the Federation of Earth to compose the second operative stage of World Government, shall immediately transfer all weapons of mass destruction and all other military weapons and equipment to the World Disarmament Agency, which shall immediately immobilize such weapons and equipment and shall proceed forthwith to dismantle, convert to peacetime uses, recycle the materials thereof, or otherwise destroy such weapons and equipment. During the second operative stage, all armed forces and para-military forces of the nations which have joined the Federation of Earth shall be completely disarmed and either disbanded or converted on a voluntary basis into elements of the non-military World Service Corps.

17.4.7   Concurrently with the reduction or elimination of such weapons, equipment and other military expenditures as can be accomplished during the second operative stage of World Government, the member nations of the Federation of Earth shall pay annually to the Treasury of the World Government amounts equal to one-half of the amounts saved from their national military budgets during the last year before joining the Federation and shall continue such payments until the full operative stage of World Government is reached. The World Government shall use fifty percent of the funds thus received to finance the work and projects of the World Economic Development Organization.

17.4.8   Upon formation of the Executive Cabinet for the second operative stage, the Presidium shall issue an invitation to the General Assembly of the United Nations Organization and to each of the specialized agencies of the United Nations, as well as to other useful international agencies, to transfer personnel, facilities, equipment, resources and allegiance to the Federation of Earth and to the World Government thereof. The agencies and functions of the United Nations Organization and of its specialized agencies and of other international agencies which may be thus transferred, shall be reconstituted as needed and integrated into the several organs, departments, offices and agencies of the World Government.

17.4.9   Near the beginning of the second operative stage, the Presidium in consultation with the Executive cabinet, shall formulate and put forward a proposed program for solving the most urgent world problems currently confronting the people of Earth.

17.4.10 The World Parliament shall proceed with legislation necessary for implementing a complete program for solving the current urgent world problems.

17.4.11 The World Parliament and the World Executive working together shall develop through the several organs, departments and agencies of the World Government whatever means shall seem appropriate and feasible to implement legislation for solving world problems; and in particular shall take certain decisive actions for the welfare of all people on Earth, including but not limited to the following:

17.4.11.1 Declaring all oceans, seas and canals having supra-national character (but not including inland seas traditionally belonging to particular nations) from twenty kilometers offshore, and all the seabeds thereof, to be under the ownership of the Federation of Earth as the common heritage of humanity, and subject to the control and management of the World Government.

17.4.11.2 Declare the polar caps and surrounding polar areas, including the continent of Antarctica but not areas which are traditionally a part of particular nations, to be world territory owned by the Federation of Earth as the common heritage of humanity, and subject to control and management by the World Government.

17.4.11.3 Outlaw the possession, stockpiling, sale and use of all nuclear weapons, all weapons of mass destruction, and all other military weapons and equipment.

17.4.11.4 Establish an ever-normal granary and food supply system for the people of Earth.

17.4.11.5 Develop and carry forward insofar as feasible all actions defined under Sec. 3.10 and 3.12 of the First Operative Stage.

17.5 Full Operative Stage of World Government

17.5.1 The full operative stage of World Government shall be implemented when this World Constitution is given either preliminary or final ratification by meeting either condition (17.5.1.1) or (17.5.1.2):

17.5.1.1 Ratification by eighty percent or more of the nations of Earth comprising at least ninety percent of the population of Earth; or

17.5.1.2 Ratification which includes ninety percent of Earth's total population, either within ratifying nations or within ratifying nations together with additional World Electoral and Administrative Districts where ratification by direct referendum has been accomplished, as provided in Article 17, Section 1.

17.5.2 When the full operative stage of World Government is reached, the following conditions shall be implemented:

17.5.2.1 Elections for Members of the House of Peoples shall be conducted in all World Electoral and Administrative Districts where elections have not already taken place; and Members of the House of Nations shall be elected or appointed by the national legislatures or national governments in all nations where this has not already been accomplished.

17.5.2.2 The terms of office for Members of the House of Peoples and of the House of Nations serving during the second operative stage, shall be continued into the full operative stage, except for those who have already served five years, in which case elections shall be held or appointments made as required.

17.5.2.3 The terms of office for all holdover Members of the House of Peoples and of the House of Nations who have served less than five years, shall be adjusted to run concurrently with those Members of the World Parliament whose terms are beginning with the full operative stage.

17.5.2.4 The second 100 Members of the House of Counselors shall be elected according to the procedure specified in Section 5 of Article 5. The terms of office for holdover Members of the House of Counselors shall run five more years after the beginning of the full operative stage, while those beginning their terms with the full operative stage shall serve ten years.

17.5.2.5 The Presidium and the Executive Cabinet shall be reconstituted in accordance with the provisions of Article 6.

17.5.2.6 All organs of the World Government shall be made fully operative, and shall be fully developed for the effective administration and implementation of world legislation, world law and the provisions of this World Constitution.

17.5.2.7 All nations which have not already done so shall immediately transfer all military weapons and equipment to the World Disarmament Agency, which shall immediately immobilize all such weapons and shall proceed forthwith to dismantle, convert to peaceful usage, recycle the materials thereof, or otherwise to destroy such weapons and equipment.

17.5.2.8 All armies and military forces of every kind shall be completely disarmed, and either disbanded or converted and integrated on a voluntary basis into the nonmilitary World Service Corps.

17.5.2.9 All viable agencies of the United Nations Organization and other viable international agencies established among national governments, together with their personnel, facilities and resources, shall be transferred to the World Government and reconstituted and integrated as may be useful into the organs, departments, offices, institutes, commissions, bureaus and agencies of the World Government.

17.5.2.10 The World Parliament and the World Executive shall continue to develop the activities and projects which are already underway from the second operative stage of World Government, with such amendments as deemed necessary; and shall proceed with a complete and full scale program to solve world problems and serve the welfare of all people on Earth, in accordance with the provisions of this World Constitution.

17.6 Costs of Ratification

The work and costs of private Citizens of Earth for the achievement of a ratified Constitution for the Federation of Earth, are recognized as legitimate costs for the establishment of constitutional world government by which present and future generations will benefit, and shall be repaid double the original amount by the World Financial Administration of the World Government when it becomes operational after 25 countries have ratified this Constitution for the Federation of Earth. Repayment specifically includes contributions to the World Government Funding Corporation and other costs and expenses recognized by standards and procedures to be established by the World Financial Administration.

ARTICLE 18: Amendments

18.1 Following completion of the first operative stage of World Government, amendments to this World Constitution may be proposed for consideration in two ways:

18.1.1 By a simple majority vote of any House of the World Parliament.

18.1.2 By petitions signed by a total of 200,000 persons eligible to vote in world elections from a total of at least twenty World Electoral and Administrative Districts where the World Constitution has received final ratification.

18.2 Passage of any amendment proposed by a House of the World Parliament shall require an absolute two-thirds majority vote of each of the three Houses of the World Parliament voting separately.

18.3 An amendment proposed by popular petition shall first require a simple majority vote of the House of Peoples, which shall be obliged to take a vote upon the proposed amendment. Passage of the amendment shall then require an absolute two-thirds majority vote of each of the three Houses of the World Parliament voting separately.

18.4   Periodically, but no later than ten years after first convening the World Parliament for the First Operative Stage of World Government, and every 20 years thereafter, the Members of the World Parliament shall meet in special session comprising a Constitutional Convention to conduct a review of this World Constitution to consider and propose possible amendments, which shall then require action as specified in Clause 2 of Article 18 for passage.

18.5   If the First Operative Stage of World Government is not reached by the year 1995, then the Provisional World Parliament, as provided under Article 19, may convene another session of the World Constituent Assembly to review the Constitution for the Federation of Earth and consider possible amendments according to procedure established by the Provisional World Parliament.

18.6   Except by following the amendment procedures specified herein, no part of this World Constitution may be set aside, suspended or subverted, neither for emergencies nor caprice nor convenience.

ARTICLE 19: Provisional World Government

19.1   **Actions to be Taken by the World Constituent Assembly**
Upon adoption of the World Constitution by the World Constituent Assembly, the Assembly and such continuing agency or agencies as it shall designate shall do the following, without being limited thereto:

19.1.1   Issue a Call to all Nations, communities and people of Earth to ratify this World Constitution for World Government.

19.1.2   Establish the following preparatory commissions:
    19.1.2.1   Ratification Commission.
    19.1.2.2   World Elections Commission.
    19.1.2.3   World Development Commission.
    19.1.2.4   World Disarmament Commission.
    19.1.2.5   World Problems Commission.
    19.1.2.6   Nominating Commission.
    19.1.2.7   Finance Commission.
    19.1.2.8   Peace Research and Education Commission.
    19.1.2.9   Special commissions on each of several of the most urgent world problems.
    19.1.2.10   Such other commissions as may be deemed desirable in order to proceed with the Provisional World Government.

19.1.3   Convene Sessions of a Provisional World Parliament when feasible under the following conditions:
    19.1.3.1   Seek the commitment of 500 or more delegates to attend, representing people in 20 countries from five continents, and having credentials defined by Article 19, Section 3.
    19.1.3.2   The minimum funds necessary to organize the sessions of the Provisional World Parliament are either on hand or firmly pledged.
    19.1.3.3   Suitable locations are confirmed at least nine months in advance, unless emergency conditions justify shorter advance notice.

19.2 Work of the Preparatory Commissions

19.2.1   The Ratification Commission shall carry out a worldwide campaign for the ratification of the World Constitution, both to obtain preliminary ratification by national governments, including national legislatures, and to obtain final ratification by people, including communities. The ratification commission shall continue its work until the full operative stage of World Government is reached.

19.2.2 The World Elections Commission shall prepare a provisional global map of World Electoral and Administrative Districts and Regions which may be revised during the first or second operative stage of World Government, and shall prepare and proceed with plans to obtain the election of Members of the World Parliament to the House of Peoples. The World Elections Commission shall in due course be converted into the World Boundaries and Elections Administration.

19.2.3 After six months, in those countries where national governments have not responded favorable to the ratification call, the Ratification Commission and the World Elections Commission may proceed jointly to accomplish both the ratification of the World Constitution by direct popular referendum and concurrently the election of Members of the World Parliament.

19.2.4 The Ratification Commission may also submit the World Constitution for ratification by universities and colleges throughout the world.

19.2.5 The World Development Commission shall prepare plans for the creation of a World Economic Development Organization to serve all nations and people ratifying the World Constitution, and in particular less developed countries, to begin functioning when the Provisional World Government is established.

19.2.6 The World Disarmament Commission shall prepare plans for the organization of a World Disarmament Agency, to begin functioning when the Provisional World Government is established.

19.2.7 The World Problems Commission shall prepare an agenda of urgent world problems, with documentation, for possible action by the Provisional World Parliament and Provisional World Government.

19.2.8 The Nominating Commission shall prepare, in advance of convening the Provisional World Parliament, a list of nominees to compose the Presidium and the Executive Cabinet for the Provisional World Government.

19.2.9 The Finance Commission shall work on ways and means for financing the Provisional World Government.

19.2.10 The several commissions on particular world problems shall work on the preparation of proposed world legislation and action on each problem, to present to the Provisional World Parliament when it convenes.

19.3 Composition of the Provisional World Parliament

19.3.1 The Provisional World Parliament shall be composed of the following members:

19.3.1.1 All those who were accredited as delegates to the 1977 and 1991 Sessions of the World Constituent Assembly, as well as to any previous Session of the Provisional World Parliament, and who re-confirm their support for the Constitution for the Federation of Earth, as amended.

19.3.1.2 Persons who obtain the required number of signatures on election petitions, or who are designated by Non-Governmental Organizations which adopt approved resolutions for this purpose, or who are otherwise accredited according to terms specified in Calls which may be issued to convene particular sessions of the Provisional World Parliament.

19.3.1.3   Members of the World Parliament to the House of Peoples who are elected from World Electoral and Administrative Districts up to the time of convening the Provisional World Parliament. Members of the World Parliament elected to the House of Peoples may continue to be added to the Provisional World Parliament until the first operative stage of World Government is reached.

19.3.1.4   Members of the World Parliament to the House of Nations who are elected by national legislatures or appointed by national governments up to the time of convening the Provisional World Parliament. Members of the World Parliament to the House of Nations may continue to be added to the Provisional World Parliament until the first operative stage of World Government is reached.

19.3.1.5   Those universities and colleges which have ratified the World Constitution may nominate persons to serve as Members of the World Parliament to the House of Counselors. The House of Peoples and House of Nations together may then elect from such nominees up to fifty Members of the World Parliament to serve in the House of Counselors of the Provisional World Government.

19.3.2   Members of the Provisional World Parliament in categories (1) and (2) as defined above, shall serve only until the first operative stage of World Government is declared, but may be duly elected to continue as Members of the World Parliament during the first operative stage.

19.4   Formation of the Provisional World Executive

19.4.1   As soon as the Provisional World Parliament next convenes, it will elect a new Presidium for the Provisional World Parliament and Provisional World Government from among the nominees submitted by the Nominating Commission.

19.4.2   Members of the Provisional World Presidium shall serve terms of three years, and may be re-elected by the Provisional World Parliament, but in any case shall serve only until the Presidium is elected under the First Operative Stage of World Government.

19.4.3   The Presidium may make additional nominations for the Executive Cabinet.

19.4.4   The Provisional World Parliament shall then elect the members of the Executive Cabinet.

19.4.5   The Presidium shall then assign ministerial posts among the members of the Executive Cabinet and of the Presidium.

19.4.6   When steps (1) through (4) of Section 19.4 are completed, the Provisional World Government shall be declared in operation to serve the welfare of humanity.

19.5   First Actions of the Provisional World Government

19.5.1   The Presidium, in consultation with the Executive Cabinet, the commissions on particular world problems and the World Parliament, shall define a program for action on urgent world problems.

19.5.2   The Provisional World Parliament shall go to work on the agenda of world problems, and shall take any and all actions it considers appropriate and feasible, in accordance with the provisions of this World Constitution.

19.5.3   Implementation of and compliance with the legislation enacted by the Provisional World Parliament shall be sought on a voluntary basis in return for the benefits to be realized, while strength of the Provisional World Government is being increased by the progressive ratification of the World Constitution.

19.5.4   Insofar as considered appropriate and feasible, the Provisional World Parliament and Provisional World Executive may undertake some of the actions specified under Section 3.12 of Article 17 for the first operative stage of World Government.

19.5.5   The World Economic Development Organization and the World Disarmament Agency shall be established, for correlated actions.

19.5.6   The World Parliament and the Executive Cabinet of the Provisional World Government shall proceed with the organization of other organs and agencies of the World Government on a provisional basis, insofar as considered desirable and feasible, in particular those specified under Section 3.10 of Article 17.

19.5.7   The several preparatory commissions on urgent world problems may be reconstituted as Administrative Departments of the Provisional World Government.

19.5.8   In all of its work and activities, the Provisional World Government shall function in accordance with the provisions of this Constitution for the Federation of Earth.

SOURCES OF EPIGRAPHS

Chapter One: Buckminster Fuller. *Operating Manual for Spaceship Earth.* New York: Pocket Books, 1972, p. 47.

Chapter Two: Albert Einstein. *Einstein on Peace.* Otto Nathan and Heinz Norden, eds. New York: Avenel Books, 1981, p. 473.

Chapter Three: Robert Ornstein and Paul Ehrlich. *New World: New Mind.* Los Altos, CA: ISHK, 2018, pp. 264-65.

Chapter Four: Al Gore. *Earth in the Balance: Ecology and the Human Spirit.* New York: Plume Books, 1993, p. 14.

Chapter Five: Joel Kovel. *The Enemy of Nature: The End of Capitalism or the End of the World?* London: Zed Books, 2007, p. 48.

Chapter Six: Jürgen Habermas. *The Postnational Constellation: Political Essays.* Max Pensky, trans. Cambridge, MA: The MIT Press, 2001, p. 60.

Chapter Seven: Errol E. Harris. *Twenty-first Century Democratic Renaissance: From Plato to Neoliberalism to Planetary Democracy.* Appomattox, VA: Institute for Economic Democracy Press, 2008, p. 145.

Chapter Eight: Hannah Arendt. *Between Past and Future.* New York: The Viking Press, 1968, pp. 169-71.

ILLUSTRATIONS

BIBLIOGRAPHY

Swami Agnivesh (2015). *Applied Spirituality: A Spiritual Vision for the Dialogue of Religions.* New York: Harper Element Books.

Akerman, Frank and Lisa Heinzerling (2004). *Priceless: On Knowing the Price of Everything and the Value of Nothing.* New York: The New Press.

Anderson, Benedict (2006). *Imagined Communities: Reflections on the Origin and Spread of Nationalism.* New York: Verso.

Angus, Ian (2016). *Facing the Anthropocene: Fossil Capitalism and the Crisis of the Earth System.* New York: Monthly Review Press.

Armstrong, Karen (2007). *The Great Transformation: The Beginning of Our Religious Traditions.* New York: Random House.

Sri Aurobindo (1973). *The Essential Aurobindo.* Robert A. McDermott, ed. New York: Schocken Books.

Bates, Albert and Kathleen Draper (2018). *Burn: Using Fire to Cool the Earth.* White River Junction, VT: Chelsea Green Publishing.

Beck, Don Edward and Christopher Cowan (2005). *Spiral Dynamics: Mastering Values, Leadership and Change.* Oxford: Blackwell Publications.

Benkler, Yochai (2006). *The Wealth of Networks: How Social Production Transforms Markets and Freedom.* New Haven, CT: Yale University Press.

Benton, Ted, ed. (1996). *The Greening of Marxism.* New York: Guilford Press.

Berdyaev, Nicolas (1960). *The Destiny of Man.* New York: Harper & Row.

Berkeley, George (1957). *A Treatise Concerning the Principles of Human Knowledge.* New York: Bobbs-Merrill Company.

Berry, Thomas, et al. (1994). *Worldviews and Ecology: Religion, Philosophy, and the Environment.* Maryknoll, NY: Orbis Books.

Berry, Wendell (2010). *What Matters? Economics for a Renewed Commonwealth.* Berkeley, CA: Counterpoint Publications.

Birch, Charles and John B. Cobb, Jr. (1990). *The Liberation of Life.* Denton, TX: Environmental Ethics Books.

Blain, Bob (2004). *Weaving Golden Threads: Integrating Social Theory.* Appomattox, VA: Institute for Economic Democracy Press.

Blum, Bill (1995). *Killing Hope: U.S. Military and CIA Interventions Since World War II.* Monroe, ME: Common Courage Press.

Blum, William (2005). *Rogue State. A Guide to the World's Only Superpower.* Monroe, ME: Common Courage Press.

Bohm, David (1980). *Wholeness and the Implicate Order.* New York: Routledge Publishers.

Bormann, Herbert F. and Stephen R. Kellert, eds. (1991). *Ecology, Economics, Ethics: The Broken Circle.* New Haven, CT: Yale University Press.

Boswell, Terry and Christopher Chase-Dunn. 2000. *The Spiral of Capitalism and Socialism: Toward Global Democracy.* Boulder, CO: Lynne Rienner Publisher.

Brecher, Jeremy and Tim Costello (1994). *Global Village or Global Pillage: Economic Reconstruction from the Bottom Up.* Boston: South End Press.

Brown, Ellen Hodgson (2008). *Web of Debt: The Shocking Truth About Our Money System–The Sleight of Hand That Has Trapped Us in Debt and How We Can Break Free.* (revised and updated). Baton Rouge, Louisiana: Third Millennium Press.

Brown, Lester R. (1973). *World Without Borders.* New York: Vintage Books.

Brown, Lester R. (2001). *Eco-Economy: Building an Economy for the Earth.* New York: W.W. Norton Publisher.

Brown, Lester R. (2011). *World on the Edge: How to Prevent Environmental and Economic Collapse.* New York: W.W. Norton.

Bucke, Richard Maurice (1974). *Cosmic Consciousness.* New York: Causeway Books.

Buell, Frederick (2003). *From Apocalypse to Way of Life: Environmental Crisis in the American Century.* New York: Routledge.

Caldicott, Helen, M.D. (1992). *If You Love this Planet: A Plan to Heal the Earth.* New York: W.W. Norton.

Capra, Fritjof (1975). *The Tao of Physics: An Exploration of the Parallels between Modern Physics and Eastern Mysticism.* Berkeley: Shambhala Press.

Capra, Fritjof (1996). *The Web of Life: A New Scientific Understanding of Living Systems.* New York: Random House.

Capra, Fritjof (2004). *The Hidden Connections: A Science for Sustainable Living.* New York: Anchor Books.

Carson, Rachel (1962). *Silent Spring.* Grenwich, CT: Fawcett Publications.

Catton, Jr., William R. (1982). *Overshoot: The Ecological Basis of Revolutionary Change.* Chicago: University of Illinois Press.

Chase-Dunn, Christopher (1998). *Global Formation: Structures of World Economy* (2nd Ed.). New York: Rowman & Littlefield.

Chomsky, Noam (1996). *What Uncle Sam Really Wants.* Tucson, AZ: Odonian Press.

Chomsky, Noam (2003). *Hegemony or Survival: America's Quest for Global Dominance.* New York: Henry Holt & Company.

Chossudovsky, Michel (1999). *The Globalization of Poverty: Impacts of IMF and World Bank Reforms.* London: Zed Books.

Chossudovsky, Michel and Andrew Gavin Marshall, eds. (2010). *Global Economic Crisis: The Great Depression of the XXI Century.* Montreal, Canada: Global Research.

Clark, Ramsey (1994). *The Fire This Time: U.S. War Crimes in the Gulf.* New York: Thunder's Mouth Press.

Cobb, Jr., John B. (1992). *Sustainability: Economics, Ecology & Justice.* Maryknoll, NY: Orbis Books.

Cohen, Joel E. (1995). *How Many People Can the Earth Support?* New York: W.W. Norton.

Constitution for the Federation of Earth. Found on-line at www.EarthConstitution.world and many other locations, in Spanish, for example, at www.ConstitucionMundial.com, in India at www.WCPAindia.org.

Corning, Peter (2003). *Nature's Magic: Synergy in Evolution and the Fate of Humankind.* Cambridge: Cambridge University Press.

Daly, Herman E., ed. (1973). *Economics, Ecology, Ethics: Essays Toward a Steady-State Economy.* San Francisco: W.H. Freeman & Company.

Daly, Herman E. (1996). *Beyond Growth: The Economics of Sustainable Development.* Boston: Beacon Press.

Daly, Herman E. (2007). *Ecological Economics and Sustainable Development: Selected Essays of Herman Daly.* Northampton, MA: Edward Elgar Publisher.

Daly, Herman E. (2014). *From Uneconomic Growth to a Steady-State Economy.* Cheltenham, UK: Edward Elgar Publishing.

Daly, Herman E. and Joshua Farley (2011). *Ecological Economics: Principles and Applications. Second Edition.* Washington, DC: Island Press.

Daly, Herman E. and Kenneth N. Townsend (1993). *Valuing the Earth: Economics, Ecology, Ethics.* Cambridge: The MIT Press.

Delio, Ilia (2013). *The Unbearable Wholeness of Being: God, Evolution, and the Power of Love.* Maryknoll, NY: Orbis Books.

De Martino, Richard (1960). "Zen Buddhism and the Human Situation," in Fromm, Suzuki, and de Martino, *Psychoanalysis and Zen Buddhism.* New York: Harper & Row.

Descartes, Rene (1975). *The Philosophical Works of Descartes.* Vol. 1. Elizabeth S. Haldane and G.R.T. Ross, trans. Cambridge: Cambridge University Press.

Devine, Pat (1988). *Democracy and Economic Planning: The Political Economy of a Self-Governing Society.* Cambridge, UK: Polity Press.

Dewey, John and James H. Tufts (1963). *Ethics: Revised Edition.* In Somerville, John and Ronald E. Santoni, eds. *Social and Political Philosophy.* Garden City, NY: Doubleday & Company.

Dewey, John (1993). *The Political Writings.* Debra Morris and Ian Shapiro, eds. Indianapolis, IN: Hackett Publishing Company.

Dickens, Peter (1992). *Society and Nature: Towards a Green Social Theory.* Philadelphia: Temple University Press.

Donnelly, Jack (2003). *Human Rights in Theory and Practice* (2nd Ed.). Ithaca, NY: Cornell University Press.

Dworkin, Ronald (1986). *Law's Empire.* Cambridge: Harvard University Press.

Earth Constitution Institute (ECI): www.EarthConstitution.world.

Ecimovic, Timi et al. (2007). *The Sustainable (Development) Future of Mankind.* Medosi, Slovania: SEM Institute for Climate Change.

Ehrlich, Paul R. and Anne H. Ehrlich (1990). *The Population Explosion.* New York: Simon & Schuster.

Eisler, Riane (1988). *The Challice & The Blade: Our History, Our Future.* San Francisco: HarperSanFrancisco.

Eisenstein, Charles (2018). *Climate: A New Story.* Berkeley, CA: North Atlantic Books.

Eisley, Loren (1959). *The Immense Journey: An Imaginative Naturalist Explores the Mystery of Man and Nature.* New York: Vintage Books.

Elgin, Duane (2000). *Promise Ahead: A Vision of Hope and Action for Humanity's Future.* New York: Harper Collins.

Ellis, Erle C. (2018). *Anthropocene: A Very Short Introduction.* Oxford: Oxford University Press.

Ellul, Jacques (1980). *The Technological System.* Joachim Neugroschel, trans. New York: Continuum Publishers.

Engdahl, F. William (2009). *Full Spectrum Dominance: Totalitarian Democracy in the New World Order.* Weisbaden: edition.engdhal.

Engdahl, F. William (2016). *The Lost Hegemon: Whom the Gods Would Destroy.* Wiesbaden, Germany: mine.Books.

Engelhardt, Tom (2014). *Shadow Government: Surveillance, Secret Wars, and a Global Security State in a Single, Superpower World.* Chicago: Haymarket Books.

Escobar, Pepe (2006). *Globalistan: How the Globalized World Is Dissolving into Liquid War.* Ann Arbor, MI: Nimble Books.

Farrell, Clare, et al. (2019). *This is Not a Drill: An Extinction Rebellion Handbook.* New York: Penguin Books.

Ferguson, Niall (2004). *Colossus: The Price of America's Empire.* New York: Penguin Press.

Finnis, John (1980). *Natural Law and Natural Rights.* Oxford: Clarendon Press.

Foster, John Bellamy, et al. (2010). *The Ecological Rift: Capitalism's War on the Earth.* New York: Monthly Review Press.

Fox, Matthew (1990). *A Spirituality Named Compassion.* San Francisco: Harper & Row.

Frank, S. L. (2020). *The Unknowable: An Ontological Introduction to the Philosophy of Religion.* Brooklyn, NY: Angelico Press.

Freire, Paulo (1974). *Pedagogy of the Oppressed.* Myra Bergman Ramos, trans. New York: Seabury Press.

Fromm, Erich (1962). *Beyond the Chains of Illusion: My Encounter with Marx and Freud.* New York: Simon & Schuster.

Fromm, Erich (1981). *On Disobedience and Other Essays.* New York: The Seabury Press.

Fromm, Erich (1996). *To Have or To Be?* New York: Continuum Books.

Fuller, R. Buckminster (1972). *Operating Manuel for Spaceship Earth.* New York: Pocket Books.

Fuller, R. Buckminister (1981). *Critical Path.* New York: St. Martin's Press.

Fuller, Lon L. (1969). *The Morality of Law* (Rev. Ed.). New Haven, CT: Yale University Press.

Gabel, Peter (2013). *Another Way of Seeing: Essays on Transforming Law, Politics, and Culture.* New Orleans: Quid Pro Books.

Gandhi, Mahatma (1972). *All Men Are Brothers.* Krishna Kripalani, ed. New York: UNESCO and Columbia University Press.

Gehring, Verna G., ed. (2003). *War After September 11.* New York: Roman & Littlefield.

Gewirth, Alan (1996). *The Community of Rights.* Chicago: University of Chicago Press.

Georgescu-Roegen, Nicholas (1971). *The Entropy Law and the Economic Process.* Cambridge: Harvard University Press.

Glover, Jonathan (1999). *Humanity: A Moral History of the Twentieth Century.* New Haven, CT: Yale University Press.

Goerner, Sally J., Robert G. Dyck, Dorothy Lagerroos (2008). *The New Science of Sustainability: Building a Foundation for Great Change.* Triangle Center for Complex Systems (n.p.)

Gore, Al (1993). *Earth in the Balance: Ecology and the Human Spirit.* New York: Plume Books.

Grandin, Greg (2007). *Empire's Workshop: Latin America, the United States, and the Rise of the New Imperialism.* New York: Henry Holt & Co.

Greco, Jr., Thomas (2009). *The End of Money and the Future of Civilization.* White River Junction, VT: Chelsea Green Publisher.

Habermas, Jürgen (1984). *Theory of Communicative Action. Volume One: Reason and the Rationalization of Society.* Thomas McCarthy, trans. Boston: Beacon Press.

Habermas, Jürgen (1987). *Theory of Communicative Action. Volume Two: Lifeworld and System: A Critique of Functionalist Reason.* Thomas McCarthy, trans. Boston: Beacon Press.

Habermas, Jürgen (1998a). *On the Pragmatics of Communication.* Maeve Cooke, ed. Cambridge: The MIT Press.

Habermas, Jürgen (1998b). *Between Facts and Norms: Contributions to a Discourse Theory of Law and Democracy.* Cambridge: The MIT Press.

Habermas, Jürgen (2003). *The Future of Human Nature.* William Rehg, et al. trans. Cambridge, MA: Polity Press.

Hackett, Ian (2004). *Transcending Terror: A History of Our Spiritual Quest and the Challenge of the New Millennium.* U.K. and New York: 0 Books.

Hardt, Michael and Antonio Negri (2000). *Empire.* Cambridge: Harvard University Press.

Harrington, Michael (1989). *Socialism: Past and Future. The Classic Text on the Role of Socialism in Modern Society.* New York: Arcade Publishing

Harris, Errol E. (1977). *Atheism and Theism.* Atlantic Highlands, NJ: Humanities Press.

Harris, Errol E. (1987). *Formal, Transcendental, and Dialectical Thinking: Logic and Reality.* Albany, NY: State University of New York Press.

Harris, Errol E. (1988). *The Reality of Time.* Albany, NY: State University of New York Press.

Harris, Errol E. (1991). *Cosmos and Anthropos: A Philosophical Interpretation of the Anthropic Cosmological Principle.* London: Humanities Press International

Harris, Errol E. (1992) *Cosmos and Theos: Ethical and Theological Implications of the Anthropic Cosmological Principle* London: Humanities Press.

Harris, Errol E. (2000a). *The Restitution of Metaphysics.* Amherst, NY: Prometheus Books.

Harris, Errol E. (2000b). *Apocalypse and Paradigm: Science and Everyday Thinking.* Westport, CT: Praeger Publishers.

Harris, Errol E. (2008). *Twenty-First Century Democratic Renaissance: From Plato to Neoliberalism to Planetary Democracy.* Appomattox, VA: The Institute for Economic Democracy.

Harris, Errol E. (2014). *Earth Federation Now! Tomorrow is Too Late.* Appomattox, VA: The Institute for Economic Democracy.

Harris, Jonathan M., ed. *Rethinking Sustainability: Power, Knowledge, and Institutions.* Ann Arbor, MI: The University of Michigan Press.

Harris, Jonathan M., et al. (2001). *A Survey of Sustainable Development: Social and Economic Dimensions.* Washington, DC: Island Press..

Hart, H.L.A. (1994). *The Concept of Law* (2nd Ed.). Oxford: Oxford University Press.

Harvey, David (2005). *The New Imperialism.* Oxford: Oxford University Press.

Hawken, Paul, ed. (2017). *Drawdown: The Most Comprehensive Plan Ever Proposed to Reverse Global Warming.* New York: Penguin Books.

Hazen, Robert M. (2012). *The Story of Earth: The First 4.5 Billion Years, From Stardust to Living Planet.* New York: Penguin Books.

Hegel, G.W.F. (1967). *The Phenomenology of Mind.* J.B. Baillie, trans. New York: Harper & Row.

Heinberg, Richard (2011). *The End of Growth: Adapting to Our New Economic Reality.* Gabriola Island, BC: New Society Publishers.

Heinberg, Richard and Daniel Lerch, eds. (2010). *The Post-Carbon Reader: Managing the 21ˢᵗ Century's Sustainability Crises.* Healdsburg, CA: Watershed Media.

Held, David (1995). *Democracy and the Global Order: From the Modern State to Cosmopolitan Governance.* Stanford, CA: Stanford University Press.

Henderson, Hazel (1988). *The Politics of the Solar Age: Alternatives to Economics.* Indianapolis, IN: Knowledge Systems, Inc.

Henderson, Hazel (2006). *Ethical Markets: Growing the Green Economy.* White River Junction, VT: Chelsea Green Publishing.

Hick, John (2004). *An Interpretation of Religion* (2ⁿᵈ Ed.). New Haven, CT: Yale University Press.

Homer-Dixon, Thomas F. (1999). *Environment, Scarcity, and Violence.* Princeton, NJ: Princeton University Press.

Hubbard, Barbara Marx (1998). *Conscious Evolution. Awakening the Power of Our Social Potential.* Novato, CA: New World Library.

Hume, David (1949). *A Treatise on Human Nature.* Volume Two. New York: E.P. Dutton & Co.

Jabbari, Mehdi, Majid Shafiepour Motlagh, Khosro Ashrafi, and Ghahreman Abdoli (2020). "Global Carbon Budget Allocation Based on Rawlsian Justice by Means of the Sustainable Development Goals Index." *Environment, Development & Sustainability* 22, no. 6 (August 2020): 5465-81.

Jacobson, Nolan Pliny (1982). "A Buddhistic-Christian Probe of Our Endangered Future." *The Eastern Buddhist.* Vol. XV, No. 1, Spring 1982.

Jacobson, Nolan Pliny (1974). *Buddhism: The Religion of Analysis.* Carbondale, IL: University of Southern Illinois Press.

Jamieson, Dale (2008). *Ethics and the Environment: An Introduction.* Cambridge: Cambridge University Press.

Jaspers, Karl (1953). *The Origin and Goal of History.* New Haven, CT: Yale University Press.

Jaspers, Karl (1957). *Man in the Modern Age.* Eden and Cedar Paul, trans. New York: Anchor Books.

Johnson, Chalmers (2004). *The Sorrows of Empire: Militarism, Secrecy, and the End of the Republic.* New York: Henry Holt & Company.

Johnson, Chalmers (2006). *Nemesis: The Last Days of the American Republic.* New York: Metropolitan Books.

Jonas, Hans (1984). *The Imperative of Responsibility: In Search of an Ethics for the Technological Age.* Chicago: University of Chicago Press.

Kafatos, Menas and Robert Nadeau (1990). *The Conscious Universe: Part and Whole in Modern Physical Theory.* Berlin: Springer-Verlag.

Kant, Immanuel (1957). *Perpetual Peace.* Louis White Beck, trans. New York: Macmillan.

Kant, Immanuel (1964). *Groundwork of the Metaphysic of Morals.* H. J. Paton, trans. New York: Harper & Row.

Kant, Immanuel (1965, orig. pub. 1781). *Critique of Pure Reason.* Norman Kemp Smith, trans. New York: St. Martin's Press.

Kent Deirdre (2006). *Healthy Money, Healthy Planet: Developing Sustainability through New Money Systems.* Nelson, New Zealand: Craig Potton Publishing.

Kitchener, Richard F., ed. (1988). *The World View of Contemporary Physics: Does it Need a New Metaphysics?* Albany, NY: State University of New York Press.

Klare, Michael T. (2008). *Rising Powers, Shrinking Planet: The New Geopolitics of Energy.* New York: Metropolitan Books.

Klein, Naomi (2007). *The Shock Doctrine: The Rise of Disaster Capitalism.* New York: Henry Holt and Company.

Klein, Naomi (2014). *This Changes Everything: Capitalism vs the Climate.* New York: Simon & Schuster.

Kohlberg, Lawrence (1984). *The Psychology of Moral Development: Volume Two, The Nature and Validity of Moral Stages.* San Francisco: Harper & Row.

Kolbert, Elizabeth (2014). *The Sixth Extinction: An Unnatural History.* New York: Henry Holt & Co.

Korten, David (2001). *When Corporations Rule the World* (2nd Ed.). San Francisco: Kumarian Press.

Korten, David (2006). *The Great Turning: From Empire to Earth Community.* San Francisco: Berrett-Koehler and Kumarian Press.

Kovel, Joel (2007). *The Enemy of Nature: The End of Capitalism or the End of the World?* London: Zed Books.

Krishnamurti, Jiddu (1989). *Think on These Things.* New York: Harper & Row.

Krishnan, Rajaram, Jonathan M. Harris, and Neva R. Goodwin, eds. (1995). *A Survey of Ecological Economics.* Washington, DC: Island Press.

Kuhn, Thomas (1970). *The Structure of Scientific Revolutions. Second Edition Enlarged.* Chicago: University of Chicago Press.

Kuzminski, Adrian (2013). *The Ecology of Money: Debt, Growth, and Sustainability.* New York: Lexington Books.

Laszlo, Ervin (2002). *The Systems View of the World: A Holistic Vision for Our Time.* Cresskill, NJ: Hampton Press.

Laszlo, Ervin (2007). *Quantum Shift in the Global Brain: How the New Scientific Reality Can Change Us and Our World.* Rochester, VT: Inner Traditions.

Laszlo, Ervin (2014). *The Self-Actualizing Cosmos: The Akasha Revolution in Science and Human Consciousness.* Rochester, VT: Inner Traditions.

Laszlo, Ervin (2017). *The Intelligence of the Cosmos. Why Are We Here? New Answers from the Frontiers of Science.* Rochester, VT: Inner Traditions.

Leech, Garry (2012). *Capitalism: A Structural Genocide.* London: Zed Books.

Lenin, V.I. (1939). *Imperialism: The Highest Stage of Capitalism.* New York: International Publishers.

Lenton, Tim (2016). *Earth System Science: A Very Short Introduction.* Oxford: Oxford University Press.

Levinas, Emmanuel (1969). *Totality and Infinity.* Alphonso Lingis, trans. Pittsburgh: Duquesne University Press.

Lietaer, Bernard (2001). *The Future of Money: Creating New Wealth, Work, and a Wiser World.* London: Century Publications.

Lifton, Robert J. (1993). *The Protean Self: Human Resilience in an Age of Fragmentation.* New York: Basic Books.

Lloyd, Jason et al., eds. (2017). *The Rightful Place of Science: Climate Pragmatism.* Washington, DC: Consortium for Science, Policy, and Outcomes.

Locke, John (1978). *An Essay Concerning Human Understanding.* A.S. Pringle-Pattison, ed. Sussex, UK: The Harvester Press.

Lokanga, Ediho (2018). *Digital Physics: The Meaning of the Holographic Universe and Its Implications Beyond Theoretical Physics.* NP: Ediho Lokanga, Buku_masapo34@yahoo.com.

Lovelock, James (1991). *Healing Gaia: Practical Medicine for the Planet.* New York: Harmony Books.

Lovins, L. Hunter, et al. (2018). *A Finer Future: Creating an Economy in Service to Life.* Gabriola Island, BC: New Society Publishers.

Macy, Joanna and Molly Brown (2014). *Coming Back to Life.* Gabriola Island, BC: New Society Publishers.

Marsh, James L. (1995). *Critique, Action, and Liberation.* Albany: State University of New York Press.

Martin, Glen T. (2005a). *Millennium Dawn: The Philosophy of Planetary Crisis and Human Liberation.* Appomattox, VA: Institute for Economic Democracy Press.

Martin, Glen T., ed. (2005b). *World Revolution Through World Law.* Appomattox, VA: Institute for Economic Democracy Press.

Martin, Glen T. (2008). *Ascent to Freedom: Practical & Philosophical Foundations of Democratic World Law.* Appomattox, VA: Institute for Economic Democracy.

Martin, Glen T. (2010a). *Constitution for the Federation of Earth: With Historical Introduction, Commentary, and Conclusion.* Appomattox, VA: Institute for Economic Democracy.

Martin, Glen T. (2010b). *Triumph of Civilization: Democracy, Nonviolence, and the Piloting of Spaceship Earth.* Appomattox, VA: Institute for Economic Democracy.

Martin, Glen T. (2013). *The Anatomy of a Sustainable World: Our Choice Between System Change or Climate Change.* Appomattox, VA: Institute for Economic Democracy Press.

Martin, Glen T. (2016). *One World Renaissance: Holistic Planetary Transformation Through a Global Social Contract.* Appomattox, VA: Institute for Economic Democracy Press.

Martin, Glen T., ed. (2016). *Our Common Human Future: The UN as an Effective Peace and Sustainability System.* Appomattox, VA: Institute for Economic Democracy Press.

Martin, Glen T. (2017). "Gandhi's Satyagraha and the Earth Constitution," chapter in *Examining Global Peacemaking in the Digital Age: A Research Handbook,* Bruce L. Cook, ed. Published by IGI Global, November 2017, pp. 361-371.

Martin, Glen T. (2018). *Global Democracy and Human Self-Transcendence: The Power of the Future for Planetary Transformation.* London Cambridge Scholars Publishing.

Martin, Glen T. (2020). "Deep Sustainability: Really Addressing Our Endangered Future," Chapter Three of *Struggles and Successes in the Pursuit of Sustainable Development.* Keong Tan, et al. eds. New York: Routledge Publisher.

Marx-Engels (1972). *The Marx-Engels Reader* (2nd Ed.). Robert C. Tucker, ed. New York: W.W. Norton & Company.

Marx, Karl (1990). *Capital: A Critique of Political Economy,* Vol. 1. Ben Fowkes, trans. Ernest Mandel, ed. New York: Penguin Books.

Maslin, Mark (2013). *Climate: A Very Short Introduction.* Oxford: Oxford University Press.

Maslow, Abraham (2014). *Toward a Psychology of Being.* Floyd, VA: Sublime Books.

Mayur, Rashmi (1996). *Earth, Man, and Future: For the Renaissance of Men and Women in the New Millennium.* Mumbai: Institute for a Sustainable Future.

McKibben, Bill (2007). *Deep Economy: The Wealth of Communities and the Durable Future.* New York: Henry Holt.

McKibben, Bill (2019). *Falter: Has the Human Game Begun to Play Itself Out?* New York: Henry Holt Publisher.

Meadows, Donna (2008). *Thinking in Systems: A Primer.* White River Junction, VT: Chelsea Green Publishing.

Meadows, Donna, Jorgen Randers, and Dennis Meadows (2004). *Limits to Growth: The 30-Year Update.* White River Junction, VT: Chelsea Green Publishing.

Meeker-Lowry (1988). *Economics as If the Earth Really Mattered.* Philadelphia: New Society Publishers.

Mies, Maria and Vandana Shiva (1993). *Ecofeminism.* London: Zed Books.

Miranda, José Porfirio (1986). *Marx Against the Marxists: The Christian Humanism of Karl Marx.* Trans. John Drury. Maryknoll, NY: Orbis Books.

Morganthau, Hans (2006). *Politic Among Nations* (7th Ed.). New York: McGraw Higher Education (orig. pub. 1948).

Munitz, Milton K. (1986). *Cosmic Understanding: Philosophy and Science of the Universe.* Princeton, NJ: Princeton University Press.

Murphy, Gardner (1975). *Human Potentialities.* New York: The Viking Press.

Nietzsche, Friedrich (1969). *On the Genealogy of Morals & Ecce Homo.* Edited with Commentary by Walter Kaufmann. New York: Vintage Books.

Nordhaus, Ted and Michael Shellenberger (2007). *Break Through: Why We Can't Leave Saving the Planet to Environmentalists.* New York: Houghton Mifflin Harcourt.

O'Connor, Martin, ed. (1994). *Is Capitalism Sustainable? Political Economy and the Politics of Ecology.* New York: The Guilford Press.

Ornstein, Robert and Paul Ehrlich (2018). *New World: New Mind.* Los Altos, CA: ISHK

Ostrom, Elinor (1990). *Governing the Commons: The Evolution of Institutions for Collective Action.* Cambridge: Cambridge University Press.

Panikkar, Raimon (2006). *The Experience of God: Icons of the Mystery.* Minneapolis, MN: Fortress Press.

Panikkar, Raimon (2013). *The Rhythm of Being: The Unbroken Trinity.* Maryknoll, NY: Orbis Books.

Parenti, Michael (1995). *Democracy for the Few* (6th Ed.). New York: St. Martin's Press.

Parenti, Michael (1995). *Against Empire: A Brilliant Exposé of the Brutal Realities of U.S. Global Domination.* San Francisco: City Lights Books.

Parenti, Michael (2011). *The Face of Imperialism.* Boulder, CO: Paradigm Publishers.

Perkins, John (2004). *Confessions of an Economic Hit Man.* San Francisco: Berrett-Kohler Publishers.

Petras, James and Henry Veltmeyer (2005). *Empire with Imperialism: The Globalizing Dynamics of Neo-liberal Capitalism.* London: Zed Books.

Piketty, Thomas (2014). *Capital in the Twenty-First Century.* Cambridge, MA: Harvard University Press.

Pinker, Steven (1995). *The Language Instinct: How the Mind Creates Language.* New York: Harper Perennial Books.

Pinker, Steven (2007). *The Stuff of Thought: Language as a Window into Human Nature.* New York: Penguin Books.

Pogge, Thomas (2015). "The Sustainable Development Goals: Brilliant Propaganda?" *Annals of the University of Bucharest,* Political Science Series 17 (2): 25-46.

Polak, Frederik L. (1967). "Utopia and Cultural Renewal." In *Utopias and Utopian Thought: A Timely Appraisal.* Frank E. Manuel, ed. Boston: Beacon Press: 281-95

Pollin, Robert (2015). *Greening the Global Economy.* Cambridge: The MIT Press.

Prugh, Thomas, Robert Costanza, and Herman Daly (2000). *The Local Politics of Global Sustainability.* Washington, DC: Island Press.

Raworth, Kate (2017). *Doughnut Economics: 7 Ways to Think Like a 21st Century Economist.* White River Junction, VT: Chelsea Green Publishing.

Redfern, Martin (2003). *The Earth: A Very Short Introduction.* Oxford: Oxford University Press.

Reves, Emery (1946). *The Anatomy of Peace.* New York: Harper & Brothers.

Rifkin, Jeremy (1989). *Entropy: Into the Greenhouse World* (Rev. Ed.). New York: Bantam Books.

Rifkin, Jeremy (2009). *The Empathic Civilization: The Race to Global Consciousness in a World in Crisis.* New York: Penguin Books.

Rifkin, Jeremy (2011). *The Third Industrial Revolution: How Lateral Power is Transforming Energy, the Economy, and the World.* New York: St. Martin's Press.

Rifkin, Jeremy (2013). *The Zero Marginal Cost Society: The Internet of Things, the Collaborative Commons, and the Eclipse of Capitalism.* New York: St. Martin's Press.

Rifkin, Jeremy (2019). *The Green New Deal.* New York: St. Martin's Press.

Robinson, William J. (2004). *A Theory of Global Capitalism: Production, Class, and State in a Transnational World.* Baltimore, MD: The Johns Hopkins University Press.

Rolston, III, Holmes (1988). *Environmental Ethics: Duties to and Values in the Natural World.* Philadelphia, PA: Temple University Press.

Romm, Joseph (2018). *Climate Change: What Everyone Needs to Know.* Oxford: Oxford University Press.

Roy, Arundhati (1999). *The Greater Common Good.* Bombay: India Book Distributers Publisher.

Sanders, Barry and Mike Davis (2009). *The Green Zone: The Environmental Costs of Militarism.* Oakland, CA: AK Press.

Schneider, Stephen H. (1989). *Global Warming: Are We Entering the Greenhouse Century?* San Francisco: Sierra Club Books.

Schumacher, E. F. (1973). *Small is Beautiful: Economics as if People Mattered.* New York: Harper & Row.

Scranton, Roy (2015). *Learning to Die in the Anthropocene: Reflections on the End of a Civilization.* San Francisco: City Lights Books.

Screpanti, Ernesto (2014). *Global Imperialism and the Great Crisis: The Uncertain Future of Capitalism.* New York: Monthly Review Press.

Shannon, Thomas Richard (1989). *An Introduction to World-System Perspective.* Boulder, CO: Westview Press.

Shiva, Vandana (2002). *Water Wars: Privatization, Pollution, and Profit.* Cambridge, MA: South End Press.

Smith, John (2016). *Imperialism in The Twenty-First Century: Globalization, Super-Exploitation, and Capitalism's Final Crisis.* New York: Monthly Review Press.

Smith, J.W. (2005). *Economic Democracy: The Political Struggle of the 21st Century.* Appomattox, VA: Institute for Economic Democracy Press.

Smith, J.W. (2013). *Cooperative Capitalism: A Blueprint for Global Peace and Prosperity.* Appomattox, VA: Institute for Economic Democracy Press.

Smith, Stephen (2011). *Environmental Economics: A Very Short Introduction.* Oxford: Oxford University Press.

Soddy, Frederick (2012). *Cartesian Economics: The Bearing of Physical Science upon State Stewardship.* New York: Cosimo Classics (orig. pub. 1921).

Speth, James Gustave (2004). *Red Sky at Morning: America and the Crisis of the Global Environment.* New Haven, CT: Yale University Press.

Speth, James Gustave (2008). *The Bridge at the Edge of the World.* New Haven, CT: Yale University Press.

Stapp, Henry P. (2011). *Mindful Universe: Quantum Mechanics and the Participating Observer* (2nd Ed.). Berlin: Springer Publishers.

St. Clair, Jeffrey and Joshua Frank (2018). *The Big Heat: Earth on the Brink.* Petrolia, CA: CounterPunch Publisher.

Strauss, Leo (1965). *Natural Right and History.* Chicago: University of Chicago Press.

Swimme, Brian and Thomas Berry (1992). *The Universe Story: From the Primordial Flaring Forth to the Ecozoic Era.* San Francisco: Harper San Francisco.

Teilhard de Chardin, Pierre (1959). *The Phenomenon of Man.* New York: Harper & Row Publishers.

Teilhard de Chardin, Pierre (1969). *Hymn of the Universe.* New York: Harper Colophon Books.

Toffler, Alvin (1970). *Future Shock.* New York: Bantam Books.

Toffler, Alvin (1990). *The Third Wave: The Classic Study of Tomorrow.* New York: Bantam Books.

Tokar, Brian (1997). *Earth for Sale: Reclaiming Ecology in the Age or Corporate Greenwash.* Boston: South End Press.

Turner, R. Kerry, David Pearce & Ian Bateman (1993). *Environmental Economics: An Elementary Introduction.* Baltimore, MD: The Johns Hopkins University Press.

U.N. Sustainable Development Goals: https://sustainabledevelopment.un.org/post2015/transformingourworld

Valentine, Douglas (2017). *The CIA as Organized Crime: How Illegal Operations Corrupt America and the World.* Atlanta, GA: Clarity Press, Inc.

Vaughan-Lee, ed. (2014). *Spiritual Ecology: The Cry of the Earth.* Point Reyes, CA: The Golden Sufi Center.

Wallace-Wells, David (2019). *The Uninhabitable Earth: Life After Warming.* New York: Penguin/Random House.

Wallerstein, Immanuel (1983). *Historical Capitalism.* London: Verso.

Ward, Barbara and Rene Dubos (1972). *Only One Earth: The Care and Maintenance of a Small Planet.* New York: W.W. Norton.

Weiss, Edith Brown (2001). "Planetary Rights," in *The Philosophy of Human Rights.* Patrick Hayden, ed. New York: Paragon House, pp. 618-636.

Wilber, Ken (1985). *No Boundary: Eastern and Western Approaches to Personal Growth.* Boston: Shambhala Publisher.

Wilber, Ken (1996). *Eye to Eye: The quest for the New Paradigm* (3rd Ed.). Boston: Shambhala Publisher.

Wilber, Ken (2006). *Integral Spirituality: A Startling New Role for Religion in the Modern and Postmodern World.* Boston: Integral Books.

Wilber, Ken (2007). *The Integral Vision: A Very Short Introduction to the Revolutionary Integral Approach to Life, God, the Universe, and Everything.* Boston: Shambhala Publisher.

Wilber, Ken (2017). *A Brief History of Everything* (20th Anniversary Ed.). Boulder, CO: Shambhala.

Williams, Chris (2010). *Ecology and Socialism: Solutions to Capitalist Ecological Crisis.* Chicago: Haymarket Books.

Wittgenstein, Ludwig (1958). *Philosophical Investigations.* G.E.M. Anscombe, trans. New York: Macmillan.

Wolin, Sheldon (2008). *Democracy Incorporated: Managed Democracy and the Spector of Inverted Totalitarianism.* Princeton, NJ: Princeton University Press.

World Constitution and Parliament Association (WCPA): www.EarthConstitution.world and www.WorldParliament-gov.org.

Worldwatch Institute (2008). *State of the World: Innovations for a Sustainable Economy 2008.* New York: W.W. Norton & Co.

INDEX

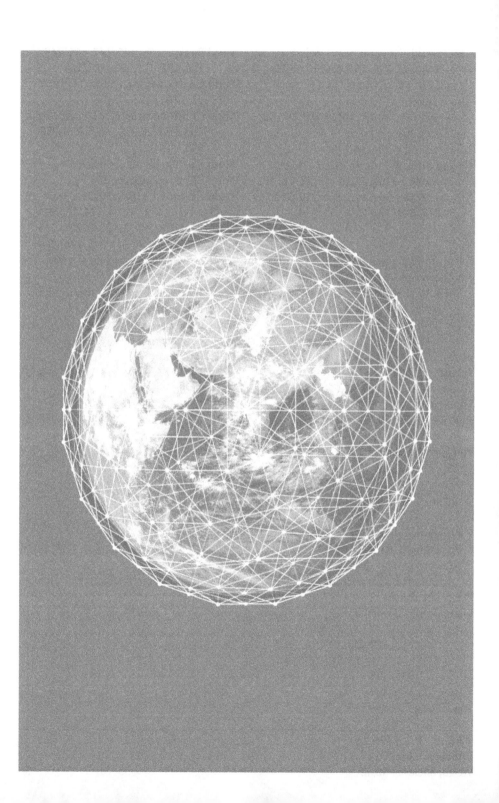

ACKNOWLEDGEMENTS

I want to thank Ms. Hannah Kocen and Ms. Lindsey Bulita, my students at Radford University, for help in creating graphs, and images, as well as the indexing, included in this book. Students like these make the profession of being a professor so rewarding.

I want to thank my wife, Phyllis Turk, for her loving support and editing contributions for my work. I also want to thank Rev. Laura George, J.D. and all the wonderful people at Oracle Institute Press for their excellent support and collaboration in the publication process for this book.

Chapters Two and Three draw from my earlier books, *The Anatomy of a Sustainable World* (2013) and *World Revolution Through World Law* (2005). I want to thank Dr. B. Sidney Smith, Director at IED Press, for his support in my appropriation of that material for the present volume.

Finally, since every book is ultimately a collaborative process, I am very grateful to the many fine thinkers, too numerous to name, some long deceased, who have contributed to this manuscript. Ultimately, of course, I am responsible for the final product, for better or for worse.

ABOUT THE AUTHOR

Glen T. Martin, Ph.D., is Professor Emeritus of Philosophy at Radford University in Virginia, USA. He is President of the International Philosophers for Peace, Executive Director of the Earth Constitution Institute (ECI), and President of the World Constitution and Parliament Association (WCPA). ECI and WCPA are cooperating worldwide organizations that sponsor the *Constitution for the Federation of Earth* and organize sessions of the Provisional World Parliament. Dr. Martin travels and lectures worldwide on behalf of the *Earth Constitution* and the philosophical foundations of emerging world law and the development of a binding world court system.

Dr. Martin was in college during the 1960s when the U.S. government was drafting young people into its imperial military machine. He declared himself a conscientious objector and refused to enter any organization dedicated to war, which he considers to be just another form of murder. Dr. Martin has long been a critic and resister of the dominant world system of militarized sovereign nation-states integrated with growth-based, exploitive capitalism. Instead, Dr. Martin has philosophically developed an alternative world federalist vision of planetary peace and justice based on the *Earth Constitution*.

Dr. Martin has received several international peace awards, including the Gusi Peace Prize International in Manila, Philippines (2013). He has published hundreds of articles concerning freedom, peace, justice, human rights, and sustainability in popular as well as academic venues, and he is the author or editor of a dozen books. For more information on the author, please visit:

ECI website: www.EarthConstitution.world
WCPA India website: www.WCPAindia.org
Dr. Martin's website: www.OneWorldRenaissance.com

ABOUT THE PUBLISHER

The Oracle Institute is a 501(c)(3) educational charity and spiritual think-tank that studies the nexus between religion, politics, human rights, and conscious evolution. Oracle's mission is to help humanity progress into the Fifth Spiritual Paradigm and adopt a Culture of Peace.

Oracle Press is our award-winning publishing house, where we create books to assist humanity's quest for Truth, Love and Light. Our texts focus on the world's great religions, wisdom traditions, social trends, and civic responsibilities. To date, our press has garnered six prestigious book awards in categories such as religion, spirituality, children's fiction, and the performing arts.

Oracle Press works closely with authors to co-create beautiful, relevant, and impactful books published under the Oracle and Peace Pentagon imprints. Our non-profit press also offers services to authors who wish to self-publish when the project is aligned with the educational mission of The Oracle Institute.

To learn more about Oracle Press, visit:
www.TheOracleInstitute.org/Press

To shop at our online Bookstore, visit:
www.TheOracleInstitute.org/Bookstore

To learn more about The Oracle Institute, our programs and events,
or our intentional community – the Valley of Light – visit us at Oracle Campus:

Peace Pentagon
88 Oracle Way
Independence, VA 24348

www.TheOracleInstitute.org
www.PeacePentagon.net

P. 60 MultiNational Corporations run
the world, not nation State

P. 69 - DAVID BOHM

P. 258 - SPACE

P. 284 - OTHER PLANETS

CPSIA information can be obtained
at www.ICGtesting.com
Printed in the USA
LVHW051358030721
691839LV00013B/1013